Kepler's Witch

AN ASTRONOMER'S DISCOVERY OF COSMIC ORDER
AMID RELIGIOUS WAR, POLITICAL INTRIGUE, AND
THE HERESY TRIAL OF HIS MOTHER

James A. Connor

with translation assistance by Petra Sabin Jung

HarperSanFrancisco
A Division of HarperCollins*Publishers*

KEPLER'S WITCH: *An Astronomer's Discovery of Cosmic Order Amid
Religious War, Political Intrigue, and the Heresy Trial of His Mother.*
Copyright © 2004 by James A. Connor. All rights reserved. Printed in
the United States of America. No part of this book may be used or
reproduced in any manner whatsoever without written permission
except in the case of brief quotations embodied in critical articles and
reviews. For information address HarperCollins Publishers, Inc., 10 East
53rd Street, New York, NY 10022.

HarperCollins books may be purchased for educational, business, or
sales promotional use. For information please write: Special Markets
Department, HarperCollins Publishers, Inc., 10 East 53rd Street, New
York, NY 10022.

HarperCollins Web site: http://www.harpercollins.com

HarperCollins®, 📖 ®, and HarperSanFrancisco™ are trademarks of
HarperCollins Publishers, Inc.

FIRST HARPERCOLLINS PAPERBACK EDITION PUBLISHED IN 2005

Library of Congress Cataloging-in-Publication Data is available.

ISBN 0–06–075049–9

Truthfully I may confess
that as often as I contemplate
the proper order, as one results from another
and becomes diminished,
it is as if
I have read a heavenly passage
not written in meaningful letters
but with the essential things in this world
which tells me: Put your reason herein
to comprehend these things.

JOHANNES KEPLER,
IN HIS CALENDAR FOR 1604

Contents

Foreword

JOHANNES KEPLER is most often remembered for his venerable three laws of planetary motion, for which he has earned the title "the father of celestial mechanics." But Kepler's accomplishments were wide-ranging. In addition to acclaim for his laws of planetary motion, he is also considered the founder of modern optics. He was the first to investigate the formation of pictures with a pinhole camera, the first to explain the process of vision by refraction within the eye, the first to formulate eyeglass designs for nearsightedness and farsightedness, and the first to explain the use of both eyes for depth perception, all of which he described in his book *Astronomiae Pars Optica*. In his book *Dioptrice* (a term coined by Kepler and still used today), he was the first to describe real, virtual, upright, and inverted images and magnification (and he created all those ray diagrams commonly used in today's optics textbooks), the first to explain the principles of how a telescope works, and the first to discover and describe the properties of total internal reflection. Galileo may have used the telescope that Johann Lippershey had invented to discover the moons of Jupiter and to see the first hints of Saturn's rings, but it was Kepler who explained how the telescope works.

Kepler probably was the first real astrophysicist, as we know the term in the modern sense, using physics to explain and interpret astronomical phenomena. He was the first to explain that the tides are caused by the moon. In his book *Astronomia Nova*, he was the first to suggest that the sun rotates about its axis. He wrote what may have been the first sci-fi story, a view of earth from the moon. In his book *Stereometrica Doliorum Vinariorum*, he developed methods for calculating the volume of irregular solids that became the basis of integral calculus. And he was the first to derive the birth year of Christ that is now universally accepted.

Today when we think about scientists, we have images of university professors in ivy-covered halls, laboratories full of elaborate instruments, cadres of graduate students, laptop computers more powerful than all those used to send men to the moon, and large government grants. Yet with all of these resources, how much of today's research will stand the test of time the way the works of Kepler have? Much has changed in four hundred years, but Kepler's laws are as exact now as they were then. How was he able to accomplish so much? Actually, given the times he lived in and the meager resources, how was he able to accomplish anything?

Context is the window to understanding. To understand Kepler and his accomplishments, one needs to understand the times in which he lived—the culture, people, places, politics, religion, and his family. Wars, witch hunts, pestilences, and death were common everyday occurrences. The potions that people used as cures were a far cry from today's medical miracle drugs. *Kepler's Witch* communicates a feeling for the hardships, difficulties, rejections, loneliness, and heartbreak that Kepler endured. He lived on the verge of poverty. His salary was almost always in arrears. His only resources were paper, pen, and one man's treasure trove of astronomical observations.

What drove Kepler? What sustained him? How could he endure? It certainly wasn't the money or even glory. He had few peers who even recognized his accomplishments. In *Kepler's Witch* readers get a feeling for the source of his strength, his vision, and his perseverance, how he was able to do so much with so little in spite of all the evils that surrounded him in life. Kepler believed in an almighty God, the creator of heaven and earth. He believed in Jesus Christ as personal Savior. He believed in the ideal of the holy catholic (universal) church. Kepler was a man of peace in search of harmony—in particular, the harmony of the heavens. Somehow he knew it was there, even if life on earth was far from harmonious.

David Koch
National Aeronautics and Space Administration
Deputy Principal Investigator for The Kepler Mission*

*The Kepler Mission is a special NASA space mission for detecting terrestrial planets—that is, rocky, earth-sized planets—around other stars. It is scheduled for launch in 2006.

With Thanks

A BOOK like this one is a river, drawing from many sources. I owe thanks to each one and will name them, in no particular order except as they come to mind. First, great thanks to Petra Sabin Jung, who helped with the difficult task of translating Kepler's German into modern English. Also, thanks to all my friends in Weil der Stadt: Dr. Manfred Fischer, the president of the *Kepler Gesellschaft*; Dr. Ernst Kühn, my sherpa guide up the steep slopes of Tübingen; Hubert and Elisabeth Bitzel, who were so kind and joined in my search for a Kepler sundae; Herr Bitzel, Hubert's brother, the curator of the Kepler Museum, and a great Kepler fan; Dr. Wolfgang Schütz, the curator of the Weil der Stadt city museum, the town historian, and a wonderful source of information. Also, thanks to all the shopkeepers in the *Marktplatz* of Weil, who gave me samples. Thanks also to Anna Madsen and her father, Dr. Madsen, a Lutheran minister from Wisconsin. I met them both in Regensburg, and we had a wonderful little "Americans in exile" moment in the foyer of the Kepler Museum. Thanks to the people at the Sweet Home Hotel in Prague, who helped me get things mailed.

Thanks also to my agent, Giles Anderson, of the Anderson Literary Agency; to my editors, Steve Hanselman and Mickey Maudlin, at Harper San Francisco; and to my publicist, Roger Freet—three men who made the Big Decisions. And thanks to Cindy DiTiberio, also of Harper San Francisco, who made all the other decisions. Thanks also to David Koch, Michael Gurian, Charles de Fanti and Leni Fuhrman, Malachy McCourt, John Anthony Connor, Margarette Alma Connor, William Craven, Sr., the New York Mets (for teaching me humility), and, of course, my wife, Beth Craven Connor, for teaching me everything else.

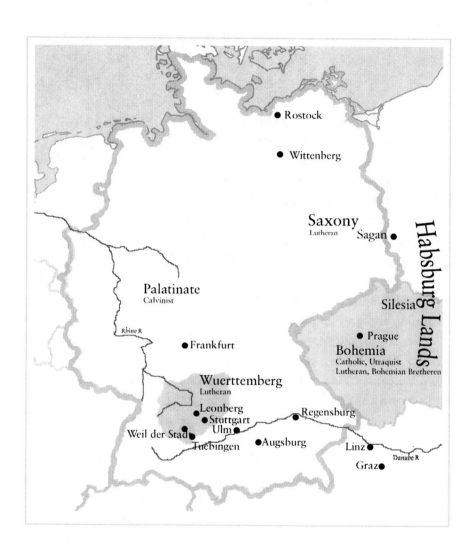

● Rostock

● Wittenberg

Saxony
Lutheran   ● Sagan

Habsburg Lands

Palatinate
Calvinist

Silesia

Rhine R

● Frankfurt

● Prague

Bohemia
Catholic, Utraquist
Lutheran, Bohemian Bretheren

Wuerttemberg
Lutheran

● Leonberg
● Stuttgart   ● Regensburg
Ulm ●

Weil der Stadt ●
Tuebingen   ● Augsburg   Linz ●
                                        Danube R
                                   Graz ●

Introduction: So Why Kepler?

AFTER HAULING MY STACK OF LUGGAGE down the platform, I finally came across the last unclaimed seat on the night train from Stuttgart to Prague. Since no one shooed me away, I heaved my luggage into the upper bins and collapsed into the seat. Beside me were two Italian men who pretended to be asleep. Opposite them by the window was a blond German woman with a sack lunch on her lap. Beside her, in the middle seat, was a Korean boy, taller and lankier than I expected. He was traveling around the world, and in his broken English he said he wanted to know everyone's story. Across from me was a short, unnaturally thin German student with a buzz cut and an excess of earrings, sitting wound into himself in a sort of existential fetal position. I was not surprised when he pulled out a packet of cigarette papers and rolled his own.

"You an American?" he asked.

"That's right."

"So what are you doing in Germany?" he went on, as if I alone had no right to be there.

"I'm writing a book about Johannes Kepler."

"What makes you think we want to know what you have to say about Kepler? You cowboy Americans and your cowboy wars. How many people have you killed this week?"

He watched me, waiting for me to bite. I wanted to explain to him about World War I and World War II, but I didn't think this was the time. "Well, we were attacked, you see," I said finally. "I was just across the river in Jersey at the time, so I saw it myself. So don't talk to me about our 'cowboy wars.'"

The temperature in the compartment dropped considerably. People fidgeted and looked at one another nervously, wondering if there was going to be a shouting match. Discussing American foreign policy is only pleasant under the most controlled conditions, especially since I didn't agree with the policy myself, though I wasn't going to tell him that. The student leaned back in his seat, muttered "cowboys," and pulled out a well-thumbed pocket edition of sayings from the Qur'an. For the next two hours, the trip rolled along pleasantly enough. The Korean student talked about his hometown and then all about the countries he had traveled through, counting them all on his fingers. The Italian men said they both came from Belluno, a city just north of Venice. The German woman spoke briefly, in German. One of the Italian men translated for the rest. She was on her way to visit her son, who was staying with his father, her estranged husband, who was in turn staying with his mother. Though they were not divorced, she worked and lived in Frankfurt, and her husband lived in a little town outside Berlin.

Suddenly, the student tucked his copy of the Qur'an back into his pocket and cocked his head at me.

"So why Kepler?" he said.

I looked at my shoes and thought about how to answer him. After a while, I looked up and said, "Because in 1620 Kepler's mother was being tried for witchcraft. Germany was well into the Thirty Years' War. Kepler had already lost his first wife and little boy to disease, and in the years following he lost three more children. In his adult life, he was chased out of one town after another by the Counter-Reformation. He was excommunicated by his own church. And yet, throughout most of these years he was writing a book called *The Harmony of the World*. This," I said, "is a man worth knowing."

The German student eyed me and sucked on his cigarette.

Since then, I have thought a good deal about this question. Great people show up now and then in this world. What makes them great is complicated. Some say Kepler was a genius, which he certainly was, but his scientific intelligence was not the source of his greatness. Johannes Kepler was one of the most powerful scientific minds of his century—he was an equal to Galileo in almost every way, a precursor to Newton, a man who had done the spadework for most of the important discoveries that defined science in the seventeenth century. And yet Kepler was also great in the way Gandhi and Martin Luther King, Jr., are great. He was a man who fought for peace and reconciliation between the Christian churches, even when it nearly cost him his life. Some people are born to greatness; some are made great by the events of their day. Some, such as King, Gandhi, and Kepler become great because they make choices full of moral courage. Kepler was a believing Lutheran and would never become a Catholic, even when it would have benefited his career to do so. People all around him were jumping from one church to another. Kepler's father-in-law did it. So did many of his acquaintances and rivals, simply to better their political or social position or not lose their earthly possessions. But Kepler believed in the Reformation; he believed in it with his soul. He stood fast with the Lutheran church, even when that church excommunicated him. When the Counter-Reformation chased the Lutherans out of their homes, he went with them, all the while fighting with the leaders of his own church in order to maintain the integrity of his conscience.

This book is a response to the question that the German student posed so succinctly—"Why Kepler?" Kepler is the man who finally confirmed Copernicus. He made a first, close attempt at defining a law of gravity. But above that, he was a man who contemplated in mathematics the glory of God. His life, his work, his mathematics were always about God. Everything he did was about God. Kepler found God in the hidden mathematical harmonies of the universe in as deep a way as he found God in the revelations of Scripture.

This is a man worth knowing. This book places him in his world, in his faith, in the events of his day. So why is Kepler not better known in our world? Everyone knows Galileo, even if all they know is that he was the guy who fought with the pope. Most people at least know the name Copernicus and that he had something to do with a revolution. They know Newton as the guy who had an apple fall on his head. But many Americans do not even know Kepler's name; it's as if he has been written out of the history of science. The National Air and Space Museum in Washington, D.C., has almost nothing about him. Why? Kepler was, as author Arthur Koestler called him, the "watershed" where the medieval world finally gave way to the modern. After Kepler's time, the scientific movement codified its method, largely following the lead of Newton's *Principia*. Newton, whether by accident or design, kept his own personal thoughts and mystical speculations, of which he had many, out of his scientific writing, a practice that later became the model for the scientific mind-set—distant, observing, uninvolved.

After the death of Tycho Brahe, Kepler and Galileo were the preeminent astronomers of their day. If Galileo was the great observer, Kepler was the great theorist, and yet their relationship was not overly cordial. I suspect that some of this came from the fact that Kepler was a Protestant and Galileo was a Catholic, and despite all the astronomical ties that bound them together, that religious difference kept them apart. Both men suffered for their religion. Both men helped to fashion the scientific world. And yet Galileo has come down to us through history as the martyr for science, while Kepler has been treated by some as a sort of embarrassment. This is largely, I believe, because Kepler did not wish to separate his science from his metaphysics or his metaphysics from his mysticism. He could not therefore fit the profile of the perfect scientist as it formed itself in the century after his death. Kepler did not have enough scientific cool— neither did Galileo or Newton for that matter, but when mythologies fashion themselves, such things don't matter.

This mythology that gathered around Newton and Galileo, this pseudo-history, pictured the scientific method, the method of the plodding empiricist, as free of metaphysical speculation and exploding from the heads of these two men, entire. But in truth, real history is always much more complicated than myth, and men like Kepler, sometimes forgotten, played a bigger role than the myth would allow. In my grumpier moments, I suspect that Galileo has been given his part in the myth because he fought with the pope, which made him a scientific Hercules. In my less grumpy moments, I suspect that there is some truth to the myth, because Kepler's mind, as it appeared in his work, was far-ranging. Was Kepler a scientist, a philosopher, or a theologian? The answer is yes to all three. Scientific work for Kepler was always grist for his theological mill, a chance to praise God.

In some ways, the whole problem comes down to Newton, who either by accident or by intent failed to give Kepler the credit he deserved. And some of Newton's own friends and supporters, including the Scottish astronomer David Gregory and the English astronomer Edmond Halley, of the comet fame, chided Newton for not giving Kepler proper recognition.[1] I suspect that Newton, the archegotist, in his darker moments knew quite well what he owed to Kepler, who had brought him right up to the doorstep of his theory of gravitation, who had laid the foundation for his work on optics, and who, as Leibniz recognized and Newton dismissed, had set the stage for the invention of calculus. He knew what he owed to Kepler but would not acknowledge it.

But this seems almost too appropriate to the rest of Kepler's life. He was a man caught between the grinding wheels of history, not only religious history, but scientific as well. In the last part of his life, he struggled through the first years of the Thirty Years' War, the war between Christians in which Reformers and Counter-Reformers tore at the body of the faith. Everyone suspected everyone, and Kepler, who would not abandon his own beliefs, suffered excommunication from his own Lutheran church on the basis of those suspicions. Sadly, much the same thing happened to him after his death at the hands of the scientific community.

If I have any mea culpas to make in this book, one is this—I did not try to give an account, except as a sketch, of Kepler's science. There are many great books about his science, and they are listed in the Source Readings. Read them, for they are more than worth the effort. This book, rather, is about Kepler's life, about his suffering and his triumphs. Perhaps if you read this book, knowing Kepler will make your own life work a little better.

I have alternated the chapters in *Kepler's Witch* with translations of Kepler's letters and journal entries. They tell the story, in Kepler's own words, of the crises he suffered. The best part about studying letters is that you find that great people in history are no longer legendary figures, but ordinary human beings caught in mundane torments. In studying Kepler, I found that all of his discoveries were made against deep opposition and were the result of tenacity. He never achieved anything easily.

The translations are keyed to the main events in Kepler's life. Some of these events are complex, for they span the length of Europe and sometimes cover a period of a century or more. This makes the translations crucial to the story. But more important, far too little of Kepler's writing has been translated into English. Mind you, a network of scholars has over the years translated most of his greater scientific works, usually from Latin, but the kitchen details, the facts of his daily existence, have been left out.

Still there have been some marvelous, informative biographies of Kepler, most notably the one by Max Caspar entitled, prosaically, *Kepler*. Every Kepler biographer owes mountains of gratitude to Caspar, and I am no exception. With the help of Martha List, Caspar collected and edited all of Kepler's sundry writings, from his scientific work to his letters to the account of his mother's witchcraft trial. This gigantic library-sized collection is still in print, and I bought a good chunk of it myself. It is called *Johannes Kepler Gesammelte Werke,* referred to in the notes as *GW.* As often as I could, I returned to the *GW*, ferreted out the original German, and translated it myself. But translating it into readable modern English was not always easy, for several reasons. Kepler wrote in both German and Latin. His Latin style was impeccable, but his German, though his native tongue, was not very good, often florid and overly ornate. In addition, he wrote in what is now called *Frühneuhochdeutsch,* a transitional form of German that is neither medieval nor modern, a form that evolved quickly into later variants that in turn evolved into modern German. It is only now being recovered by scholars.

All of this to dig out the life of a man worth knowing. The rest of that train trip, the German student and I talked about Kepler. When he left the train, he shook my hand and said, "Good day." A beginning.

LETTER FROM KEPLER TO THE SENATE OF LEONBERG
JANUARY 1, 1616

⟨∞⟩

Earnest, caring, wise, and especially benevolent Gentlemen, to
whom I am devoted to the best of my ability:
  I wish you a joyful new year.
  On December 29, with unspeakable sadness, I read a letter
my sister, Margaretha Binder, sent me dated October 22. As I
understand it, there is a case before you concerning several
people accused by the court, based purely on the imaginative
ranting of your dear darling housewife and sister, Ursula
Reinbold. Everyone knows that until this day, this woman has
lived frivolously, and now, according to you, she has become
mentally ill. Caught in the middle of this depressing web of
suspicion, my own dear mother, who has lived honorably into
her seventieth year, has been accused by you of giving this
same crazy person some silly magic potion, which you say
caused her insanity.
  But apparently even the dung heap of suspicion, slander,
and gossip that these people have been spreading around
town has not been enough. These same people have let them-
selves be blinded by the devil, the master of all misunderstand-
ing, superstition, and darkness, and have let themselves be
deceived. Forsaking God, they thought they could help their
dear darling kinswoman by enlisting the aid of the devil. They
forced my poor mother to perform some stupid, superstitious
magical ritual, a ritual that they would have you believe was
meant to assuage their fear and terror of the accused person,

7

namely, my poor mother. As you know, this use of witchcraft against witchcraft, a superstitious cure at best, is highly illegal. This is well known among the jurists, who widely reject it as a pactum tactitum cum diabolo, a tactical pact with the devil, and therefore an ungodly remedy. Indeed, many of you would consider it an indicium ad torturam, an indication for torture, if an accuser were to choose such a course and afterwards claim that the devil was conjured during this highly dangerous procedure against innocent persons, all because of this superstitious ritual.

I am writing not only in reaction to my sister's letter, but also in reaction to word from other trustworthy sources whose reports break my heart. My own dear mother, in her old age, has been more and more abandoned by time, and now they tell me she is to be dishonored and stripped of her possessions? Also, secretly, her son, my own brother, is to lose his estate as well? I have also heard that my mother was threatened with prison by the local constabulary and strongly urged by men who were armed when they questioned her, as if they meant to kill her on the spot, but still she resisted. Finally, these men used kind words and deceitful promises to persuade a poor old woman, making her think that nothing would happen to her. They used every other devilish trick you can imagine, demanding that she perform this forbidden ritual, supposedly to help the crazy woman, Ursula Reinbold, whom she did no harm to in the first place and could not help. From what I have heard, after she finally performed the ritual demanded of her and cured the woman, you claimed that this was done with the help of the devil's magic. So now, they say, my mother deserves the death penalty, when in fact the people who forced her into such actions deserve it more.

In summary, it is not surprising to me that this whole situation has left my mother feeling very afraid and unjustly attacked. All she wanted to do was to save her life, to still her grief, and to appease her accusers by giving them what they wanted, even though the blessed Almighty forbade this. Because of this, she was sentenced to torture by a judge who is not very knowledgeable in the law, and she could have suffered a terrible death. So, even when God has mercy on a sick woman and lets her regain her health, these devils would use this blessing to slander my mother's good and honest name as if they had driven a knife into her neck. And these same fiends would leave my poor, innocent mother with the feeling that they had actually helped her by first instilling fear in her, and then, by using the above mentioned superstitions, banned one devil with another.

Now as I understand from my sister's correspondence, both her landlord and her brother, who also endured this demonic injury and this terrible danger, are to be charged in a jury trial! This was a warning for me. Though it was carelessly mentioned in my sister's letter, other people have advised me as well that two legal actions have already been put forward. I do not know if my name appears on the list of the accused, or if they plan to include claims on my assets or income. I do not know if I will need to secure my holdings and my good name, in case these may be jeopardized, particularly if the accusers want to claim that I too practice the forbidden arts. I have no idea if this ridiculous situation, which has been blown all out of proportion, will also blow away my fifteen years of imperial service. This would break my mother's heart entirely (which is, of course, my chief concern, far more important to me than any of my personal sorrows).

Based on this and because of my dual interest, I address the honorable court with my well-founded request to immediately forward to me copies of all documents received by the court from both sides to date. My own courier will bring the documents to the Cantstatt Post Office, where they have orders to forward them to me in Prague, where, if my health allows, I will apply for permission to travel to my home.

In the meantime, I want to respectfully remind and request that right honorable, wise, and steadfast Gentlemen give my cause the attention it deserves, so that the law can follow its proper course with due process. Although these are my kin who are accused and not I (doubtless though we have the same name), I want to have my protest noted, and that nothing be omitted, so that my judgments about the case might be considered, granted, and approved or contractually engaged. I also desire that all judgments rendered against either party, to the limits of the jurisdiction of Leonberg, take account of my rights and obligations, which I have not surrendered. Rather, I maintain those lawful rights for the sake of my deserving, widowed mother, a law-abiding and commendable woman, and because of my desire to protect her and her assets. Let it be known that I will also seek the help of my friends and mentors, and that I will gain favors from such well-known and respected persons as I am acquainted with. I intend to contest this matter and bring to bear the full extent of my powers until it is finally remedied in accordance with the written laws.

Herewith, right honorable, provident, and wise, especially gracious Lords, putting myself and my kin into your protection and awaiting the necessary action.

Kepler's house on the Leonberg market square

I

With Unspeakable Sadness

*Where Kepler's mother, Katharina, is accused of
witchcraft by a former friend, which the gossip of
the townspeople whips into a fury against her.*

ON SEPTEMBER 28, 1620, the Feast of St. Wenceslas, the executioner
showed Katharina Kepler the instruments of torture, the pricking needles,
the rack, the branding irons.[1] Her son Johannes Kepler was nearby, fum-
ing, praying for it to be *over*. He was forty-nine and, with Galileo Galilei,
one of the greatest astronomers of the age—the emperor's mathematician,
the genius who had calculated the true orbits of the planets and revealed
the laws of optics to the world. Dukes listened to him. Barons asked his
advice. And yet when the town gossips of Leonberg set their will against
him, determined to take the life of his mother on trumped-up charges of
witchcraft, he could not stop them. Still, he never gave up trying, and in
that he was a good deal like his mother.

It was five years into the trial, and the difficult old woman would not
bend—she admitted nothing. Not surprising, for if truth be told, Katharina
Kepler was a stubborn, cranky, hickory stick of a woman who suffered
from insomnia, had an excess of curiosity, and simply couldn't keep her

nose out of other people's business. She was known to be *zänkisch*—quarrelsome—and nearly everyone said she had a wicked tongue. Perhaps that was why her old friends and neighbors were so willing to accuse her of witchcraft, why five years before they had forced her at sword point to perform an illegal magical ritual just to gather evidence that she was indeed a witch, and why they eventually handed her over to the magistrate for trial.

The ordeal consisted of two years of accusations and five years of court action, from 1613, when the accusations of handing out poison potions were first made, to 1620, when they convicted Katharina and sentenced her to the *territio verbalis,* the terrorization by word,[2] despite all Johannes could do. There were tidal forces at work in this little town. The events around the duchy of Württemberg would gather into themselves all the violent changes of the day, for by their conviction of Katharina, the consistory (the duke's council), the magistrates, and the Lutheran church authorities had bundled together their fear of Copernicus and their anger against Johannes, a man they had already convicted of heresy. The Reformation, like an earthquake, had cracked Western Christianity, stable since the fifth century, into Catholics and Protestants, and the Protestants into Lutherans, Zwinglians, Calvinists, Anglicans, and Anabaptists, with the many camps drifting apart like tectonic plates. Even the heavens had begun changing, and Kepler had been a part of that change. Copernicus, an obscure Polish priest, had published his *On the Revolutions of the Heavenly Spheres,* which had dethroned the earth from its place at the universe center and sent it spinning through the heavens like a top revolving around the sun. Fear ruled Europe—fear of difference, fear of change. And there, in one corner of Swabia in southern Germany, the mother of a famous man, a mathematician and scientist, a respected, pious Lutheran, nearly paid with her life.

Like his mother, Johannes was willing to fight. He had taken a hand in her defense, writing much of the brief himself. He was not present at the sentencing, though, for he would not have been permitted to accompany her to the *territio.* But only a few days before, Kepler had petitioned the *Vogt,* the magistrate, of Güglingen, the town where the trial had taken place, to get on with it, so when it was over old Katharina could finally have some peace.

Early that morning, she was led to the torturer by Aulber, the bailiff of Güglingen, who was accompanied by a scribe for recording her confession, and three court representatives. The torturer, with the bailiff standing to one side, then shouted at her for a long time, commanding her to repent and tell the truth and threatening her if she didn't. He showed her each instrument and described in detail all that it would do to her body— the prickers, the long needles for picking at the flesh; the hot irons for branding; the pincers for pulling and tearing at the body; the rack; the garrote; and the gallows for hanging, drawing, and quartering. He adjured her to repent, to confess her crimes, so that even if she would not survive in this world, she could at least go to God with a clear conscience.

Meanwhile Johannes, almost insane with rage and fear, waited in town for the ordeal to be over. Kepler was a slight man with a jaunty goatee and a dark suit with a starched ruff collar; he was slightly stooped from bending over his calculations and he squinted from bad eyesight, a parting shot from a childhood bout with smallpox. His hands were gnarled and ugly, again a result of the pox. Perhaps he paced as he waited for news, shook his fists at the empty room. Essentially a peaceful man, he was given to rages when he knew an injustice was being done. After all, these were his neighbors, his childhood friends, not strangers, who had forced this trial. The accusation, the trial, the conviction, and the sentence were all the work of hateful people, people who had wanted some petty vengeance, people who had seen their chance to get their hands on his mother's small estate. It was the work of a fraudulent magistrate, a good friend of the accusers, and of a judicial system gone mad.

Being imperial mathematician meant that the courts in Leonberg couldn't touch him, but they could do as they liked with his mother. Imperial protections went only so far. In the end, no mere scientist could expect that much security. Thirteen years later, the other great astronomer, Galileo, would face charges of heresy before the Inquisition in Rome. The executioners at that time would perform the same *territio* on him.

Such things happened all too often at that time, because people were afraid. In the seventeenth century, mystery tolled like a bell in people's lives, disturbing their dreams. They lived in fear of unseen forces and anything beyond their understanding terrorized them. Like her son Johannes, Katharina was more intelligent than most people, and so the way her mind ran in oblique ways set people's teeth on edge. Unlike her son, however, who had the best education Germany offered at the time, she was illiterate. Her mind, forever restless, had no proper expression. Her formidable intelligence had been stuffed back inside her, always moving, always seeking something, but with no way out. It is likely that she had been born the child of a rape. Unlike everyone else around her, she was short, thin, and dark, just like her son Johannes. Less than a year before she was born, the Spanish had invaded the region, and typical of armies, their soldiers had raped every woman they could find. Katharina's mother may have been one of them.

Katharina owned a house in Kirchgasse and some fields that she rented out to local farmers, which together provided for a modest living. She was always clever in the making of money, always finding new ways to increase her "little estate."³ At seventy-four, what she missed was a purpose to her life. Her children had grown up and moved away. Her husband, Heinrich, had deserted her years before, rushing off to fight one war after another until he died out there somewhere. She knew something of herbs, knowledge she had gained from an aunt who had raised her and who had later been burned as a witch. Understandably, Katharina wanted to make herself useful and so offered medicines, salves, and healing potions for the sick as well as herbal tonics for the healthy. She walked from farm to farm, speaking benedictions and offering medicines for both livestock and people:

Bid Welcome to God
Sun and Sunny Day.
You come riding along—
Here is a person,
Let us pray to you O God—

Father, Son and Holy Spirit,
And the Holy Trinity.
Give this person blood and flesh
And also good health.[4]

Her son Johannes lived with his wife and children in far-off Linz, in Upper Austria, where he lived in what everyone in her circle must have thought was imperial elegance. They had just had a new baby, and so there was joy in the house, even though one of the other babies had already died, the other was sick, and Kepler worried constantly about money. Little did anyone in Leonberg, Kepler's old neighborhood, understand, that for all his high position, Kepler had to scramble for money just like they did. The emperor owed him over 10,000 gulden—a fortune—in back pay, but like all the other court advisers, Kepler had to stand in line to get a few scraps of what was owed him. Only the emperor and his top lieutenants, it seemed, actually lived in imperial splendor.

As difficult a woman as Katharina could be, her neighbors often came to her with their medical problems—when the medicine worked, she was a hero; when it didn't, she was a villain. They assumed that she had intended it to fail, which meant that she was malevolent, a witch, like her aunt before her. But as suspect as Katharina was, to some her son Johannes was far worse. He must have been crazy, and a witch himself, a follower of Copernicus and a heretic. The small local court in Württemberg couldn't understand Kepler's science, so they supplemented their prejudices with church dogmas, both Lutheran and Catholic, to argue against his mother.

Kepler knew that the town of Leonberg had already executed six other women for witchcraft by that year—what had once been a vague anxiety over the spirits of darkness had boiled up into a national mania. Not that he would have disagreed with the idea that there were witches. No one would have. It was just that he was certain his mother was not one of them. Admittedly, the times were uncertain, the forces of darkness on the move. Germany was at the crumbling edge of the Thirty Years' War—Catholics, Lutherans, and Calvinists rampaged through the countryside on one wave or another of Reformation or Counter-Reformation. Jesuits were everywhere, whispering into the ears of kings. And, admittedly, Johannes Kepler had tried to bring the warring factions together, and for his pains was excommunicated by his own Lutheran church. Is it surprising then that, caught in these tidal forces and shackled by her neighbors' petty fears, Katharina Kepler ended up in prison for over a year, sometimes in chains, sometimes tortured, and almost lost her life on the gallows or at the stake?

Katharina's fate had finally been decided, not in Leonberg or Güglingen, but in Tübingen, at the famous university there, decided by the law faculty, who reviewed the court case and determined the sentence. This was Kepler's old university, where he had studied, but that had not made him their favorite. In fact, quite the opposite.

The year of Katharina Kepler's trial, the summer heat had not yet completely dissipated; the fall colors were appearing as the professors at Tübingen, meeting in solemn conclave, decided that the evidence against the Kepler woman was mostly circumstantial, and that they could not in good conscience condemn her to death or even actually torture her, though the law permitted them to do so. However, since she had been convicted and sentenced to the *territio* by the Duke of Württemberg himself, they could not in good conscience set a convicted witch free without punishment, even after the duke had, in his own uncertainty, asked them to review the case. Meanwhile, Kepler was writing letters and sending petitions to nearly everyone, trying to head off that dreadful day. Still, he failed.

But Katharina would not bend. Even after the executioner had done his worst, after he had shouted and commanded and adjured himself hoarse, after he had shown her all his tools and explained each one's purpose and had described how she would suffer most horribly under them if she did not confess her evil and renounce her lord, the devil, Katharina, unbowed, said: "Do what you want to me. Even if you were to pull one vein after another out of my body, I would have nothing to admit." Then she dropped to her knees and prayed a fervent Pater Noster. God, she said, would bring the truth to light, and after her death he would reveal the terrible violence that had been done to her. She knew that God would not call his Holy Ghost from her nor would he abandon her in her suffering.

⟨⟩

Article 6:
The witness heard the same story (about the potion) from the
accuser herself, because he had once been sick and stayed in
the same hospital as she did. There, the accuser, Ursula
Reinbold, asked him how he got well.

The witness did not respond to any of the interrogational
questions, except to say that after the glazier's wife had told
him how she had received a potion from the Kepler woman
and had tasted it, she said immediately, "Good Devil, what is
this?[1] *What did you give me to drink? It is as bitter as gall!"*
The witness did not respond to what Ursula said about this
drink the Kepler woman had given her, because she should
have known what kind of a woman Katharina was. The
Kepler woman later found out what the witness was saying
about her and sent Michel to him to ask why he had bad-
mouthed her, which the witness, however, did not admit. In-
stead, he asked Michel not to bother him with those sorts of
things any more. Soon after, the Kepler woman stopped the
witness on the market square and asked him personally if the
good Michel had said anything to him. If so, what was his an-
swer? He told her that she understood the good Michel cor-
rectly, that he had indeed said those things. Then the accused
admitted that, yes, she had given a drink to the glazier's wife,

but that she had mixed up the jugs. She had two little jugs sitting on the sill, and in one of them she had prepared a potion out of herbs. Witness admitted to the interrogator that he should have made this information known to the magistrate earlier. Even so, he wanted it noted that, because of the length of time, he might have forgotten several details.

Article 9:
Witness said that even if Mt. Engelberg were made of gold and was given to him, he would not speak a falsehood.

Testimony of Benedict Beutelsbacher, German Schoolmaster of Leonberg
1620

Article 6:
The witness (Beutelsbacher) said that the accuser, Ursula Reinbold, suffered great agony, especially when the moon changed. What the cause of this pain was, he did not know.

Article 21:
Several years ago, the Kepler woman often visited the witness, either to relay regards from her son, who lived in Linz and was once a schoolmate, or to ask him to read a letter for her and other things of that nature. But more important, once at the end of a long summer day, the witness came home after working in the field and locked the two doors of his house himself and therefore felt all was safe. Suddenly, as the witness was eating his supper, the Kepler woman came into the room to visit him and his wife, right through the locked doors! The two of them were startled and frightened. The Kepler woman asked him to write a letter for her to her son in Linz. The witness refused to do this with a variety of excuses, but finally, against his will, wrote the letter for her. He does not remem-

ber the content of the letter, which surprises him a great deal.
After all, the accused had come into his house with both doors
locked! Last, on a Sunday about ten years ago—the witness
cannot remember the specific date—he was called to her house
to read and write several letters. After finishing the task, be-
tween two and three o'clock in the afternoon, the witness
asked to leave to have a light meal. She told him no, that he
had gone to so much trouble for her, she wanted to offer him
something to drink. She had a rather good wine in her cellar,
and he had to try it. The witness refused repeatedly, because,
to tell the truth, he was not thirsty. Finally, against his will he
had to wait while she fetched the wine and then gave him
some in a pewter cup. He only tried a little bit. At this, the ac-
cused asked him why he drank so little. Then she handed an-
other cup to Margretta, Bastian Meyer's housewife, who was
also there at the time. The Kepler woman encouraged her to
drink up. She said that she knew quite well that often enough
not one drink of wine turns out good in an entire week, and
that a good drink of wine happens rarely, therefore they
should drink up, because this bottle of wine turned out well.
After that, the Kepler woman persuaded the witness into
drinking a little more and convinced the other woman,
Bastian Meyer's housewife, to empty the cup entirely. Very
soon after, Margretta began to feel ill. She never recovered and
ultimately died. The witness, however, experienced a slight
pain in his thigh, a pain that increased as time went on, so
that eventually he had to use a single cane at first, and after-
wards, two canes. These days the witness has pains in his
thighs all the time, so that he cannot move at all. The pain he
experienced lingered so long and grew so intense that his man-
hood was taken from him entirely. If it turned out that his in-
jury was not the result of the potion the Kepler woman
brewed up and gave him, the witness would die from shock.

The comet of 1577

II

Appeired a Terrible Comet

Where Kepler is born in Weil der Stadt, near
Leonberg, including a description of the town, the
Kepler family, and Johannes's early childhood.

KATHARINA KEPLER GAVE BIRTH to Johannes, her first child, on December 27, 1571. It was the middle of the afternoon, two-thirty precisely, on a Thursday, the feast of St. John the Baptist—a comfortable omen, for he was born as the sun was still high in the sky. At the time, Katharina and her husband lived with his parents in Weil der Stadt, one of the smaller *freie Reichsstädte,* the free imperial cities, which owed their allegiance to the emperor himself, rather than to a local duke or prince. Weil was a small town even by sixteenth-century standards, located ten miles to the south of Leonberg and about twenty miles southwest of Stuttgart.

Germany was not a separate nation then, but part of the Holy Roman Empire, which included modern-day Germany, Austria, Hungary, Slovakia, and the Republic of Czechoslovakia with bits and pieces of other countries thrown in. At the time, it consisted of a series of little principalities, duchies, counties, and baronies, ruled by princes, dukes, counts, and barons, all of whom had the right to rule autonomously in their own territories, to determine the religion of their people, and to determine the form of governance. The emperor, traditionally with no fixed land of his own, traveled about the empire, taking his court with him and staying in the imperial cities as he traveled. The many dukes and counts and princes were loyal to him after a fashion. They supplied his taxes and his armies and supported him in other ways when they saw fit. For this, they received rewards, and some even possessed the coveted right to vote on the next emperor, which made them electors.

As an imperial free city, Weil der Stadt had certain advantages in terms of trade and taxes. Citizens of a free city were usually much better off than subjects of a duke or baron. The local ruler usually lived close by and tightly managed his people's lives, while the emperor was far away and most often left the free cities to their own devices. Imperial Weil der Stadt was loyal to the Catholic Habsburgs, the imperial family, and was therefore Catholic. All around Weil der Stadt, including nearby Leonberg, the town that Kepler grew up in and considered his home, was the Protestant duchy of Württemberg, the most contentedly Lutheran territory in the empire and a central place from which Lutheranism had spread. After the Peace of Augsburg, signed in 1555, only sixteen years before Kepler's birth, the religion of the ruler determined the religion of the people according to the formula *cuius regio, eius religio* ("whose the land, his the religion"); therefore because the emperor was a Catholic, so was Weil der Stadt. As the Reformation heated up, partly out of religious zeal and partly out of pique, the Duke of Württemberg put increasing economic pressure on Weil. But then, as the Counter-Reformation gathered strength in return, anyone who didn't want to be Catholic had to leave the city, and no one was excluded. Soon the town became a tiny island of Catholicism in Lutheran Württemberg, which caused no end of trouble.

Johannes's family was Lutheran—completely, utterly, and without reservation—the most prominent Protestants in Weil der Stadt. They were a pious family, but troubled, and sometimes their piety got in the way of their Christianity. The Kepler family had lived in Weil der Stadt for about fifty years, in a narrow house tucked into one corner of the market square, next to the city hall. His family had been the first to become Lutherans in the town, but they were respected enough that even becoming Protestants did not affect their standing. Catty-corner to Kepler's grandfather's house, now the Kepler Museum, was the Gasthaus zum Engel, the Inn of the Angel, also owned by Johannes's grandfather Sebald. He was an overbearing old patriarch with a red face that ran to fat, a quick temper, and a dignified beard that made him look important. He controlled just about everything in the lives of the people around him, starting with his family. Johannes's grandmother, another Katharina, was much the same as his grandfather—quarrelsome, intensely pious, restless. She rarely forgot old wounds. Kepler describes her as "fidgety, clever, given to telling lies, but pious in all matters of religion. She was slender, fiery, lively, always moving, jealous, vengeful, and full of resentments."[2]

Kepler's horoscope and his memoir are the only traces left of his youth and family life, and the dark picture he painted there is accurate enough in the details, though the melancholy tone of it can be partially understood as the sadness of youth and a bright young man's jaundiced view of his family. No one knows what his mother thought of his success or fame, for her thoughts were never written down, and time has washed them away. She was not a particularly good mother, and more than likely she never understood his life and his love of the heavens, though in many ways she was responsible for both. In 1577, a great comet haunted the sky, one of the most brilliant ever seen, which was observed and described by astronomers from England to Japan. Tycho Brahe first noticed it while fishing. In Scotland, James Melvill described it this way:

This yeir, in the winter, appeired a terrible Comet, the stern [star, i.e., nucleus] wharof was verie grait, and proceiding from it toward the est a lang teall. In appeirance, of an ell and a halff, like unto a bissom or scurge maid of wands all fyrie. It rease nightlie in the south weast, nocht above a degree and a halff ascending above the horizon, and continowed about a sax oukes [weeks], or two moneth, and piece and piece weir away. The graittest effects wharof that out of our contrey we hard was a grait and mightie battell in Barbaria in Afric, wharin thrie kings war slean, with a hudge multitud of peiple. And within the contrey, the chasing away of the Hamiltones.[3]

Of all the great minds to see this comet, Johannes Kepler, then six years old, was one. His mother took him by the hand and led him up a hill out-side of town where the two of them watched through the long evening. Tycho Brahe, whose life would cross so importantly with Kepler's, watched the same scene from beside his personal fish pond five hundred miles away. For Kepler, it was one of the few pleasant memories he had of his mother.

Johannes describes his father, Heinrich, as a rough and beastly man, "a vicious man, with an inflexible nature, a quarrelsome man, who was doomed to a bad end," a soldier to his bones, perhaps the resurrected image of the long-dead family knights. The urge to go to war was strong in him, and he spent his adult life sniffing for battle. Kepler blames this on the planets, saying that Saturn was in trigon to Mars in the seventh house, which of course explained everything.

Nevertheless, Heinrich almost certainly abused his wife and perhaps his children as well, and once tried to sell his sickly young son Heinrich, his own namesake and one of Johannes's younger brothers, into servitude. An adventurer, he had no skills to speak of except gunnery, which he had picked up on one of his soldiering forays. Once he got into a fistfight in Weil der Stadt and had to pay a fine. Another time he lost an inn he owned because he got into a brawl. He fought in several wars, once in Belgium, in the pay of the emperor fighting against Protestant rebels, which did not endear him to his family. When the urge came on him he would disappear off to some war; he went twice to the Netherlands to fight as a mercenary, the first time in the pay of the emperor and the sec-ond time in the pay of the Duke of Alba.

On one such occasion, he was gone so long that when Johannes was merely three years old and baby Heinrich was merely an infant, Katharina left them with their grandparents and set off after him. Just as she was leaving, Johannes took sick with smallpox and nearly died. Katharina, intent on finding her husband, left anyway, handing care of Johannes over to her in-laws, who wanted nothing to do with him. Angry with their son Heinrich for running away and with Katharina for dropping their sick son on them, they tended the boy without much enthusiasm. Surprisingly, the boy recovered, but his health was shattered. The pox had weakened his eyesight, and for years he suffered from sores, scabs, and putrid wounds, possibly because his immune system had been damaged. His hands were also deformed and he moved in a clumsy, jerky way, as if the virus had also affected his nervous system. Like his younger brother Heinrich, he was accident-prone.

Nevertheless, Katharina found her husband. One can imagine the meeting: Heinrich the runaway, in his cups or at camp staring dully at a plate of stew, looking up and seeing Katharina, his partner in a loveless marriage, marching at him down the row of tents, so furious that electricity sparked around her head. If he could have run, he would have, but there was nothing else to do but follow her home. Finally, after moving his family to Leonberg and after trying his hand at innkeeping in Ellmendingen, he forced his family back to Leonberg and then disappeared altogether. Some said he died in Augsburg after fighting for the Neapolitan navy.

Johannes's mother, Katharina, née Guldenmann, was as restless as her husband. There were problems in her marriage right from the beginning. Because Katharina gave birth to Johannes only seven months after her wedding to Heinrich, all the old women in town were busy counting the months on their fingers. Certainly it is possible Johannes was born prematurely, for he was small and sickly most of his life. Kepler no doubt believed this himself. Still, one has doubts, because while Katharina was pregnant, she was regularly beaten by her parents, as if they were trying to make her lose the baby, though both mother and son survived.[4] The image of a hurried, forced marriage between an angry Heinrich, a military straight who was cold and distant, and a pregnant Katharina explains a great deal about the family's history.

Only a mother whose eccentricities hid a vast intelligence could imagine turning her own father's skull into a drinking cup just because she had heard it was an ancient pagan custom. She possessed the kind of intelligence that could either blossom into genius, as with Johannes, or fester into madness. All her life she struggled against her illiteracy. In that time, few women could read, and through the years Katharina felt humiliated that she could not even read her son's letters and was forced to rely on Beutelsbacher, the schoolteacher, who would later turn against her in her witch trial. To make up for it, she collected herbs and mixed potions from them, and it is possible that she did in fact poison Beutelsbacher and the wife of Bastian Meyer, but almost certainly without meaning to. Her little tin jug often sat in the corner for days, unwashed and untended. Who knows what kinds of bacteria were growing in there?

Her children were a mixed lot, however. And although Johannes's youngest brother, Christoph, who became a pewterer, and his sister, Margaretha, who married a clergyman, turned out relatively well, the middle brother, Heinrich, teetered on the edge of insanity. Possibly a borderline schizophrenic, he was accident-prone, was constantly being beaten by other children and bitten by animals, nearly drowned, and was almost burned alive. Eventually he wandered off when his father tried to sell him, only to show up in Prague as a palace guard when Johannes was there as imperial mathematician, and then to return to live with his mother years later, much abused by life.

The Keplers, all boxed into that little house, argued eternally. They were not poor, however, and they could even claim some echoes of nobility. The angel in the name of the inn referred to the Kepler coat of arms, which may or may not have been a last remnant of some long-lost nobility. Grandfather Sebald acted as if there were no question of his knightly roots, since he could trace them back to Sebald Kepner, a nobleman who became a bookbinder after a sudden descent into poverty and then moved to Weil der Stadt. Four generations later, the Sebald who was the grandfather of Johannes Kepler the astronomer, no longer obviously noble, ended up as the mayor of Weil der Stadt. He served for ten years before Johannes came into the world and remained a prominent burgher in town throughout the boy's youth.

Towns like Weil could easily maintain their independence simply by paying their taxes and helping the emperor out when he needed them. The emperor asked little more than this, because the *freie Reichsstädte* were not his property, but were individual republics, each with its own representative in the Imperial Diet. Weil had its own magistrate, either the mayor of that year or the one of the previous year; its own police force; its own laws; and its own justice system.

The mayor, in his day old Sebald, reported directly to the emperor, wherever he happened to be, and thus had direct contact with the imperial court. The republics all had trading rights and privileges and enjoyed economic benefits stemming from the fact that the town paid no tithes to the local prince or duke, but only the standard imperial tax. For Americans, this would be like paying federal, but no state income tax. The mayor of Weil was elected by a select group of the people. Only burghers (citizens, men with land), merchants, and craftsmen—goldsmiths and tinsmiths— could vote. The men who sat on the town council were those with enough money and leisure time to concern themselves with government. These were the people who produced Johannes Kepler, and this was the town that fashioned the character of his family and, through them, his own character as well. What feistiness he showed in his later years derived at least partly from that soil. What genius he showed derived, at least partly, from there as well.

Because of its position in the empire, there was a great sense of independence in Weil der Stadt. But this feeling of republican independence flowered at a time when the world itself was changing, a time when people in Germany had begun to ask for a deeper understanding of the world, a deeper understanding of the Word of God, the book of Scripture, as well as a deeper understanding of creation and the natural world in general, the Book of Nature. Printers had set up shop in various towns around Swabia, so books in German were becoming more available. Soon after the Reformation began, Luther's Bible was everywhere. A new consciousness had begun to emerge in the towns, a consciousness that called for independent thought, a consciousness that would eventually lead from the Reformation to the first glimmers of the scientific revolution. The citizens of Weil der Stadt were more a part of this than they knew.

ᏈᎳᎵᎣ

*This person was born with the destiny to spend most of his
time working on the difficult things others shirk from. As a
boy, before his time, he already studied prosody and poetic
meter. He attempted to write comedies and chose the lengthi-
est psalms to commit to memory. He tried to learn by heart all
the examples in Crusius's grammar book. In his poems, he ini-
tially troubled himself with acrostics, griphens, and anagrams.*

*Later, however, when his growing judgment let him dis-
dainfully see the true meaning of such things, he tried even
more difficult forms of lyric. He once wrote a Pindaric melos,
a Greek chorus. Another time, he became interested in unusual subjects, like the immobility of the sun, the creation of
rivers, and watching the fog from Mt. Atlas. He enjoyed
riddles and searched for the most acrid jokes. He played with
allegory by following the strands through to the smallest detail and then pulling them out by the hairs. When imitating,
he sought to stick with the exact text and then to apply it to
his own material.*

*When writing down his work, he liked paradoxical state-
ments; for example, that one should learn French rather than
Greek. As an opponent [in a debate], he never said anything
he did not mean. When he wrote down his ideas, the final version always contained something other than the draft. But
more than all other studies, he loved mathematics.*

*In every type of learning he immersed himself by challeng-
ing each idea, and he critically interpreted everything he read.*

So he held on to insignificant notes that he wrote himself and stubbornly kept borrowed books, as if they could be of use to him at a later time.

It was unbearable for him to let even a short period of time pass unproductively; despite his strong desire for human company, he stayed away from it. In monetary matters stubborn, when budgeting tough, critically pursuing petty details, all things with which he wasted time. At the same time he is not opposed to work, so much so, that solely the desire for knowledge keeps him at it. And still there are all the beautiful things he aspired to, and in most cases he grasped the truth.

Mercury in the seventh house means haste and an aversion to work, because he is also swift; the sun in Saturn's sixth means conscientiousness and perseverance. These two things are in conflict: to continuously feel regret about lost time while still willingly losing it again and again. Because Mercury affects a tendency for play and fun, this person enjoys the spirit of lighter things. As a child, he was devoted to play; as he grew older he found enjoyment in other things, and he therefore turned to other things; to find out what brings a person joy, therefore, remains subject to opinion. Since being stingy with money deters one from play, he often plays by himself. One has to note the following here: holding on to money does not have the goal of wealth, but rather the alleviation of the fear of poverty. Of course, most greediness grows out of unfounded worry. Or perhaps not; rather, the love of money possesses many. His eyes are fixed upon gain and reputation. Perhaps it is the fear of poverty that can be blamed for much. Because he is presumptuous and contemptuous of mass opinion, he tends to be hard.

By nature he is very well suited for pretense of all kinds. There is also a tendency toward disguise, deceit, and lies. It has its root in common with the jokes and jest. Mercury does this, instigated by Mars. But one thing prevented these dis-

guises: The fear for his reputation. Because foremost he yearns for true recognition, and every type of defamation is unbearable to him. He would pay very good money to buy himself free of even harmless, but wicked gossip, and poverty frightens him only because of the shame.

The astronomical clock in Prague's Old Town Square

III

Born with a Destiny

Where Kepler receives his education
as a scholarship student under the care
of the Duke of Württemberg.

IN KEPLER'S OWN WORDS:

From the beginning of his life, this person [the subject of this horoscope, Kepler himself] had enemies.[1] The first I can remember was
Holp. The rest, indeed, were all my fellow pupils, especially Molitor
and Wieland. In Maulbronn and in Tübingen it was Köllin; in
Bebenhausen it was Braunbaum; and in Maulbronn it was Ziegel-
heuser. I am listing only the long-term enemies, of course. In Tübingen, it was Huldenreich, Seiffer, and Ortolph, while in Adelberg it
was Lendlin; in Maulbronn it was Spannenberg, while in Tübingen
it was Kleber; in Maulbronn it was Rebstock and Husel, while in
Tübingen it was Dauber, Lorhard, Jaeger, a relative of mine, Joh,
Regius, Murr, Speidel, Zeiler, another Joh, and Molitor, the brother,
and An. Krell, the father-in-law [presumably of Molitor]. Mostly,

35

these were people his own age. Some of the others not his own age were merely casual acquaintances.

This person harbors dark thoughts about his enemies. And why would he do this? Could it be because his enemies compete with him for industry, success, distinction, and fortune? Or could it be because the Sun and Mercury are in the seventh house?[2]

The fact that Kepler listed his enemies but not his friends is telling. In his self-study, which is often brutal, he confesses to a difficult personality. Like a monk cruciform on the chapel floor, he lists his faults. He says that he is deceitful, overanxious to please, yet quick-tempered. "This person has the nature of a dog. He is just like a spoiled little pet. . . . He likes to gnaw on bones and chew on hard crusts of bread. He is voracious, without discipline. When something is put before him, he snatches it up."

Kepler was an odd boy, intense yet withdrawn. He perceived with the clarity of a child that he was not wanted in his family—his grandparents, saddled with a sickly grandchild by their own irresponsible son, had treated him roughly; his mother had been cold all his life; his father was distant and brutal. He defended himself when he could and tried to appease the great adult powers when he could not. His only comfort was in his own mind and in his thoughts of the Lutheran religion, which he had picked up from his grandmother. He was an intensely pious boy, and at times his piety twisted him about.

Perhaps it was the discomfort he felt in his own life or the sense of responsibility he felt for his intelligence, but he spent much of his youth searching his soul for forgotten sins. When he found one—a vengeful thought, a moment of uncharity, an error in his thinking—he assigned himself a penance. Once while at school he fell asleep and missed the evening prayer, and the next morning he assigned himself the task of repeating a number of sermons he had heard over and over, as if that would somehow placate God. Nevertheless, as hard as he tried, he was certain that he would have otherwise received the gift of prophecy, had it not been for his wicked life, his worldliness, and his unremembered sins. If only he could be a better person. If only he could be a saint.

By the age of ten, Johannes had learned to read Bible stories. One of his favorites was the story of Jacob and Rebecca, for he saw in them an example of perfect love, and with the fervor that only a child could maintain for long he resolved that one day, should he ever find his own Rebecca, the two of them would practice the perfect life together by strictly following the Mosaic law. Never mind that he wasn't Jewish, and never mind that the only girls he was likely to meet weren't Jewish either.

He was, from his earliest days, at war both with the world and with himself. Sometimes, in order to fit in, he allowed himself to be seduced to evil. He once joined in the general hatred of a boy named Seiffer because everyone else hated him too. At the urging of his teachers, he turned informer. In return, the boys insulted him because of his father's reputation. He had a wicked sense of humor, one that was often misunderstood and often kindled smoldering hatred among his acquaintances, hatred that lasted for years. Worst of all, many of his classmates envied his industry and his success.

He was a small boy—sallow-faced with a dark glance. On top of that, he was often sick, sometimes with real illnesses, sometimes with imagined ones. Nevertheless, he threw himself into playing games. He would fight when called upon, though he was often beaten for it, and if he believed that he was in the right, which was most of the time, he would never give in. This last trait followed him throughout his life, and although it helped him to survive the troubles of the Thirty Years' War, it brought him endless grief with his own Lutheran church.

What Kepler endured, however, was not only a product of his personality. His time on earth boiled with struggle, not only for him, but also for the Holy Roman Empire. Dark change was in the air. When Kepler was born in 1571, his parents had him baptized in the local Catholic church, Sts. Peter and Paul. Yet, his family were Lutherans, and Johannes was therefore raised a Lutheran. So, on the ground at least, where ordinary people lived out their lives, the membrane between Lutherans and Catholics was still permeable. For example, in 1535, when the local Catholic priest married, the town council didn't object, but by 1572, the year after Kepler's birth, when a second priest married, the town council dismissed him. The gulf between Lutherans and Catholics was widening every day. Eventually, a Lutheran child could not have been baptized in a Catholic church and would have had to travel to Leonberg for the sacrament. By the time Johannes was a young man, the easy relationship between believers in the new faith and believers in the old had become nearly impossible.

Soon after the Keplers moved to Leonberg, they enrolled Johannes at the German school, but throughout the year there were always interruptions because Heinrich kept leaving to fight in a war and the family had to make do without him. In 1578, because of Johannes's obvious intelligence, the teachers at the German school convinced Katharina to send him on to the Latin school, where he could learn to be a scholar or, better yet, a Lutheran pastor. By the end of that year, however, when Kepler was only six years old, his father bought an inn at Ellmendingen, near Pforzheim, and moved the whole family out there. The inn wasn't very successful, because Heinrich didn't really want to be there. Instead, an unhappy man, he drank, beat his wife, trumpeted around the house, and threatened his children. Eventually, running short of money, Heinrich sent Johannes out to work on a farm, something he wasn't suited for, so that his education in the Latin school took him five years rather than three. In the meantime, the boy was always coming down with something.

At the Latin school, Kepler used Philipp Melanchthon's grammar book as his guide. Melanchthon had been Luther's chief collaborator in the Reformation, the man responsible for the humanistic voice of Lutheranism and the founder of Lutheran education. Part of studying Latin meant studying the classics: Cato, parts of Cicero's letters, and the comedies of Terence.[3] Every day there were times for prayer and studying the Lutheran catechism. On Sundays Johannes went to church with his classmates and sang in the choir. During either the third or the fourth year, those students singled out by their teachers as likely candidates studied for the *Landesexamen,* a sort of standardized test, in Stuttgart. They went not only with good grades, but also with letters of recommendation from both their pastor and their schoolmaster, speaking of their good qualities in writing and erudition, their high level of intelligence, and their bright Christian character. The master of the *Pädagodium*, along with one of the teachers from the school and several church leaders, gave the examination. Each step was therefore carefully watched over by representatives of the church.[4] After they accepted Kepler into the scholarship system, he swore an oath before God that he would follow the rules of the monastery schools to which he would be going, that he would continue his education in theology at the *Stift* in Tübingen and would complete his studies there, and that he would serve the duke for the rest of his life or as long as the duke desired his services.

On May 17, 1583, Johannes traveled to Stuttgart to take the *Landesexamen.* He was eleven years old at the time, and his score on the examination was so high it gave him a place in the duke's scholarship system even though he came from an undistinguished family. As soon as his parents gave permission, the Duke of Württemberg assumed complete financial responsibility for young Johannes. The duchy would until the end of his school days supply his tuition, food, and clothing, and although he could never leave the system without repaying the duchy for his study, he could be expelled for misconduct or theologically suspicious beliefs.

Thirteen months later, after the failure of his inn, Heinrich dragged the family back to Leonberg, where Johannes could complete his Latin school education, and where Katharina gave birth to Kepler's sister, Margaretha, on June 26. Margaretha would become the only one of his siblings that Johannes truly cared for.

It was in Leonberg that Kepler's intellectual personality took shape. If any such emergence could be pinpointed to one day, it would have to be that Sunday when Kepler, aged twelve, heard a fresh young deacon in his church preach a long, violent sermon against the Calvinists. The growing dissension between the Christian denominations had been preying on Kepler's mind for some time, though it hadn't come to full consciousness until that day. It troubled him terribly that Christians could be so vicious to one another. From that point on, he questioned what the preachers said and resolved to test them. Whenever one glossed a biblical text in a way that he found troubling, he pulled himself off to a corner with the Luther Bible and, in good Lutheran fashion, consulted the text himself. With time, he realized that the very people that the preachers were attacking so vehemently had their good points too.

To understand Kepler the man, the philosopher, the scientist, however, one must first understand Kepler the Lutheran. The new faith was in his marrow, and all his science was at heart a prayer. In his childhood and through his school days, Lutheranism surrounded him, even in Catholic Weil der Stadt, largely because of old Sebald and his grandmother Katharina. Although his parents might possibly have baptized him in the old faith in the baptistery of the Catholic parish of Sts. Peter and Paul, he spent his life utterly fixed in the new one. It is important to remember, however, that by the time that Kepler was in school, Württemberg Lutheranism had become profoundly conservative, as conservative in its own way as Roman Catholicism, and in spite of its position as the new way of Christianity, it was closer to what we would call fundamentalism than it was to liberalism. This would cause Kepler no end of trouble in the years to come. But in spite of everything, Lutheran he was, and Lutheran he would remain, even when his livelihood and his career would have been greatly enhanced by conversion to Catholicism.

Once in Leonberg, he bathed in Lutheranism—in church, in his home, even on the street. Street vendors, innkeepers, soldiers, old widows, and laughing children all daily peppered their speech with blessings and curses, calling on God to support, to witness, to punish. It is impossible for us to understand a world so permeated with religion, for the daily life of ordinary people was forever filled with God—and the devil. Good and evil in a harvest dance, with human souls caught between.

Grounding this was the Bible, newly available to all. For Luther, God's divine law was revealed only in Scripture and not through observation of either nature or society. Although creation was for Luther the most beautiful work of all, where the Creator had left vestiges of the divine like broken twigs along a forest path, the study of nature could never reveal the wholeness of truth. Only the Word of God could do this, for human reason alone did not have the power. Astronomy must therefore be separate from theology, for human reason could never come to understand God's will for the world. *Sola scriptura*—the Scriptures alone reveal the mind of God and should alone guide the affairs of human beings.

On the other hand, Philipp Melanchthon, one of Luther's counterparts in the Reformation in Wittenberg, saw a strong connection between natural law and moral law and maintained that an understanding of nature could indeed inform our understanding of God and creation. As the reformer of education in the new faith, Melanchthon saw philosophy as essential to a trained Christian mind and mathematics as essential to philosophy. Like Plato, he set forth mathematics as the preeminent subject in the philosophy curriculum, for not only did mathematics have practical importance in people's lives, but it also refined the mind in logic. Arithmetic and geometry, Melanchthon said, reveal the order of things and teach the mind to recognize and separate things, for the "soul is a reasoning being which understands things and observes order."

For Melanchthon, as for Kepler, the order of the world was a shadow of the mind of God. The human mind vibrated with this order, felt it, and reacted to it. In Scholastic language, the mind was co-natural with the order of the world—they were made for each other and could be tuned to each other like radios. Philosophy was essential, because it trained the mind to this divine order, and astronomy was central to philosophy, because the most perfect order, the only true harmony, hummed in the celestial spheres. These heavens, this order, had been established by God to bring human beings to perfection. Just as the heavens move in perfect circles, gliding through the night sky, stately, unhurried, noble, like kings in procession, so the trained mind moves in rational patterns, argument to argument, from truth to greater truth, even to the greatest truth of all, the truth of God. Of all the mathematical sciences, astronomy was the most important. The whole of mathematics, arithmetic as well as geometry, was important for Melanchthon only because it revealed the order of the heavens.

Thus, the heavens contain a celestial light that the ordered mind contemplates and, in doing so, plumbs the secret places of God, revealing even the Creator to the observing mind, for only by design, a divine design, could such heavenly order come about. And this design is not only wonderful in itself, but also useful to human beings, for the order of the heavens is also a moral lesson. Its perfection, regularity, and harmony teach the divine law of righteousness to the human soul, fallen and beset by chaos. Nothing in the heavens occurred by chance for Melanchthon. It is God's will that human beings know the divine, and the study of the perfection of the sky is one of the ways into that knowledge, a knowledge and a truth that will set you free. Those who believe that the heavens exist only by chance wage war against the human soul.

For Melanchthon, unlike Luther, the order of the heavens revealed God's mind, both as Creator and as Father of the human race. The movements of the sun and moon, stars and planets are used to regulate human action, setting forth times for planting and for harvesting, times to buy and sell, and times for rest. By ordering human lives, God reveals the truth—that order is of God and that chaos is of the Evil One. Those who stand on the side of the angels support God's order on earth as it is revealed in the heavens. The stately movement of the celestial spheres, therefore, becomes the template for human morality.

The stars also carry prophecy, portents and signs of events to come, most commonly disasters. God reveals the future in the stars, for the stars reveal the mind of God. Events in this world come from the movements of the heavens as effect follows cause. Thus, Melanchthon, like most thinkers of his day, trusted in the astrological sciences, for the stars in their perfection were closer to God, to the First Mover of Aristotle, who moves all things without Himself being moved. Therefore, a person can read the stars and predict the future, at least in a general way. Nevertheless, not all readings of the heavens are equally valid. Any attempt to observe the sky in order to predict specific events or to read the particulars of the future was superstition for Melanchthon. True astrology is the science of the subtle influences that the stars have over the inclinations of human souls and human societies.[5] This was not too far from Kepler's own view, in which the stars helped to shape the general flow of the world, its tendencies and its limitations, but did not control individual events. (Contemporary science no longer believes this of the stars, but still there are influences. Our own genes have taken the place of

Melanchthon's and Kepler's stars, and the inclinations of our hearts are influenced by strings of proteins.)

It is this embrace of astrology more than anything else that puts Kepler at a distance from our age. Astrology lost its credibility in the century after Kepler, when Newton, who still believed ferociously in alchemy, abandoned it for a more thoroughgoing mechanical cosmology. What Newton did not acknowledge, however, for which he was chided by other astronomers in his time, was that this mechanical insight had first been Kepler's. In Kepler's own day, however, astrology was still queen and the larger towns all across Europe, from Stuttgart to Leonberg, from Tübingen to Prague all built astronomical clocks, not to tick off the seconds of a person's life—*Click!* Then gone forever—but to map out the heavenly realms, to give insight into the events of the day. Kepler's attachment to astrology followed Melanchthon's and was Lutheran to the bone. Kepler's warnings about using astrological predictions as a reliable guide mirrored Melanchthon's concerns.

Oddly enough, this ease with astrology puts the sensibilities of the seventeenth century at some distance even from those of modern Christians. Although liberal Christians follow Newton and wave off astrology as empty pseudo-science, conservative and fundamentalist Christians fear it as a manifestation of the occult, as dark superstition. How different this is from the ideas held by the founders of Protestantism and seventeenth-century Christians in general, who saw the stars and their movements as divinely created sacraments, windows into the mind of God. Still, the stars were incomplete windows. Astronomy can reveal the order of God built into the world and open the eyes of human beings to God's good government, but it cannot achieve the revelation of Christ and the story of the salvation of humanity. Astronomy can make visible something of the intentions God had for the world, but not the story of God's relationship with the human race. The mind that shaped the world has left its imprint in the world and has given us access to it. Partly, according to Melanchthon, this is because the human mind originated in the heavens and is a direct creation of God. But the human mind has fallen, been corrupted by sin, and the story of God's generosity cannot, therefore, be written in the stars.

43

On October 16, 1584, after Kepler graduated from the Latin school, his parents sent him off to the lower seminary, the school at Adelberg, once a monastery of the Premonstratensian Order near Mt. Hohenstaufen. Two years later he went on to the upper seminary in the Cistercian abbey at Maulbronn. The children of the wealthy who attended the Latin school afterward studied at the *Pädagodium,* a college in Stuttgart or in Tübingen, which was meant to prepare them for entry into the university. The monastery schools, on the other hand, were an alternative route into the university reserved mainly for the gifted children of the lower classes, which finally led to an education at the *Stift,* the Lutheran seminary.[6] All told, there were thirteen monastery schools in the duchy of Württemberg, forming a system that was unique throughout the empire, and with the *Stift* they formed a separate school system from the *Pädagodium* and the normal university.

It was in Adelberg that Kepler formed his ideas on the ubiquity doctrine that got him into so much trouble in his later life. The ubiquity doctrine was Luther's response to Thomas Aquinas's doctrine of transubstantiation, whereby the hidden substance, the underlying reality, of the bread and wine were transformed during the Mass into the Body and Blood of Christ. Luther disagreed with this, as did all the Reformers, and presented the alternative idea that the communicant does indeed receive the Body and Blood, because, by his existence as God, Christ's Body and Blood became universal and were everywhere. A person could receive the true presence of Christ because Christ was everywhere. The Calvinists denied this idea, claiming that the bread and wine remain bread and wine and that the communicant receives special assistance from Christ, who is in heaven, during Communion. It was this idea that Kepler leaned toward, since he could not find any mention of the ubiquity doctrine in either Scripture or the church fathers. (The Lutheran church eventually abandoned the ubiquity doctrine altogether.)

Time and again, the preachers railed against the Calvinists. They took special note of the Calvinist doctrine of Communion and shot as many holes in it as they could. But Kepler, true to form, studied the scriptural texts mentioned, meditated on them, and finally concluded that the

Calvinists were right all along. But that was as far as he would go with them. He could never accept the doctrine of predestination, for he thought it barbaric that God would condemn people for no fault of their own, just because they weren't "chosen." One of his fellow seminarians at the time teased him about his constant doubts, saying "Freshman, do you want to contest the predestination as well?" As it turned out, he did. After a great internal debate, he decided that he could not accept that pagans would be damned by a loving God, just because they had never learned of Christ.[7]

On March 5, 1587, his brother Christoph was born. Two years later, in 1589, after Heinrich the father tried to sell sixteen-year-old Heinrich the son into slavery, Heinrich the son ran away from home. Soon after, Heinrich the father also disappeared, for the last time, running off to fight for the king of Naples and then to die somewhere far away. All indications are that he was not much missed.

Meanwhile, young Johannes, off in the seminary, thrived. Up at four in the morning on summer days and five in the winter, Johannes lived the life of a monk, singing psalms as the sun rose, scrubbing hallways and classrooms, then studying the rest of the day. The boys ate and worked and studied and prayed from waking to sleeping, with no time for play. They ate in silence, and all of them dressed in the same knee-length coat without sleeves. They read, studied, and even disputed in Latin, so that they became more adept in that language than in their own workaday German. Along the way, they read Xenophon, Demosthenes, Virgil, and Cicero, but no Catullus or Ovid. They studied rhetoric, dialectics, and music. Later on, once they reached the upper seminary, they would study geometry and arithmetic.[8]

He loved the study and pursued it with the same intensity he pursued everything. But then he started having trouble. His doubts about the ubiquity doctrine and his typically aggressive defense of his own views set both his teachers' and his schoolmates' teeth on edge. Preachers from Tübingen visited Adelberg while Kepler was there, and they preached Lutheran doctrine with great fire and vehemence. But Kepler, true to form, his mind always toeing out new paths, didn't quite agree with any of them, because he was always looking for the other side, the place where those vilified

45

enemies of the true faith were right. In particular, he didn't like the manner of the sermons, the viciousness of them. And typically he argued his positions forcefully and with wry humor, which led to arguments and sometimes to fistfights. "Be careful," a friend of his told him. "Take such stands in the classroom only. If you speak like that in public, you could be called a heretic."

"My beliefs are my beliefs," Kepler told him. "I will make no secret of them."[9] After the first flush of Reformation, new ideas were no longer the fashion, and young Kepler the prodigy produced new ideas like sparks from a rocket. Denied health and strength and stature, he found his one compensation in his mind, forever questing, forever seeking the answers to the mysteries. As a boy, in his few spare hours he studied prosody and wrote his own comedies; he memorized the longest psalms in the Bible, just because they were the longest. His poems, he said, were mostly word games, acrostics, anagrams, and griphens. He loved paradoxes. Night after night, this intense, small dark-haired boy sat by the fire with the family Bible in his lap, his finger tracing out the lines, his mouth silently moving with the words, shoving each line into his memory.

This intensity carried over into school, into his study, his prayers, and his battles. When he fought with his classmates, he meant everything he said. He was always in control of his words, and when his words burned, he meant them to. Perhaps this was why in later life he berated himself so for his quick rages, why one minute he was searing and vicious and the next apologetic. He was, though he never admitted it, a haughty genius as a young man, with that blind and simple arrogance of the young. Every day, he practiced in his life what he practiced in his science—he attacked, he ridiculed, he challenged his opponents as he waged war on the world. His violent tempers were gradually reformed, though they plagued him from time to time throughout his life, and where he once demanded, he later offered. He was more the Christian gentleman when he left the seminary than when he entered.[10]

∽ꝏꝏ∾

*When, for the first time in my life, I tasted the sweetness of
philosophy, I was taken by a forceful passion for it in general,
not yet for astronomy in particular. I had a certain talent, and
it was not hard for me to comprehend the geometrical and
astronomical concepts, supported by figures, numbers, and
relationships sufficient for educational standards. But those
were necessary exercises and nothing that would have
revealed a very strong inclination toward astronomy.*

*Since I was supported by the Duke of Württemberg, I had
to watch my fellow students often bristle out of love for their
homeland as the duke, upon request, sent them to foreign
countries. I was stronger and decided early, wherever I was to
be sent, to follow willingly. My first task was an astronomical
exercise, and I was actually sent there [to Graz] by order of
the faculty. The distance didn't bother me, as I have said I
condemned this fear in others, but rather the unexpected and
disdainful type of task and also my limited education in this
area of philosophy. So I tackled it more supported by spirit
than by science and promised myself not to abstain from my
right to a way of life that appeared brighter.*

The astronomical clock in Tübingen

IV

Taken by a Forceful Passion

Where Kepler enters Tübingen University
and prepares for his calling as a priest of the
Book of Nature.

JOHANNES KEPLER STUDIED AT TÜBINGEN from 1588 to 1594. During
that time, Elizabeth I was still queen of England and would remain so
until the earliest days of the seventeenth century. On July 19, 1588, one
month before Kepler arrived, the Most Fortunate Spanish Armada sailed
within striking distance of the English coast. All that year, astrologers had
been predicting disaster for Spain, but no one had listened, for who would
dare speak against the Armada, with its hundred galleons, its thousands
of brave soldiers, the hope and power of the nation? Soon after setting out
from Cadiz, however, with horns blaring and banners flying, the Armada
ran into trouble. Strong storms out of the North Sea pounded the Spanish
ships one after another, decimating the fleet. Then the British sent a single
fire ship out among the Spaniards who, seared by panic, cut their anchors
and drifted away. When all this was over, only a few crippled, ragged-
bone vessels from the Most Fortunate Armada returned home.[1]

That year, Kepler's future master Tycho Brahe, in Denmark, declared
that, because of his observation of the comet in 1577, the comet that

Kepler had watched with his mother, Katharina, he had concluded that the crystalline spheres of ether in the Aristotelian cosmology did not exist. A few years later, Shakespeare received his first review, a bad one, from Robert Greene, a rival playwright and pamphleteer. Pope Sixtus V presided over the Counter-Reformation in Rome, while Jesuit colleges sprouted up throughout Europe like overnight mushrooms. The Turks advanced into Austria close to Vienna. On the other side of the world, the Jamestown colony was still thirteen years away, while the landing at Plymouth Rock wouldn't happen until 1620, the year of Katharina Kepler's trial for witchcraft. Galileo was then teaching mathematics in Pisa, but would later move to Padua. In Prague, the young Habsburg prince Ferdinand II returned from his studies at a Jesuit college in Bavaria and, still full of fervor, prepared for a journey to Rome to see the pope. He had one purpose in mind—to bring the full weight of the Counter-Reformation to the empire. Almost invisibly, Europe inched toward the Thirty Years' War.

On October 4, 1587, Johannes Kepler registered at Tübingen University. The registration was called the *depositio*, because the students had to "deposit" their "horns," or *cornua*. In essence, this was a leftover medieval university ritual, presided over by the older students, in which the incoming freshmen had to dress up as billy goats and prance around as a rite of initiation. Moreover, the freshmen had to pay for the honor of it all.[2] In September 1588, Johannes took and passed his baccalaureate examination. Everyone agreed that he had done brilliantly, but because there were no spaces available at the *Stift* (the seminary), the university turned him back to Maulbronn for one more year. Finally, on September 17, 1589, he returned to Tübingen, possibly on foot, emerging from the wooded area of the Schönbuch and onto the twisted streets of the lower town.

From the bridge across the Neckar, Tübingen rises like music, a crescendo swelling from the riverside up through the town to the *obere Stadt,* the upper town, to the Stiftkirche, the seminary church, and then winds back along the ridge to the Hohentübingen, the fortress on the crest. Tübingen is a tall city, cramped into the narrow space between the rivers, so that the buildings, even the meanest half-beam houses, seem to rise up and up like trees. The town was old even in Kepler's day, starting with a few narrow streets and a few thatched houses built by the Ale-manni, the proto-Germans who had lived there after the fall of Rome. Sometime in the eleventh century, the counts of Tübingen built themselves a fortress on the high hill overlooking the old town and then expanded the village into a city, with a new marketplace, a new parish church, and new city walls. Pilgrims on their way to Santiago de Compostela, the great pilgrimage site dedicated to St. James, stopped at the church in Tübingen to rest and prepare for the push first into France and then into Spain. Across the pilgrim road from the church they built a hostel to house the pilgrims, which the city later used to warehouse victims of the Black Death.

In 1342, the counts of Tübingen ran out of money and sold the city to the counts of Württemberg, making Tübingen part of a larger and more powerful political order. In 1477, Count Everhard the Bearded of Württemberg founded the university. An etching of him standing languid in full armor, holding an unsheathed sword beside him, its tip resting on the ground as a sign of peace, forms the centerpiece of the grisaille artwork on the Tübingen town-hall façade. Above his head, on the brick face of a wide dormer, the astronomical clock still ticks away the seasons and charts the motions of the skies.

In the late sixteenth century, astronomical clocks, which tracked the position of the sun as it traveled through the zodiac, were more than pretty showpieces used to bolster civic pride. They were practical tools for farmers and merchants, charting not only the time of day, but also the phases of the moon, the seasons of the year, and the general comings and goings of the heavens. At the end of the sixteenth century, people throughout Europe ordered their world astrologically. The swirling motions of the planets, the sun, and the moon meant fortune or failure to both peasants and kings alike. Kepler spent much of his life writing horoscopes—it was a lucrative business—and in an age before psychology and economics, it was the main way in which people mapped the troubles of their lives. One can imagine Kepler, newly arrived in town, pack on his back, standing within a knot of farmers, merchants, and students, staring at the clock as if to read God's plan for the day. Farmers planted and harvested according to the clock, while merchants bought and sold according to it. As a ministerial student,

Kepler was deeply aware of the influence that the heavens had on earthly affairs, all of which was mapped out every day, every week, and every month by the clock. The clock was a window into God's mind.

Throughout the late sixteenth century, Tübingen's population was relatively small. However, in all of southwest Germany, it was second only to Stuttgart. The town fluctuated between three thousand and thirty-five hundred inhabitants, but only a few of those were citizens. The university too was small, with four or five hundred students, a hundred of which belonged to the *Stift*.[3] The university was the heart of the town's social and cultural life. The students often staged plays in the marketplace and gave lectures to the townspeople. During the carnival of 1591, on Ash Wednesday, Kepler played Herodias, the wife of Herod Antipas, in a play entitled *Ioannes Decollatus,* concerning the beheading of John the Baptist. He was flush that year, because the town council of Weil der Stadt had voted him a scholarship, the *Stipendium Ruffinum,* out of a fund endowed by a priest named Rudolf Ruff in 1494. They gave it to him because of his *"Fürtrefflich und herlich Ingenium,"* his "extraordinary and glorious ingenuity." This tripled his pocket money. He was suddenly so rich that he lost a quarter of a *Reichstaler,* otherwise known as a "taler" or "dollar," while gambling with the boys.

Most of the university students were the sons of wealthy merchants, landed gentry, and minor nobility, and they were a rowdy lot. The university had its own police force, its own magistrate, and even its own jail, although because the authorities expected the town's magistrate to keep order, they gave him jurisdiction over the students. He could turn them over to the university magistrate for trial, but because most students were the sons of important men, the university rarely did anything. All too often, regular students meandered through town, drinking and brawling heroically, when they had a mind to. Not a few young women were raped in the streets, but there was little justice for them.

> Thunderbolt! How these drunken wenches march on
> My Lord brother, come. Let us escort them.

A strong beer, a smarting pipe
And a maid in her finery—that's to my taste.[4]
Goethe, Faust

Meanwhile, the seminarians at the *Stift* scurried to class and busied themselves with study, a bright contrast to the rest of the students. Students at the university wore gowns with hems colored to indicate their field of study: medicine, law, theology. The quality of the gowns varied widely, and the poorest were usually those of the seminary students—dark monastic robes. The seminarians were by all accounts the serious ones, because they attended by the duke's good graces and studied at his command, and because they were the sons of obscure parents, who had no protection from the consequences of their misdeeds. Years before, when they first entered the monastery schools, they had sworn themselves to the duke's service, promising lifelong fealty, to serve at his discretion and to leave at his discretion. They slept in unheated cells, rising in the predawn darkness at four or five to recite their morning prayers while the sunlight gathered in the stained-glass windows and suffused red-gold throughout the nave.[5] After prayers, the rector of the seminary lectured on the theological point of the day. The sermons were often charged with sectarian politics, with hot brimstone against the pope and with sly warnings against the Calvinists. While the students listened, beneath their feet the long-dead Dukes of Württemberg slept on to resurrection day. Here Kepler listened to the Word of God proclaimed to the students. Here, his Lutheranism, held at some cost by his family, took seed, sent down roots, and blossomed.

Sermons and lectures at the university were public entertainment then as much as public instruction, and people often walked miles to hear a good preacher. The Stiftkirche was a "hall church," specifically designed without pillars so that everyone in the congregation could see the preacher. As with all things Lutheran, each lecture, each sermon was intensely scriptural. The boys gathering in the seminary church every morning, shivering from the cold, yawning, rubbing sleep from their eyes,

53

attended as best they could as each preacher fought to make Scripture come alive. The great characters of the Old and New Testaments walked before them—Gideon, Samuel, David, Daniel, John the Baptist, Mary Magdalene, and all about them danced St. Paul, whose words permeated the writings of Dr. Luther like water permeates soil.[6]

Lutheranism was more than just the religion of his parents and grandparents for Kepler. It was a religion that made sense to him. He saw it as a religion that never asked him to submit his reason to any other authority than God. In religion, one need not consult any other authority than the Scriptures and the fathers of the church. Even when princes commanded, they could not violate his conscience. Neither could Satan, the Prince of Darkness, or any other demon. For Kepler, God not only preserved the heavens above, but also the reasoning mind that contemplated them. Kepler would sit in the stone church, his breath floating visibly before him, reciting Luther's Morning Prayer:

> My Heavenly Father,
> I thank You, through Jesus Christ, Your Beloved Son,
> that You kept me safe from all evil and danger last night.
> Save me, I pray, today as well, from every evil and sin,
> so that all I do and the way I live will please You.
> I put myself in Your care, body and soul and all that I have.
> Let Your holy Angels be with me,
> so that the evil enemy will not gain power over me.
> Amen.[7]

A prayer for inner peace. A prayer for salvation. A prayer for freedom from evil. A vital prayer, a necessary prayer, for the intellectual environment in which the faculty preached was a turbulent one. Throughout his life, Kepler remained an Augsburg Lutheran because, he believed, it allowed him freedom of conscience. There was no church authority standing between him and the Scriptures or between him and the church fathers. He could follow his own path, think his own thoughts, and find

God in his own way, without pope or bishops standing between him and his Redeemer.

The Counter-Reformation, meanwhile, gathered like storm clouds in the distance. But for the Württemberg orthodoxy, the immediate threat came from the followers of other Reformers, from Calvin and Zwingli, who from the Lutheran point of view had taken the Reformation along paths that God had not intended and were therefore heretics. Lutheranism was the middle way between conservative Catholics and radical Anabaptists and Calvinists, those rebaptizers and predestinarians. Controversies that nearly came to blows within the Lutheran confession earlier in the century had finally settled themselves by the promulgation of the Book of Concord in 1580, eight years before Kepler's arrival at Tübingen. While Catholics looked to the Council of Trent for guidance, Lutherans looked to the Book of Concord—and to the university theologians who preached it, men such as Jakob Heerbrand, who taught theology to Kepler, and Matthias Hafenreffer, who taught Scripture. After the heady years of the early Reformation and the subsequent years of turbulence with both Rome and Geneva, the first islands of Lutheran orthodoxy had appeared above the flood, thanks to the work of university theologians who were streamlining the faith for Lutheran unity and for the creation of a pure doctrine. But repression follows orthodoxy like a jackal. The *Stift* quickly developed a culture of denunciation, with students denouncing each other for minor infractions of the rules, suspicious talk, and potential heresy. Kepler received his share of denunciations, lying spread-eagle on the chapel floor as he listened to the catalog of his faults and misdemeanors. Now and then he returned the favor.

Twenty-first-century people most often think of Kepler as a scientist, but that was not his intention when he arrived at Tübingen. More than anything else, he desired a pulpit and longed for the life of a Lutheran preacher. He was born a Lutheran and would die a Lutheran, and for all his later troubles with the church, when his people suffered, Kepler suffered with them. Eventually he would be chased from one town to the next by the Counter-Reformation. After his death, the Lutheran cemetery in which he was buried became a battlefield; soldiers died on top of the dead, destroying Kepler's marker stone and losing his burial place forever. All in the name of God and a faith that justifies.

Kepler's studies began with two years in the Faculty of Arts in preparation for three years of study in theology. The faculty lectured in Latin, allowing Kepler to hone his Latin, to learn Greek and Hebrew, and to study dialectics, mathematics, and rhetoric. At the end, he received his master's degree, coming in second out of fourteen. Because of his record, the chancellor of the university assigned him to study with Martin Crucius, the great classicist, and with Michael Mästlin, the great astronomer and mathematician.

In a short time, he developed a reputation as a dutiful scholar and a good fellow, if a bit of an intense one. He took part in the school plays, where he sang in the chorus and once or twice, because of his slight build, played a woman. He was also a bit of a prankster. One student, Zimmerman, the son of a Lutheran pastor, didn't like Kepler much, and Kepler didn't like him. The two young men were the antitheses of each other. Kepler was intense, chewing on ideas like they were meat; he didn't have much tact or patience for those who couldn't keep up. Zimmerman, too inert for study, too undisciplined for scholarship, simply wanted to sail through the university. So Kepler and a few of his friends sewed the arms of Zimmerman's gown shut, forcing Zimmerman to come to Matthias Hafenreffer's class armless. Hafenreffer told Superintendent Gerlack, who called in the pranksters. Kepler admitted it, denounced himself, and took his punishment, but that didn't stop the pranks.[8]

The other reputation he had developed was as an astrologer. Astrology was for the seventeenth century what economics is for the twenty-first. Astrology tried to form predictions about an uncertain future based on strict mathematical calculation, just as economics does with the laws of the market. Both are wrong about as often as they are right. Astrologers assumed that the heavens were never mute, but full of meaning, and that their meaning could be read as a text. Oddly enough, postmodernists say something similar about the world, that all things have sign value, and that nothing merely exists: a cigar is never simply a cigar. Stars and planets in their complex relationships formed an alphabet. The fact that Mars was red, for example, said that it was hot and dry and warlike. Its place in the heavens meant that it had its allies and its enemies, and where it stood in the zodiac, in agreement with some planets and in opposition to others, had its effect on the world of human beings, an effect that could be read like a book.[9]

Because his love for puzzles and acrostics had started when he was a child, Kepler was particularly good at reading signs. He soon learned, however, that being a good astrologer required more than just math skills. One student, Rebstock, a fellow with a red face and beer breath, accosted Kepler in the hallway and demanded a horoscope. Kepler reluctantly agreed and, after obtaining the man's birth date, set to calculating his chart. What Kepler learned that day, however, is how dangerous it is to read all the signs. Rebstock's noisy drinking habits had to be taken into account, so Kepler predicted that the fellow would one day become a drunk, which wasn't much of a stretch. The stars tell all, but so does beer breath. Rebstock didn't like the report and forced his way into Kepler's room, where the two duked it out. The next day, Kepler asked Mästlin for advice. What should he do? If he was going to be an astrologer, he had to read all the available signs, and that included a beer breath, because the stars were so often hard to read. Sometimes his predictions worked and sometimes they didn't, so what could he do to make them more secure? Mästlin told him to just predict disaster. That would be bound to come true sooner or later.[10]

In Lutheran education, the Bible was foremost, and the method of biblical exegesis, the close reading and interpretation of original texts, informed all other disciplines. The method stripped away layers of accumulated interpretation to arrive at the original texts in the original languages and then built a new interpretation out of that. Kepler read Aristotle in the same way, especially the *Analytica Posteriora,* which set out Aristotle's logic, and the *Physica* and *Meteorologica,* which set out his ideas of motion and change in the world and sky inside the sphere of the moon. Oddly enough, Kepler says nothing about having read the *Ethics* or the *Topics.*[11] His inclinations as a scientist may well have surfaced even at this time, in his reading choices in Aristotle. Early on, he was "taken by a forceful passion" for philosophy in general. He loved the fisticuffs of argument, the development of clear ideas, and the defense of them. In fact, this is the main way that Kepler complicated his life. The medieval tradition in education that Tübingen inherited included the staging of public disputations, debates in which students took opposite sides of an issue and argued it. Kepler learned to do this in Maulbronn and developed a liking for it that carried on into Tübingen. But Kepler could not take a position he didn't believe in, and his mind, ever active, was always searching for a new angle to an old question, a new solution to an old problem.

What he didn't understand was the wider world—that Tübingen and the Lutheran world in general felt themselves to be under siege, that in their growing orthodoxy the faculty of the *Stift* were not particularly interested in new approaches. New ideas are always dangerous in unsettled times, and for every good idea that comes along, an army of resistance emerges from the dust to squash it back down. Kepler's refusal to condemn the Calvinists over their doctrine of Communion made him suspect. His questions about the Lutheran ubiquity doctrine, which held that believers could receive the Body and Blood of Christ in Communion because Christ, being God, was everywhere, an idea that was a central part of the Formula of Concord, turned into a problem. Then he discovered Copernicus.

His philosophy professor at the university was Vitus Müller, who along with Martin Crucius taught him Aristotle and staged many of the students' public disputations. In his first years at Tübingen, Kepler read the works of Nicholas of Cusa and found the Catholic heretic's geometrical mysticism to be similar to his own. Crucius eventually led Kepler through the labyrinth, one by one introducing him to the philosophers who would dominate his thinking throughout his life—to Plato and the Neoplatonists and through them to Pythagoras. The secrets of the universe were in the shapes of things, in the geometry of the universe, and in that geometry there were harmony, order, and perfect reason. Throughout his life, Kepler would seek that harmony, and if he could never find it on earth, he would find it in the sky. He was impelled to do this, for it was central to his intellectual life. A boy who had grown up in a chaotic family in a world where Christians preached hell to one another from their churches on opposite sides of the town square, where old women were regularly accused of witchcraft, and where emperors and princes sent armies tramping across people's farms and bloodying innocent peasants' wheatfields could either despair of all order and all civility or spend his life looking, as Kepler did, for a place where the universe made music.

Crucius was enamored of the Greeks, because they were the first, the inventors, the originators of the Western intellectual world. Whatever we think, they thought it first. For Crucius, anything not found in Aristotle could not be true.[12] Kepler accepted this for some time, so much so that Crucius asked him to collaborate on his magnum opus, a commentary on Homer, to help him interpret the astrological and astronomical allusions in the poems, but the work was never quite satisfying to Kepler. His own inclinations were too different from his teacher's. Although both men were industrious and addicted to detail, Crucius was a gleaner, a gatherer of notions, while Kepler was a separator, splitting ideas into their purist form, skimming off the dross to find the gold.

Eventually this led Kepler closer to his final mentor, Michael Mästlin, who of all of Kepler's teachers had the greatest influence on his life. Mästlin introduced Kepler to the Copernican universe, and for the young Kepler it seemed as if a new window had opened in his mind. Typical of Kepler, he instantly began taking the Copernican position in his university debates and getting himself into trouble:

> In Tübingen, as I listened attentively to the lectures of the famous
> Magister Michael Mästlin, I saw how awkward in so many ways
> the customary notion of the structure of the universe had become. I
> was delighted, therefore, by Copernicus, whom Mästlin often men-
> tioned in his talks, and I not only frequently promoted his views in
> the students' debates, but also wrote a careful disputation concern-
> ing the thesis that the first motion [the revolution of the sphere of
> the fixed stars] comes from the rotation of the earth. I also set to
> work assigning to the earth on grounds of physics, or perhaps meta-
> physics, the motion of the sun across the sky, just as Copernicus

had done on grounds of mathematics. To this end, I have bit by bit—in part out of Mästlin's lecture, and in part out of my own thoughts—gathered together all the mathematical advantages that Copernicus has over Ptolemy.[13]

Meanwhile, as Kepler gushed about his new discovery and defended Copernicus as if he were a besieged city, Mästlin, the man who started Kepler along this track, stood by nonplussed. Mästlin was a short man, taciturn and introverted, who kept his more dramatic emotions calmly sealed away. He rarely smiled and yet was rarely angry. He had a high forehead and a narrow jaw, which made him look slightly bulbous. His hair was black, worn short, and he sported a bushy goatee, ubiquitous in the seventeenth century. He had made his fame by showing that a new star, a nova that appeared in 1572, had actually been a distant object and not some trick of the atmosphere. Conservative by nature, he was unwilling to part with Ptolemy completely, though he was aware of the growing complexity of the geocentric system, and he knew that even before Copernicus pressure had been building against the system for some time.

And yet no one had any proof. The weight of observation had been growing, seeding dissatisfaction among a number of astronomers, but such a momentous change would require a wealth of evidence. Copernicus himself had been afraid to publish until after his death. By Kepler's day, there were at least four distinct models of the universe floating through the intellectual air that were whispered about or heatedly discussed by clusters of students in the smoky dark corners of beer halls. First, there was the official cosmos, the geocentric, finite universe of Aristotle and Ptolemy, reiterated by St. Thomas Aquinas. Then there was the infinite cosmos of Nicholas of Cusa, with God at the everlasting, omnipresent center. Third, there was the "heliostatic" universe of Copernicus, in which the planets, including the earth, orbited the sun, which was fixed in place. And finally there was the model resurrected by Tycho Brahe, first discussed by Plato's student Heracleides Ponticus, in which the sun orbited the earth and the planets orbited the sun.[14] Each of these systems had its supporters, and each had its detractors.

Surprisingly, the difficulty in adapting to the new placement of the earth in relation to the other heavenly bodies was not primarily that it spelled the downfall of human .dignity. Later generations seemed to think that the geocentric model promoted the dignity of humanity's place in the universe, as the apple of God's eye, while the Copernican system turned this around and set the earth spinning meaninglessly through a meaningless universe. This is not quite accurate, for Aristotle never thought of the earth as a special place or the apple of anybody's eye. The earth occupied the lowest position in the cosmos, where all things chaotic and all things corruptible eventually settled. The world beneath the sphere of the moon was the privy of the universe, where living things came into existence and then died away, where sooner or later all life returned to rot.[15] Only the heavens were eternal; only the heavens were divine. Redefining the earth as a planet, as Copernicus did, actually set the earth into the heavens with the other planets and raised property values all around.

Modern people, however, cannot fully understand or appreciate the Aristotelian model without also seeing it within its wider metaphysical framework. When Plato looked at the human person, he saw a paradox. First, there was the body—material, corruptible, an instrument for use on the material earth. Then there was the soul—immaterial, incorruptible, sharing in some fashion with the eternal and divine Ideas through its capacity for reason. What it could know, even vaguely, it could share in. These two halves of the human person were incommensurable, blind to one another. They spoke different languages, perceived the universe in different ways. Aristotle accepted this dichotomy and accepted their union within the human person, but admitted that it was one of the greatest mysteries of all. Somehow, these two parts were joined together by an intermediary substance, an apparatus so subtle that it was corporeal and incorporeal at the same time, so subtle that it approached the immaterial soul in essence, shimmering in the dark, and yet was still a body, able to join with the flesh and pass along the commands of reason. This substance, this astral body, the *prŏton organon*, the primary instrument relating soul to body, was made of the same stuff, the same spirit, *pneuma*, that made the stars. The spirit that moved the heavens, that moved the stars across the sky, was the same spirit that raised the human arm in greeting or moved the human leg in walking. The vast sky was not dead, then, but subtly, luminously alive, and we in our gross flesh were in some small way cosmic beings.[16]

Perhaps finally the Aristotelian universe was simply familiar. And it was beautiful, if a bit creaky. It did a good job synthesizing the appearances, the phenomena that people actually saw when they looked into the sky, and it looked reasonably similar to the Babylonian cosmos of Genesis. It was comfortable. It worked well enough for weekdays and sometimes on Sunday. The problem was that it was getting too complicated, and no mathematician liked that.

Twenty-first-century people often imagine that the Copernican controversy was about science against the church, but the reality was far more complex. Science as we know it did not yet exist, and the church, Protestant and Catholic alike, was in fact the normal place for intellectual discussion. Nevertheless, society itself was changing, all too fast for most people. As if under a magnet, complex social forces aligned, at once pushing Copernicus forward and shoving him back. Friends and enemies of the sun-centered universe gathered in colleges and cathedrals across Europe and shouted at one another, something that Copernicus himself described: "Since the newness of the hypotheses of this work—which sets the earth in motion and puts an immovable sun at the center of the universe—has already received a great deal of publicity, I have no doubt that certain of the savants have taken grave offense and think it wrong to raise any disturbance among liberal disciplines which have had the right set-up for a long time now."[17]

The changing cosmology scared people, affronted their sense of reason, a sense that derived its rationality, its worthiness for belief, from the single idea that God had set an order to things, an order to the heavens and an order to the earth. In the heavens, the stars and planets sailed on by force of the perfect divine will, the perfect divine law, which constrained them into circular orbits, the most perfect of all shapes. This was so, because it was the way things ought to be, the best of all possible worlds. On earth, the world was likewise ordered by divine will into church and state, into hierarchies of religious and political power. Even when Luther rebelled against the order of the church, he did so by relying on the order of the state, by turning the princes and dukes into "emergency bishops," so that the order of the whole might be maintained. Beyond this was chaos.

Who could imagine such a state? The medieval world that lingered in Kepler's day, although ancient and creaking, leaking oil and all too often rolling over the lives of peasants, overflowed with providence, with God's special care for trudging mortals doomed to die. What could possibly replace that? No wonder astronomers stepped lightly, not only out of fear of the church, though that fear was well founded. They stepped lightly out of fear of unraveling the cosmos, of pulling on the wrong string and having the whole order of the universe collapse at their feet. Nevertheless, the generation of astronomers before Kepler and Galileo, the generation of Michael Mästlin and Christoph Clavius, could see the value of the Copernican system. Its elegance and simplicity appealed to their reason as much as it did to Kepler's and Galileo's. Although Clavius was more conservative and more philosophically subtle, basing his resistance to Copernicus on the differences between reasonable hypotheses and proven fact, Mästlin was more willing to dig at the root of the problem. His criticisms of Aristotle were dangerous and he knew it, so he broadcast them carefully. Aristotle was the systematizer who had set Greek reason to the Babylonian cosmology of the Bible, and long use of his work to build a Christian cosmology had nearly identified him with Scripture itself. To reject Aristotle and his astronomical interpreter Ptolemy was to remove the cotter pins holding together the structure of rational Christianity. Mästlin never challenged Aristotle's value head on, but picked at the philosopher's mistakes from behind, showing how, contrary to common belief, his system did not support observation, but rather complicated it.

In Mästlin's lectures, the lectures that so influenced the life of Johannes Kepler, he traced the path that Aristotle had taken to create his system. "What happens to fire when it is lit?" he asked. "It rises." In this, Aristotle too was following a long tradition set down by the Pre-Socratics, that heavy things fall and light things rise, just as air and water and earth mixed in a glass jar eventually settle with the earth at the bottom, the air at the top, and the water in between. Heavy things separate from light things and wet things from dry things, hot things from cool things, and light things from dark things. Therefore, said Aristotle, the earth must be at the center of the universe, for, as anyone can see, the earth is heavy and the air, which extends outward into the sky, is light. In Aristotle's universe, and therefore in Ptolemy's, the heavens are made of clear crystal globes that gear against each other and groan in mystical harmonies. The spheres sing a song that perhaps only God can hear properly.

We are steeped in hundreds of years of Copernicanism, so we cannot see another way. In this, we are like the churchmen who condemned Galileo and the theological faculty at Tübingen who worried over Mästlin's lectures and later shook their heads over Kepler. But Ptolemy and his followers were not fools. They never tried to explain the heavens mechanically, but only to come up with a systematic account of all the phenomena that any observer could see by looking. The earth had to be the center of the universe, they argued, because otherwise the sky would seem different looking north and looking south or looking east and looking west, but it isn't—it is the same in all directions.[18] The stars are different, of course, but the spheres appear to be the same distance from the earth no matter which way you look. Ptolemy assumed that the sky was an actual physical sphere, and not endless space, but that assumption came to him from Aristotle. Moreover, Ptolemy argued, the earth cannot move from its central position, because all things move toward the center of the universe as the heavy things sort themselves out from the light things. The earth, therefore, is a massive, bulky, weighty thing, while the heavens are light and ephemeral. So which makes more sense—that the adipose earth move from place to place or even spin, or that the aerie heavens, luminous and eternal, sailing through the sky as they do by the action of heavenly spirits, move? What later seemed so costly to Copernicus— that the entire universe would turn around the earth—seemed reasonable and natural to the followers of Ptolemy, based upon the relative weight of the two systems.

Aristotle and Ptolemy had invented a system anyone could see at work in a glass jar, watching light things separate out from heavy. Then they could step outside and watch the heavens waltz across the sky like gods in evening wear, and they would say, "Of course!" But then there was this problem—the dances of the planets, the "wanderers" that didn't follow the expected pattern. They were eccentric, sweeping across the dance floor in arcs of perfect circles as expected, but during the time of their "opposition," when they were on the other side of the sun from the earth, they stopped, hand-jived backward, then reversed course and waltzed on. Ptolemy explained this strange retrograde motion by inventing circles, *epicycles,* whose center moves along a larger circle, called the *deferent,* whose center was the earth. Moderns are often stymied at this idea, wondering what the planets are orbiting in their epicycles, what points of gravity were holding the epicycles together, keeping the planets moving round and round like dancers, but although everyone could see gravity at work on the earth—heavy things fall and light things rise—they could not imagine it for the heavens. These little circles were descriptive of the appearances, each planet circling through its epicycles, all inside the planet's crystalline sphere, in the space between the inner and outer walls. The spheres themselves rotated, moving the planets across the sky, but inside the sphere the planets circled on. The presence of epicycles in the system alone wasn't the problem, however. The problem surfaced when, with increased observation, Ptolemaic astronomers needed to add more and more epicycles to the system to keep it working. For a long time, they required only twenty-seven epicycles, but by Kepler's day, they needed nearly seventy—far too complicated.

Kepler saw this at once. Though he loved theology and Scripture, though he could read in four languages (German, Latin, Hebrew, and Greek) and write in two (German and Latin), he found himself drawn toward the study of the heavens. He had always been good at mathematics, and since the day his mother took him to see the comet back in 1577, his mind and heart had been pulled toward the stars. There was beauty, majesty, and grace there. There was harmony—he could feel it—and he longed for simplicity. Real harmony, he reasoned, can never be that complicated.

Eventually, Mästlin loaned him his own copy of Copernicus's *De Revolutionibus Orbium Coelestium, On the Revolutions of Heavenly Spheres.* This was quite an act of trust, for the book was rare and a little dangerous. Mästlin told Kepler how Copernicus had delayed publication, worried about the effect of his own ideas, and how on May 24, 1543, a friend put a newly arrived copy of the book into Copernicus's hand as he lay on his deathbed. In the preface, Mästlin said, Copernicus had stated that his ideas were merely a useful means for calculating celestial events, a mathematical hypothesis. This was the same argument that the Jesuit Clavius used when discussing Copernicus with Galileo. But that is not how the rest of the book reads. Many astronomers and mathematicians suspected that Copernicus believed what he wrote and that he wrote the preface out of timidity.[19] In truth, Andreas Osiander, whom Copernicus put in charge of the book's publication, wrote the preface, and he did this without the permission of the author, which would likely have angered Copernicus, had he not been on his deathbed. The preface had the positive effect, however, of making Copernicus's book more palatable, so that men like Clavius, Mästlin, and Brahe could study it without fear, since the preface itself said that Copernicus's system was meant only as a device for calculation.[20]

Device for calculation or not, Copernicus's theory set Kepler's mind on fire. Here was the simplicity, the elegance of thought he longed for. Kepler was enough of a Platonist to believe that the universe was simple elegance and was best described by simple, elegant mathematics. Could a good and loving God, a rational God, all-wise and all-knowing, have created the epicyclic nightmare that the Ptolemaic system had become? Kepler doubted it, and yet he fretted. For all his excitement about Copernicus's cosmos, with its simplicity and elegance, the thought that the earth was no longer the center of the universe worried him. Even if Aristotle didn't think much of the earth, Christians did. Shouldn't the place where the Son of God had been born, had lived out his life, and died, the place where he was raised and from which he ascended to the Father be the center of things? Shouldn't the world of human beings, built in God's image, be the center? The answers he needed to these questions could not be found in mathematics, but in metaphysics and theology. Like Melanchthon, Kepler expected that the human mind and God's mind worked in roughly the same way, because God created humans in God's own image and likeness.

For all his doubts, however, Kepler was convinced enough to defend Copernicus to his fellow students, writing one disputation after another, first making physical arguments and then metaphysical ones. In all the emotional turmoil, in 1593 his health problems returned. He still struggled with poor eyesight, and his clumsy and somewhat malformed hands pained him. But now he had headaches, possibly migraines, as well. Emotionally, he felt that honors owed to him were being held back.

Resistance to the Copernican system was growing throughout Europe, in Catholic and Protestant countries alike. Except for Mästlin, the faculty at Tübingen was decidedly anti-Copernican. One of the favorite arguments of the objectors was that if the earth really moved, why couldn't the people living on the earth feel it? There should be some perceptual evidence of this motion, one way or another.

In 1593, Kepler wrote a short dissertation, supported by his friend Christopher Besold, imagining what the earth would look like to people living on the moon. This would be revised several times in his life and finally published as Kepler's *Somnium,* his *Dream,* after his death. The purpose of this dissertation was to demonstrate Copernicus's idea that the earth moves very rapidly, rotating and revolving around the sun, but the people living on the earth cannot see or feel this. Anyone who looks up at the sky from the earth can see the moon and its motions, so what would happen if someone were standing on the moon? Would they not see the motions of the earth, just as we see the motions of the moon?

Kepler wanted to present it as a disputation, but Vitus Müller wouldn't have it. Müller hated Copernicus's ideas and would not listen, nor would he let the thesis be heard. Though Müller never referred the incident to the general faculty, some of the professors had already begun to wonder if Kepler was suited to the ministry and whether he might be happier doing something else. Some people suspected him of being a crypto-Calvinist. Why did the boy have to go his own way all the time? And this Copernican nonsense of his! Even Magister Mästlin did not go so far. Could the Kepler boy become an embarrassment?

Then, lucky day, a solution. The Lutherans in Graz were looking for a mathematics teacher. All the way in Austria. People would be sad to see him go. Many people liked the boy and wished him well. But, all the way to Austria.

꙰

*The position of mathematician that is offered to me is in many
respects so honorable that I could not decline it, since my
parents—I am still under their supervision—are too far away
to consult in person, I want to report what my nearby
relatives think I should attend to. They asked me to gratefully
recognize the great goodwill of the Herr Chancellor and to
continue getting him on my good side by diligence in my
studies and by an irreproachable lifestyle. Additionally, they
agreed with the offer at hand, because so many advisers are of
the opinion that I should undertake this journey. There is one
thing that is rather dear to them, but that they don't want to
determine themselves and would rather leave to the decision
of the theological faculty. They would prefer if I dedicated
myself to theology, as have my classmates who have thus far
been encouraged to this end, and once the work is completed
devote myself to the church. I am not speaking particularly
about studies of the holy sciences, which I have been blessed
to enjoy up to now. And still, whatever shall become of me, if
the good Lord warrants me a healthy mind and freedom, I
would never consider interrupting this. Rather, my relatives
imagine, since my age and stature are not yet fit for a pulpit,
that I would easily, by way of a letter written by the Herr
Chancellor to Pastor Zimmerman in Graz, have the
opportunity to practice church services and by reading of the
Holy Scriptures and other authors to further my studies.*

Portrait of Michael Mästlin, who introduced Kepler to the Copernican system

V

In Many Respects So Honorable

Where a position as a mathematics teacher opens
at the Lutheran school in Graz, and Kepler takes
the position with some fear.

THE FACULTY AT TÜBINGEN wished him well, even while they fumed about him. Who could deny the boy's sincerity? Even so, their patience was wearing thin. Poor Kepler was naïve and had little idea about the trouble his enthusiasm had caused. He could not have understood the hand-wringing he induced in the orthodox people around him by his willingness to soldier on with new ideas while others held back. If only he hadn't insisted on arguing publicly for Copernicus, they said. If only, like Mästlin, he had been more circumspect. From the day Kepler left the university, therefore, he would never find his way home again. Kepler's far-ranging mind had slipped the narrow bonds of Tübingen's orthodoxy, in both theology and science.

Unfortunately, Kepler was a genius born half a century too late, at a time when the reformation was finding its feet. Perhaps it was his incessant support of Copernicus, or perhaps it was his scruples about the Formula of Concord, but either way, when Georg Stadius, the teacher of mathematics at the new Lutheran high school in Graz, in the foothills of Austria, died suddenly, the faculty at the Tübingen *Stift* found a solution to the Kepler problem. Kepler had already received his master's degree and was three and a half years into his theological studies with one half year to go, when the local authorities in Graz sent word requesting a new mathematics teacher. The Tübingen faculty held a secret meeting, and they agreed—this was perfect. Everyone knew how adept Kepler was in that subject, how solid his horoscopes were, and if this worked out, his new appointment would keep him out of the pulpit.[1] In his journal for that year, Kepler mentioned his hopes for another appointment then being negotiated for him in Württemberg, but then suddenly, on January 18, a few weeks past the Feast of Epiphany, like a new star the Graz appointment appeared above him. By February, the university faculty had decided.

Because Kepler was a scholarship student, he had to get the permission of the duke to take any assignment at all, let alone one far from home. For some time, he had watched other students at the *Stift* bristle at the thought of leaving the duchy for assignments in far-off countries, and as he wrote in his journal several years later, he had decided that if such a position were offered to him, he would accept it, and gratefully. But now they were asking him to give up the ministry, to take a position he had never expected. Certainly he liked mathematics but all his life he had set his course on service to God and to the church. Was this new job a failure?

When Kepler waffled about his new good fortune, Hafenreffer pushed him. The appointment had to be accepted as soon as possible. Was Kepler going to complain about the distance, as others had? Was Kepler afraid? Perhaps there was some pride in this? A mathematics teacher did not have the prestige that a preacher had, so this new appointment may have been a disappointment. Still, what they offered was an honorable position, and he could not decline it without people thinking he was arrogant. What should he do? He could not easily consult his mother or his grandparents, because they lived too far away. Weil der Stadt was at least a fast two-day walk, and Leonberg was farther than that. His relatives who lived nearby—Kepler does not specify—advised him to stay on the good side of the Herr Chancellor, to do whatever the Chancellor wanted him to do, and to live an exemplary life. They advised him to take the position in Graz because so many smart people seemed to think he should go there. Perhaps, they said, he was too young for a pulpit, and maybe even too short. And probably too thin, not imposing enough, as a preacher should be.

By early March, Kepler walked to Stuttgart to see the duke, and on the way stopped at Weil der Stadt to see his grandfather Sebald, who listened kindly, nodded, gave him good travel advice, and sent him on to Stuttgart, where on March 5, the duke received him. Duke Friedrich was a plain man, round and stubby in the face, with hound-dog eyes, a high forehead, and a crown of curly blond hair cut close to his head. He received Kepler in the palace kindly, asked a few questions about the position, gave his permission along with promises of support, and then sent him on his way. After all, what else could the duke do? No ruler should second-guess his own staff.

Back in Tübingen, Superintendent Gerlack, along with some good friends, loaned Kepler 50 gulden for the journey. "I cannot hide the fact that, after my appointment to come to Graz, as I was preparing for the journey from Tübingen, I had to borrow 50 gulden from the university as well as from some good friends to pay for the necessary travel costs. I promised to faithfully return such money by way of my cousin Jaeger, who accompanied me to Graz."[2] A few days later, on March 13, 1594, when Kepler was twenty-two years old, he bought a horse and he and his cousin set out for Ulm, where they would catch a barge heading down the Danube River to Linz (where years later he would live while his mother was on trial for witchcraft), and then travel overland once again, by wagon this time, to Graz.

Travel was better than it had been. Throughout the long feudal period, rule was always local—duchies, counties, cities, towns, villages—with local constabularies and local men at arms. In the wide-open lands in between there were bandits, cast-off men, and even women preying on those unfortunate enough to travel. Often, highway robbery was a family business. For safety, people moved about in bands, with hired bodyguards if they had any money, with knives and cudgels if they did not. Pilgrims traveled in large groups for mutual protection. Some larger towns and cities provided escort services along major roads, at least to the end of their jurisdiction. After that, people were on their own. At the end of the High Middle Ages and into that part of history we call the Renaissance, which is really the tail end of the medieval world, duchies and counties and principalities widened, became grander and more complex, and the spaces in between shrank. The number of bandits shrank accordingly, but still waxed and waned with the changing political tides. And there were always gypsies about. Even in Kepler's day, travel was not something people did lightly. Even with his cousin along, a young man, especially a short, slight young man, the kind of young man who played girls' parts in school plays, setting out with 50 gulden in his pocket would be an easy mark. Best to rely on the grace of God and a fast horse.

This particular young man also brought along a troubled mind. Kepler and Jaeger arrived in early April, and Kepler was counting his money. "The index of my travel expenses incurred thus far for me and my travel companion from March 24 to date: From Tübingen to here for meals, exchange, tolls, and other necessities total $31^1/_2$ gulden. Also, my lodging with Stephan Kirschner, in Schmidgasse, cost 5 gulden. Since I do not know how long my cousin will have to stay here in order to learn the necessary information to make his report back home, I want to include my cousin's expenses as well."[3] By April 21, the school paid him 60 gulden for his travel expenses, which allowed him to send the 50 back with his cousin. And then, on April 25, he suffered a bout of Hungarian fever and was in bed for two weeks. All the time, his mind churned. Would he ever return to the ministry? How could he find his purpose, his calling by God, in this new twist of his life? What's more, in his pocket he carried a letter from friends and the leaders of the Lutheran community back home, a letter that might cause problems for him in the future, though he kept his feelings in check. He wrote in his journal for 1594: "At the same time, I had brought along a Uriah letter that determined a bride for me. I, however, had brought my love along, and for a while, I was calm."[4]

What was this Uriah letter, and what was this love? In the Bible Uriah the Hittite was the husband of Bathsheba and an officer in the army of

David. David betrayed Uriah by sending a letter along with him, supposedly a commendation or recommendation, but David's general, according to the king's instruction in the letter, had Uriah placed on the front lines, where he was killed. Apparently, Uriah could not read. In Kepler's time, a Uriah letter meant either bad news, or a recommendation letter damning the bearer with faint praise, or, as in Kepler's case, a letter that could be useful to the bearer, but not really welcomed by him. Perhaps he perceived that his friends, teachers, and family were setting him up, that they wanted him to settle in Graz, be a good citizen, and make no trouble. It would not have been out of custom for them to do so. A young man needed guidance, and just about everyone who knew him believed they were the right ones to provide it.

What, then, was this love he brought with him, this love that calmed him, at least for a time? Perhaps it was his passion for philosophy, mathematics, and astronomy, his passion for the life of the mind. Perhaps again it was simply his human capacity to love, to find the hand of God in the twists and turns of his life. If given a wife by the Lutheran community, a good Lutheran woman, a helpmate in this world, and a companion throughout his preparation for the next, then he would have brought his love with him, his capacity to love her, whoever she might be, even if their marriage meant that he would never return home, never finish his training for the ministry, never find a pulpit or a church or a congregation in his life.

Graz, the capital of Styria, or Upper Austria, was an old town. Like so many others, it started off as a ford across a river, the Mur River, where Roman armies once crossed on campaigns into Germany, and where during the *Volkerwänderung*, the wandering about of the people, the Teutonic tribes eventually returned the favor. The town was a prehistoric place, a place for trade, if not always a place for a quiet life. After the Romans, the Slavs and the Bavarians in their turn occupied both the ford and the town. Sometime in the early Middle Ages, the people built a small fort on top of a dolomite hill, partly to keep out of the yearly floods and partly to keep away from the next set of invaders. The Slavs were probably the ones who built it, for they eventually named the place—Graz comes from *gradec,* which means "little fortress." Eventually, a castle—the Schlossberg—replaced the fortress. Somewhere in the twelfth century, someone mentioned Graz in the imperial chronicles for the first time.[5]

Then, in the fifteenth century, Friedrich III, the Habsburg Holy Roman Emperor, abandoned Vienna to the Hungarians and made Graz his new capital. Friedrich was a devotee of the occult and believed with nearly mystical certainty in the power of his own royalty. Through much genealogical sleight of hand, he traced his bloodlines back to Augustus Caesar and to Priam of Troy and was the first to engrave the letters *A.E.I.O.U.* on everything he owned. For most people, these are just the vowels in order, but for Friedrich, they meant *Austriae Est Imperare Orbi Universo,* or "Austria (i.e., the Habsburgs) Is Destined to Rule the Whole World." For all his mystical bravado, however, things had not gone well for him in Vienna. Right from the day of his coronation in 1452, he had been at war with Matthias Corvinus, the king of Hungary, and with his own brother as well. While besieged at the Hofburg by the Viennese people, Friedrich and his family ate their pets, along with a few hapless vultures. While in Graz, Friedrich built up the town, the cathedral, the city center. Little Graz became a city, a fine city but then in 1480, however, perhaps with the coming of the Habsburgs, Graz was devastated by one disaster after another, the "Plagues of God." First locusts, then the Black Death, then the Turks, and finally, hot on the heels of Friedrich, the Hungarians. For all the initial promise, these were not good times.

Then came the Reformation, and most of Austria followed the "new way," from the peasants, who had some stake in change, any change, to the local aristocracy, who wanted out from under the iron hand of the Habsburgs. Lutheranism had set the European imagination on fire with a new kind of fervor, a new way of seeing God, the world, and their place in it. The problem in Graz, in all of Austria, was that the Habsburgs had not changed, and by the terms of the Peace of Augsburg, *cuius regio, eius religio*—"whose the land, his the religion"—the Habsburgs were legally entitled to set the official religion of the country. Still, the law was one thing, and the ability to enforce it was another.[6] If most of the people and most of the aristocracy were followers of the new way, it would have been hard for any ruler, however autocratic, to impose the old. When Kepler arrived in Graz on April 11, 1594, Ferdinand II was grand duke, the same Ferdinand who, while Kepler had been studying in Tübingen, had studied with the Jesuits in Bavaria along with the Duke of Bavaria, and the two had determined to bring the True Faith back to their respective homelands.

Sixteen years before, in 1578, under the Pacification of Bruck, Ferdinand's father, Charles, had granted the Protestant nobility the right to practice their faith as they saw fit. This was a forced concession, and one that must have galled him. Wasn't the archduke God's own appointed representative, blessed and approved by the pope himself? Archduke Charles had to give the nobles and the people what they wanted just to prevent more uprisings. When the old duke died, his widow, the archduchess Maria, of the House of Wittelsbach and the mother of Ferdinand II, decided to take up the banner for the old faith and to prepare her son to do the same. Ferdinand was more than eager to do so. So, when Kepler arrived, he found a world divided, so different from his homeland in Württemberg, where everyone was a Lutheran from the duke to the basest peasant. After his family left Weil der Stadt, Kepler had been surrounded by people who believed as he did, but now there were Lutherans and Catholics all mixed together, balkanized by their respective churches and yet gamely trying to be one people. Instead of peace, the Habsburgs had found that they were sitting on top of a landslide, and the rocks, seemingly so solid, were shifting beneath them.

The *Stiftschule,* the Lutheran seminary school that had hired Kepler, was part of the landslide. It had been founded to compete with the new Jesuit college that the old Archduke Charles had set up in 1574. The Lutheran school therefore played a central role in the Lutheran resistance and was the principal meeting ground for the Lutheran party. When Kepler arrived, the rector of the school, Johann Papius, quickly befriended him, but Papius would last only a few more months before he left to teach medicine at Tübingen. His replacement, Johannes Regius, would not be so pleasant. At the time, there were twelve to fourteen teachers in charge of two levels, an upper and a lower, with four preachers above them keeping one eye on the archduke and the other on the teachers. The school inspectors included Pastor Wilhelm Zimmerman, the father of one of Kepler's schoolmates, the one who had been asked to leave Tübingen for lack of academic progress. He was also a friend of Mästlin. The good pastor quickly put Kepler on the spot and asked him why his son had been expelled from the university, and Kepler, typically, blurted out that it was because his son was spoiled by his mother and didn't know how to apply himself. Pastor Zimmerman, understandably, was not pleased. Kepler did his best to smooth things over, and all indications are that Zimmerman was big enough to accept his apology.

The school paid Kepler 150 gulden annually, 50 gulden less than his predecessor. Also, the administration felt a bit uncertain about the young man, so they decided to give him a probationary period of two or three months to see if he would work out. They assigned him to teach mathematics in the philosophical component of the upper level of the school, part of the second tier among the students; the first tier was for preachers and theologians, the second for those studying law. Part of that section included advanced mathematics, including astronomy, which had been important in Melanchthon's vision of education. The problem for Kepler was that not many students wanted to study mathematics—it was too hard and not very practical, except for astrology. The first year he had only a few students, and the second year he had none. No one blamed Kepler, because they knew that mathematics had never been popular, so the school administration assigned him to teach basic arithmetic, history, ethics, the poetry of Virgil, and even rhetoric as a substitute.

Kepler and the new rector, Regius, quickly found that they did not get along. "The Rector is hostile to me, and toward the end of the year, this became dangerous."[7] Regius got his nose twisted by the feeling that Kepler, that young, untried fellow from Württemberg, did not give him enough respect. The boy was brash, plain-spoken, even blunt, and would not listen to his instructions. For all this, Kepler got high marks from the school inspectors, who said that he had distinguished himself first as a speaker, then as a teacher, and finally as a debater. That second year was a year of good and bad omens. Kepler wrote that the stars indicated some future problems with his mother and her bad standing in the community and the suppression of her inheritance.

The other part of his assignment, the part that did not involve students, was as district mathematician for Graz and the surrounding region, which meant that the authorities expected Kepler to write a series of calendars, astrological forecasts giving them all a heads-up on the weather, the political situation, the crops, and any other possible disasters that might befall them. Kepler took to this, because there was extra money in it.

Personally, he was lukewarm about astrology. He referred to it often enough to make sense of the strange turns in his life, the character of each of his family members, and even some general things about the future of the nation, but he constantly worried that he might be "nourishing the superstition of fatheads."[8] Still, he practiced it, as did every other astronomer of the day. Tycho Brahe wrote horoscopes. So did Galileo, though he had an even worse opinion of astrology than Kepler did. This is what kings and emperors wanted from stargazers, to cast some light into the future effects of their royal actions. Taking a middle position on the subject, Kepler thought that astrology could cast some light into the world, but that light was diffuse, vague, and full of shadows.

In 1595, partly from his calculations and partly from his commonsense reading of the times, Kepler made three predictions: one, a terrible winter, with bitter cold weather that would damage fruit trees and cause hardship all around; two, an attack by the Turks from the south; and three, a peasant uprising. All three came true. That winter was so bad, they said, that anytime a shepherd in the mountains blew his nose, it would pop off.[9] The Turks did attack, which wasn't all that surprising, and there was a peasant revolt, again, not all that surprising. Suddenly, Kepler was a celebrity. "On October 27, we had a visit from a baron from Tschernembl, who was sent from the *Stände* (body of representatives) in Austria 'above the Enns' [north of the River Enns], to call for help against the farmers. The same called upon me and started out by mentioning the correctness of the predictions of my astrological calendar in the matter of these revolts."[10]

Another peasant revolt had broken out north of the River Enns, and the local lord, Baron Georg Erasmus von Tschernembl, later a major Protestant leader in Bohemian revolt against the Habsburgs, came by. He also mentioned that the ruler of the territories south of the Enns wanted Kepler to draw a map of his territory. Excited about the idea, Kepler agreed and, with the permission of his superiors at the school, constructed his first scientific instrument, a wooden double right-angled triangle with movable parts. It was about ten feet in length and five in height with a crossed base so that it could stand perpendicular. Soon, between the map and his calendars, Kepler had developed a second career, moving him deeper into the world of numbers, measurements, and all things mathematical.

His career as an astronomer, however, began in that same year, on July 19, 1595. He says in his journal from February 27, 1596: "The Almighty revealed a central discovery in astronomy to me last summer, after long, exhaustive work and diligence. I explained this in a special treatise which I want to publish soon. The whole work and its demonstration could, with the utmost sophistication, fit nicely into a serving cup with the diameter of a *Werkschuh*.[11] This would be a true reflection of the design of creation, as far as human reason may reach, and at the same time, no man had either seen or heard the like."[12]

The idea hit him in the middle of a lecture on the great conjunctions of Jupiter and Saturn. One can almost imagine him in the middle of a sentence, stopping, staring into the air above his students' heads for a long, excruciating moment, then stepping aside to write a furious note to himself. In spite of what the inspectors had said about him, Kepler had a reputation as a muddled lecturer, his mind hopping from one idea to the next, faster than the students could follow. For all his students could see, his mind had just made one more leap. For Kepler, however, the moment was decisive. He had put together the pieces, seen that the orbits of Jupiter and Saturn, within the Copernican system, could be drawn on the inside and then on the outside of an equilateral triangle, with Jupiter on the inside and Saturn on the outside. An equilateral triangle is the first regular polygon, one of the simplest archetypes of geometry. What if he could find similar relations between the other planets? Wouldn't this be an indication of a geometric regularity in the cosmos? For Kepler, this regularity was chock-full of meaning. It was a peek into the mind of God, into the cosmic template used to create the universe.

The problem was that it didn't quite work. He tried to draw a square inside the circle of Jupiter, and then another circle inside this square to indicate the orbit of Mars, and then another regular polygon inside the circle of Mars, and then on through the rest of the planets, but the circles he inscribed inside the polygons didn't quite match up with the known distances to the other planets. The regularity that he found in the orbits of Jupiter and Saturn couldn't be extended to the other planets, so it wasn't a universal principle. Besides, it didn't really explain why there were only six planets and not seven, or twelve, or sixteen. With the discovery of Uranus, Neptune, and Pluto (though there is some doubt about Pluto), we have since learned that there are indeed more than six planets, and that Kepler's scruples about the number was unnecessary, if in fact his system had explained anything.

But Kepler didn't stop there. He reasoned that two-dimensional figures were by nature inappropriate for explaining a three-dimensional universe with three-dimensional planets, so he looked into ways of nesting three-dimensional figures into each other to explain the distances between the planets, and that led him to the five Platonic solids. Since the time of the Greeks, the Platonic solids had been a mystery in mathematics bordering on mysticism.

There are only five such solids—the tetrahedron, with four triangular faces, four vertices, and six edges; the cube, with six square faces, eight vertices, and twelve edges; the octahedron, with eight triangular faces, six vertices, and twelve edges; the dodecahedron, with twelve five-sided faces, twenty vertices, and thirty edges; and the icosahedron, with twenty triangular faces, twelve vertices, and thirty edges. There are only five, and there can never be more than five. Plato connected these with the atoms of nature, the building blocks of everything. In this, he followed the earlier philosopher Empedocles: fire for the tetrahedron, earth for the cube, air for the octahedron, and water for the icosahedron. The dodecahedron he connected to the element *cosmos,* the stuff from which the stars and planets are made.[13] Following Plato, Kepler thought that he had found another place where these solids appeared, and since nothing that God fashions is created without a plan, he believed that he had found a way to calculate the distances between the planets in an *a priori* manner, that is, before any observation takes place. The planets were at these distances not because they happened to be there; they were there because God meant them to be there.

On August 2, 1595, Kepler wrote to Mästlin in Tübingen, remarking how he had discovered a way of calculating the planetary orbits *a priori.* He wrote again on September 14, outlined his polyhedral hypothesis, and then went on to explain why the planets moved as they did. The polyhedral hypothesis constructed an account of the number, order, and distances of the planets, but not why they moved. In his letter, Kepler suggested that an *anima movens,* a spirit of movement, existed in the sun and that its power, a *vigor movens,* weakened as one moved farther away from the sun. In this one short comment, he set the stage for his own discovery of the area law and the elliptical shape of the planetary orbits. He also set the stage for Newton's law of gravitation, including the idea that the effect of the moving action decreased with distance. Later in his life, Kepler changed the idea of a moving spirit to a mechanical movement, a force, leaving Newton with the task of defining this mechanism, this action at a distance, in terms of postulates, theorems, and universal laws.

From this point on in his life, Kepler began to search for this divine plan with grand hopes and grand visions. His life as a scientist was as theological as it was scientific. He wanted to find out what God really intended, God's final cause of the universe. He wanted to find out what was in God's mind, actually, really. To do this, he had to break with the astronomical tradition, since Ptolemy, that astronomers used mathematics to create plausible accounts of the appearances in the sky. Like Copernicus, Brahe, and Galileo, he believed that his theories represented the structure of the universe as it really is, accurately and completely. If observation didn't support his theories, and it eventually did not, he was willing to cast them aside and go on looking. The next seven months after his discovery, he set his ideas into a book he ponderously titled *Prodomus Dissertationum Cosmographicarum, Continens Mysterium Cosmographicum,* the *Forerunner of the Cosmological Essays, Which Contains the Secret of the Universe.* The subtitle was *On the Marvelous Proportion of the Celestial Spheres, and on the True and Particular Causes of the Number, Size, and Periodic Motions of the Heavens, Established by Means of the Five Regular Geometric Solids.*[14] Arguably, no book title in the history of Western civilization has ever claimed more for itself than this one. Cosmologists talk like this from time to time, however. Books on "the theory of everything" are part of a long-standing tradition. Ultimately, Kepler's theory was largely speculative, and he never could get it to fit the data. Today, it has been disproven, and in spite of what Kepler thought about it, it does not generally stand among his greater achievements.

What Kepler believed was that he had found the map to the shape of the universe, and that his "little book," as he called it, was a book of cosmography, a picture of that map. The truth of the universe, he maintained, can be found in the mysteries of geometry and not in the properties of pure numbers. Three centuries later, Einstein returned to this idea when developing his general theory of relativity, setting geometry once again at the head of the math class. Kepler considered the properties of pure numbers to be accidental, with the sole exception of the Trinity, the number of God, because God is of course the exception to everything. In saying this, he abandoned numerology. The numbers one, two, four, sixteen, the twelve tribes of Israel, forty days and forty nights had no significance in themselves, but gained significance only when they were used to calculate geometrical relations, which were grounded in nature, which measured real things, land, cities, stars and planets, and were not mere abstractions. Following Nicholas of Cusa, Kepler said that the most fundamental and most important distinction in geometry was between the straight and the curved, the curved representing God and the straight humans. There was no strict line, no impassable wall, between Kepler's astronomy and his theology. Such walls and divisions developed in later years, possibly by the tidal shifts in Western culture initiated by the Thirty Years' War, the war that caused Kepler himself so much grief.

Kepler's search of the heavens was a search for God. It was, he believed, God's will. God wanted us to find the plan of creation for ourselves and thereby share in the divine thoughts. The world did not come out as it had for no reason; distinctions like the curved and the straight were inherent in the universe from the beginning. The world as God intended it had to be the best and most beautiful possible, with perfect divine order underlying all the seeming chaos of sinful humans. On April 9, 1597, he wrote to Mästlin and argued, as Plato had done in the *Timaeus,* that just as the eye is apt to seeing light and the ear to hearing sound, so the intellect is apt to the knowledge of quantity.

Kepler finished the *Mysterium Cosmographicum* in 1596 and set about looking for a publisher. He wrote to Mästlin about this, sent him a copy of the manuscript, and asked for his opinion. Mästlin was deeply impressed with the book and offered to help get it published. He had only one criticism, that Kepler had not taken the epicycles into account, which led to problems about the thickness of the spheres, but Kepler pointed to Tycho Brahe's work on the comet of 1577, in which he showed that the spheres were not real at all, but mathematical fictions.

To get a book published, even a book on astronomy, perhaps especially a book on astronomy, it had to pass muster with the theologians, the guardians of tradition and orthodoxy. Just as Galileo would have trouble with the Inquisition a few decades later, Kepler would have trouble with the faculty at Tübingen, and for many of the same reasons. As a believing man, Kepler wrote a chapter in the book that tried to show that the Copernican universe was not in disagreement with Scripture, as Galileo would later do in his *Letter to Grand Duchess Cristina.* Hafenreffer and others gave Kepler the same advice that would be given to Galileo—keep this theory, conjecture, hypothesis, not reality. Don't try to make overreaching claims for your ideas. Otherwise, you could get in trouble.

ᏩᏤᏌᎧ

No mortal, I believe, is alive whose destiny is farther from foresight than mine. Already, I was hoping for happiness and believed in its enjoyment, when it got away from me. Contrarily, when I feared the worst and saw it coming, the good appeared. And the more my destiny is in my favor, the more I fear it will once more turn against me, as it is not constant, that it may stay that way for a long time. In the year 1596, I chose a wife and didn't think of anything else for a half year, since letters from exceedingly serious men encouraged me. Happily, I returned to Steiermark [Styria]. When no one congratulated me upon my arrival, I was taken aside and told that I had lost my bride. Since the hope for marriage had taken root over a half year, it took another half year to get over it and to convince myself entirely that it is nothing and embark on a new path of life. After all hope was gone and the outcome already reported to the authorities, another change occurred. The people [the bride, Barbara, and her father] were moved by the governing authority and by ridicule whenever they showed themselves. Then one [of my supporters] worked harder than the next for the disposition of the widow and her father; they were successful and thus prepared another marriage for me. Because of this turmoil, all my decisions for a change of path were canceled. That is how little control each man has over his future.

Marriage portraits of Barbara Müller and Johannes Kepler

VI

Married under Pernicious Skies

Where Kepler publishes the Mysterium
Cosmographicum, *and in the following year
marries Barbara Müller von Mühleck, a widow
twice over with one daughter, which
marriage is complicated.*

IN THE ORIGINAL DEDICATION of his *Mysterium Cosmographicum,* Kepler wrote a short defense of his science for the practical people of the world. Ever since the servant girl laughed at Thales of Miletus, who forgot his feet and fell into a ditch while gazing at the stars, practical people have scoffed at intellectuals, who spend their lives in pursuits far from the farm, the merchant ship, and the counting house. The patrons of the arts and the sciences, however, have generally been practical people themselves, but blessed with money, power, and the unexplainable itch for something more, so intellectuals have traditionally penned dedications in the fronts of their books to convince the powerful that their money was well spent.

"Do you want something bulky?" Kepler asks. "Nothing in the whole universe is greater or more ample than this. Do you require something important? Nothing is more precious, nothing more splendid than this in the brilliant temple of God. Do you wish to know something secret? Nothing in the nature of things is or has been more closely concealed. The only thing in which it does not satisfy everybody is that its usefulness is not clear to the unreflecting."[1]

It was the unreflecting that Kepler had to convince. The dedication of his book reads: "To the Illustrious, Eminent, Most Noble and Energetic Lord Sigismund Frederick, Free Baron of Herberstein, Neuberg, and Güttenhaag, Lord of Lankowitz, Hereditary Chamberlain and Steward of Carinthia, Counselor to His Imperial Majesty; And to the Most Serene Archduke of Austria, Ferdinand, Captain of the Province of Styria; And to Their Lordships the Most Noble Five Commissioners of the Illustrious Orders of Styria, Most Generous of Men, My Kindly and Liberal Lords, Greetings and Homage."[2] At least one of these men listed in the dedication would be responsible for his eventual exile from the province. Beyond its obsequiousness, which was the style then, the dedication set out in plain words what Kepler really thought about the book. Philosophers and scientists from the beginning of Greek civilization have sought the secret, the truth of everything, the one scheme that would encompass it all. Mathematicians call these solutions "elegant," for they are composed of complicated threads laced together into a beautiful simplicity. Kepler believed that such a solution would enable him to read the Book of Nature, and that following that road was for him a new kind of ministry: he was *sacerdos libri naturae,* "a priest of the Book of Nature." In his dedication he refers to St. Paul, who says that we should contemplate God like the sun in water or in a mirror. A little later he quotes Psalm 8:

When I see your heavens
The work of your fingers,
The moon and the stars that
You set in place . . .

While dressing himself in his room on Schmidgasse Street, while walking to and from the school, while in the classroom, while dining, while in the latrine, while preparing himself for bed, Kepler contemplated the heavens. His mind dug through geometric forms, triangles and squares, hexagons and octagons, cubes and dodecahedrons, as if they were gilded treasures hidden away in an old chest. Without practical concerns to keep him anchored in this world, he would have abandoned himself to the sky and, like Thales, would have lost his footing and fallen into a ditch. Fear of poverty alone drew him back, for he knew that a poor man is only a step away from slavery. For most of his life, he would fret over practical things, over making money, over paying the bills; his letters show this. Yet God's glory in the heavens tickled his mind, whispered in his ear, pulled him back to his desk, away from little Graz and Austria into the wider universe once more. Therefore, his friends and associates in the Lutheran community, both in Württemberg and in Graz, were determined to find him a bride.

The day he left Tübingen, he carried the Uriah letter with him, probably from the church officials in Württemberg to the church officials in Graz asking that he be introduced to some likely young woman bent on marriage in order to quickly transform him into a proper, satisfied burgher. In December 1595, they introduced him to Barbara Müller, the daughter of Jobst Müller of Gössendorf, the mill owner, who also owned an estate called Mühleck, south of Graz.[3] Kepler was entirely smitten, for Barbara was young, round, and pretty in a country way. And the fact that she brought her own fortune along didn't hurt.

In January, two friends of his, Dr. Johannes Oberdorfer, an inspector at the *Stiftschule,* and Heinrich Osius, a former professor, now deacon at the college church, approached Barbara's father in the role of "gentlemen delegates" to recommend Kepler to him and to begin negotiations leading to an eventual marriage. Barbara and her father, however, were quintessentially practical people who thought little about the orbits of the planets and who had built a tidy fortune out of hard work, good business sense, and a touch of luck. They were on their way up and yearned for noble connections. The Kepler pedigree, with its long-dead knightly ancestors, was one reason the Müllers even considered Johannes, who was otherwise a low-level teacher of mathematics with no fortune of his own.

When they met, Barbara was merely twenty-three years old, but she had already been married twice to older men, both of whom had died. Her first husband was Wolf Lorenz, a cabinetmaker from Graz with a good career and a tidy sum. She bore him a daughter, Regina. Then, after only two years of marriage, he inconveniently died. Soon after, Barbara married Marx Müller, a district paymaster at a local estate, which was a solid, respectable position. Unfortunately, he brought several loutish children with him into the marriage and then involved Barbara in a touch of scandal. Marx had been light-fingered with the receipts, something that surfaced only after his death. Fortunately, he was along in years, sickly, and by 1595 he too had died. By the end of that year, Barbara had been introduced to young Kepler.

Her father, Jobst, unfortunately, was not convinced. Who was this young man to court his daughter? Everyone knows what an impoverished young man wants from a girl with a fortune of her own. This Kepler, this stargazer, this dreamer, this son of an adventurer, this young man who had reached for the ministry and ended up a mathematics teacher. Who was he? Jobst was a doer, not a dreamer. He was a man of means. Not content to be a mill owner, he kept his hand in several businesses at once and over the years watched his wealth grow. But as a man on his way up, the only thing left to acquire was nobility. The Keplers had that, vague and distant as it was. It was all right for Barbara to have married her previous husbands, because they too were men of means. They had added to the family fortune, and if they had, by the grace of God, died all too soon, Barbara's fortunes had not died with them. But this Kepler fellow, what could he offer his daughter? Only a noble name.[4]

Kepler pressed his suit, but the suit was not well received. The arrangements floated in the air—the church authorities, including the rector at the *Stiftschule*, were for the match; the father was not. But Jobst had help. One Stephan Speidel, the district secretary, egged Jobst on in his fears, because he had plans of his own for Barbara. Not for himself, of course, but for a friend, a friend who would owe him some political favor or other for securing a match for him with a wealthy bride.

Meanwhile, Kepler had finished his book and was looking for a publisher. He was certain that he had discovered a singularly important truth, the blueprint of the universe, the most wondrous secret of all. The praise of God was his subject. What voice, he said, do the stars have to give praise as humans do? And if his little work, *The Secret of the Universe,* should clarify that voice, praising God in response to the majesty of the heavens, the moon, and the stars, who could charge him with vanity? Divine things should not be counted up like cash, like money paid for food or clothing. Therefore, sensible people should not abandon astronomy in spite of the fact that it will never feed their bellies. This statement alone would have set his prospective father-in-law's teeth on edge, had he read it. But artists give delight to the eye, Kepler went on to say, as if in retort, and musicians give delight to the ear. If astronomers give delight to the mind, is that of any less value? Would God, who does nothing without a plan, without foreknowledge of the future, leave the mind alone without delight? A shabby existence, indeed.

For Kepler, God had planted truth in nature to act as a kind of wordless Scripture, a companion to the Bible.[5] He saw himself, the priest of the Book of Nature, deciphering that book much as a minister of the church deciphers the book of Scripture. All his training at Tübingen taught him to interpret the original texts, to mine them for divine meaning, and the same divinity could be found in the world, just as it was found in the Bible. When he left the university, however, he felt his life changing, the sudden turn uncertain. The longer he stayed in Graz, the wider the rift between his new life and his old. With the completion of *Mysterium Cosmographicum,* he had realized a new vocation, a new way to serve God. Writing to Mästlin in 1595, he claimed that astronomy should be a practice done in the service of God, to give God glory and honor. Such thinking was not only in line with Mästlin's, but also with Philip Melanchthon's as well.

Following Melanchthon, Kepler argued that astronomy is natural to humanity, as natural as singing is to songbirds. We don't ask why birds sing other than to say that they delight in it, so we shouldn't ask why the human mind searches the heavens to plumb God's secrets. The study of the heavens is something we do for its own sake, because it is a part of us, natural to us. Ultimately, we study the sky to glorify God, the Creator of the stars and the Source of all perfection, for that too is a part of us. Moreover, there are lessons to be learned. God teaches us how to live a moral life as we contemplate the perfect heavens. We see the angelic stars, the peace, the royal order, the noble predictability of the constellations, with only a wandering comet now and then to give us pause and to teach us humility. The skies, therefore, speak to us of God. This is because, for Kepler, there is a strong correlation between the way that the human mind thinks and the way that God thinks. The Scholastics, after Thomas Aquinas, called this "connaturality," a resonance, an aptitude between the mind and the world, as if the knowing mind and the existing universe were made for each other. The human mind is apt to the world, and the world is apt to the mind.

More than anything else, it was this idea that was lost to us after the discoveries of quantum physics came to light, what with particles that can be in two places at once along a sliding scale of probable existence. The human imagination grows more alienated from the universe with each new day, while mathematical reasoning pieces together a wider landscape for the deep places that matter inhabits, a landscape that few can understand and none can completely imagine. Postmodernists chart this change by denying the one thing that made Kepler's universe what it was—a cosmos. Kepler's secret of the universe, his geometric hypothesis of nested Platonic solids, was possible for him only because he believed in a divine order to things, a "theory of everything" that was a direct projection of the mind of God.

The sciences we know in the twenty-first century are the grandchildren of the cosmology of Copernicus, Kepler, Galileo, and Newton, each a man of faith in his own way, who sought the stars for this cosmos, this photograph of God. Even as the West shuffled toward materialism, the yearning for mystical truth was burrowing into the mental fabric of science—for what we find beautiful we must also find true. In the twentieth century, when James Watson and Francis Crick, who were galaxies away from Kepler's Lutheran faith, discovered the structure of DNA, they knew they were on the right track because the structure they had found was beautiful. It had to be the right one, they said. The Good, the True, and the Beautiful—these are the deep properties of Plato's Ideas, and the yardstick by which Kepler measured the universe.

For Kepler, astronomy paralleled biblical studies because the heavens held a special position in creation. As for Aristotle, they were insusceptible to corruption and were therefore a cleaner revelation of God. The astronomer sees the truth because humans are created in the image of God, and perceiving the truth is how we share in God's mind, which created the truth. The bridges between the mind of God and the human mind are the archetypes, the geometrical absolutes, the perfect forms that built the cosmos out of infinite unformed, primal matter. It was out of the archetypes, the triangles and squares, the hexagons and octagons, the dodecahedrons and icosahedrons, that God made all things, for God is the ultimate mathematician.

When Kepler had first made his discovery in July 1595, he felt as if an oracular voice had spoken to him from heaven. He burst into fits of weeping. The thought that he, sinful as he was, should receive such a revelation astounded him. "Now I no longer bemoaned the lost time; I no longer became weary at work; I shunned no calculation, no matter how difficult. Days and nights I passed in calculating until I saw if the sentence formulated in words agreed with the orbits of Copernicus, or if the winds carried away my joy." His discovery needed to be published, not for his own glory, but for God's, for he "vowed to God the Omnipotent and All-Merciful that at the first opportunity I would make public in print this wonderful example of His wisdom."[6]

Now the problem was getting the book published. Kepler wrote several times to Mästlin for advice. Mästlin was enthusiastic, with a touch of caution, but he agreed to help in whatever way he could. In February 1596, Kepler's negotiations with Barbara and her father were making progress. Work on his book had finished, for the time being, and Mästlin had offered his help both in editing the work and in getting it published. Things were coming together. Then word came that his two grandfathers, aged and sick, had taken a turn for the worse and were calling for him. They wanted to see him again before they died. Kepler took leave from his teaching for two months and traveled back to Weil der Stadt and Eltingen. Old Sebald died soon after, and Melchior Guldenmann, Katharina's father, grew sicker by the day.

After Sebald's funeral, Kepler traveled on to Tübingen to speak with Mästlin face-to-face. Mästlin was generally happy with the book, though he had reservations about Kepler's introduction of an *anima movens*, a spirit of movement, or a force, in the sun. He worried that this blurred the line between astronomy and physics. Traditionally, astronomy was about geometry, about creating hypotheses that accounted for the appearances, for what people saw in the sky, while physics was about explaining movement on earth or in the atmosphere just above the earth, in the world under the sphere of the moon. Explaining the heavens in terms of earthly motion, that is, in terms of Aristotle's four causes, blurred the line in a suspicious way. What Mästlin did not realize, as twenty-first-century people do, is that this one idea was a watershed, a turning point in science leading to the modern world. Kepler's *anima movens* eventually became Newton's law of gravity, one of the grounding pillars of science.

Kepler's response to Mästlin, however, was more metaphysical than physical. "Of all the bodies in the universe," he wrote, "the most excellent is the sun, whose whole essence is nothing else but the purest light. Than it there is no greater star; singly and alone it is the producer, conserver, and warmer of all things. It is a fountain of light, rich in fruitful heat, most fair, limpid, and pure to the sight. It is the source of vision and the portrayer of all colours, though itself devoid of colour."[7] The sun was like God the Father, who created the universe. The stars were like God the Son, forever constant. The space between, the space of the moon and the planets, was like the Holy Spirit, the upholder and conserver of all things.

In spite of his concerns over some of Kepler's ideas, Mästlin applied to the Senate of Tübingen for approval of the book, the first necessary step for publication. The Senate's response was tepid. They recognized the young man's obvious talent, and they trusted Mästlin's word about its scientific value. Still, they worried about Kepler's attempts to reconcile

Copernicus with the Bible. How could anyone accept both the descriptions of creation in Genesis and a moving earth at the same time? Moreover, the book of Joshua clearly stated that the sun stopped moving in its course for fifteen minutes in order to give Israel the victory in battle. A moving earth simply did not make sense. Also, they worried that the average reader, who did not have Kepler's knowledge of Copernicus, could be led astray, and they wanted him to write a series of elucidations to explain the most difficult Copernican ideas to the newcomer.

In spite of the Senate's concerns, however, Kepler had become a minor celebrity in Tübingen. His discovery had made the rounds, and the same faculty who had fretted about him while he was a student celebrated him now that he had become an author. No one had yet come up with a scheme to calculate the number of the planets, their arrangement, and the size of their orbits or to add to that an explanation of their motion through the sky. The idea of an *a priori* cosmography, what moderns might call a theoretical cosmology, was entirely new. To tease out the thoughts of the Creator in pure mathematics was audacious, brilliant, and wonderful. After a dinner party, Martin Crucius, Kepler's old classics teacher, wrote a note about him in his otherwise pedestrian journal: *"Pulcher iuvenis,"* he said, "Charming boy!"[8]

Meanwhile, Kepler traveled on to Stuttgart, where he joined the ranks of the lower and middle managers of the duchy in the *Trippeltisch,* the place in the ducal castle where such managers resided and ate their meals. While there, he petitioned the duke to have a model of his cosmography fashioned by a local goldsmith. He envisioned a goblet about the size of a drinking cup with the various Platonic solids—all hollow—nested inside one another like a matrushka. This idea caught the duke's imagination and he commissioned it at once, but the artisan he chose to do the job was slow and not all that excited about the goblet, so the project seemed to drag on forever. Meanwhile, Mästlin had continuing trouble negotiating with the Senate, and then with the publisher about who would pay for what, how it would be paid, and this, that, and the other. Because of these delays, Kepler stayed on in the duchy for nearly seven months, even though his leave of absence from the school was for no more than two.

The *Mysterium Cosmographicum* finally came out in December 1596, and by February 1597 Kepler had returned home and had found out the awful truth—while he was away, he let his bride slip through his fingers. He shouldn't have been away so long. What does it say about the ardor of a suitor who is off traveling in foreign lands for seven months? Old Jobst had had second thoughts about the boy. While Kepler was in Swabia, his gentlemen delegates had continued their negotiations and had finally been successful. They sent word on to Swabia. While in Tübingen, Kepler heard from Papius, his old friend who had been rector at the Lutheran *Stiftschule* in Graz when he arrived, that the family had agreed and that all was well. The bride was his.

He had been looking forward to this marriage for more than half a year, but then, on his arrival in Graz, he noticed that no one had stepped up to congratulate him on his upcoming marriage. He wondered about this and feared some disaster, when a friend pulled him aside and told him that it was true, the Müllers had canceled the marriage. In spite of Kepler's nobility and his newfound notoriety, he simply didn't make enough money to satisfy old Jobst. Kepler took sick with an irritated gall-bladder and depression. "Rage, audacity," he wrote in his journal. "The position of the stars is powerful, calling out passion, hurt, and fever."[9]

In the seventeenth century, marriage was never an affair between the bride and groom alone, but a complex political dance between two families. And not just them—the whole community got involved. Public sentiment was so strongly in favor of Kepler, the jilted bridegroom, that whenever the Müllers showed their faces in public, they had to put up with sniggers and ridicule, satiric verses and unkind humor. Moreover, the church authorities paid a visit to the Müllers and, in good Lutheran fashion, instructed them on their duties. Eventually, old Jobst caved. The marriage was on once again. Poor Kepler, caught in the whirlwind of an off-and-on-again love, was turned round, first by joy, then by depression, then by joy once again. Just as he had come to accept the fact that he would not be married to Barbara, the marriage was back on. "That is how little control each man has over his future," he wrote to Mästlin.

Then something happened that added a touch of vinegar to Kepler's relationship with his old teacher. In March, just a month before Kepler's wedding, Mästlin sent a dyspeptic letter to Kepler, complaining about how much time and effort he had given to the publication of *Mysterium Cosmographicum,* so much so that he had neglected his own work, his fiery critique of the new calendar, that papist invention designed to subvert the spiritual independence of good Lutherans everywhere. Kepler didn't agree with his old teacher and told him so in a return letter, for Kepler was not only a mathematician and astronomer, but also a historian, and he knew the history of the Gregorian calendar better than his teacher. His view was practical to the core.

The European calendar used by Christians throughout the Middle Ages had first been commissioned by Julius Caesar, whose astronomers and soothsayers had informed him that the year was $365^1/_4$ days long. The problem was that Caesar's astronomers and soothsayers were wrong. The year is not $365^1/_4$ days long, but a few minutes shorter than that. The Council of Nicea (325) set down the rules for determining Easter based on the first day of spring, the equinox, but then later, in the middle of the Dark Ages, St. Gregory of Tours (544–595) wrote that these calculations were becoming a problem, because the equinox kept creeping backward, moving earlier one day every 128 years. In 750, the Venerable Bede reported that the equinox was now occurring three days earlier than the date set by the Council of Nicea. Eventually, said Bede, they would be celebrating Easter in January. Finally, the great medieval astronomer John Holywood, otherwise known as Sacrobosco, whose treatise, later called *The Sphere of Sacrobosco,* became the touchstone for all medieval astronomy, calculated that the observed year was 11 minutes and 14 seconds shorter than the Julian year.

Various solutions were put forth, but nothing actually happened until the sixteenth century, when Pope Gregory XIII handed the entire mess to Father Christoph Clavius, the famous Jesuit astronomer, commentator on *The Sphere of Sacrobosco,* and later adviser to Galileo Galilei. In papal circles Clavius was the local hero and certainly one of the great astronomical minds of the day. Clavius took an earlier solution proposed by Aloysius Lilius and simplified it, saying that the last year of each century should be a leap year only if that year was divisible by 400, years such as 1600 and 2000. This still left standing a morsel of error in the calendar, but since this came to only one day in 3,333 years, Clavius thought it was good enough. The pope agreed, and published his bull *Inter Gravissimas,* which decreed that October 4, 1582, would then jump to October 15 to reset the calendar, which would henceforth, in perpetuity, follow Clavius's worthy scheme.

Mästlin hated the idea. He was not alone in this. All over Europe, people belittled this reform, and Mästlin headed the pack. Suddenly, Protestant and Catholic differences became all-important. Mästlin was not about to let the pope in Rome tell good Lutherans how to count time. He took to the intellectual streets, writing four diatribes against the calendar. Since the first one helped him get his professorship at Tübingen, his course was set. Intellectual fur flew all over Europe between Mästlin and various Jesuits defending Clavius, including Clavius himself. Attacking the calendar reform had become a bit of a cottage industry for Mästlin, and he was on his fifth diatribe when he turned aside for a time to work on Kepler's book. Too much time, according to Mästlin. After *Mysterium Cosmographicum* came out, he wrote that letter to Kepler, and Kepler, in typically undiplomatic fashion, wrote back:

> What is half Germany doing (I ask)? How long does it mean to
> hold aloof from the rest of Europe? For what are we waiting? . . . It
> is 150 years since astronomers demanded legislation for some cor-
> rection, and Luther himself demanded it. . . . Now one correction
> has been made; no one can easily introduce another into a small
> part of Europe without great disturbance. Therefore, either the old
> form must be retained or the Gregorian accepted. But which? . . .
> The states have proved their independence of the pope for almost
> twenty years; let that suffice. He already sees that we may, if we
> wish, retain the old calendar. If we choose to emend it in the same
> was as he did, it is not because we are forced to do so, but because
> it seems good to us to do so. . . . It is a disgrace to Germany that

they who discovered the art of reformation should alone remain un-
reformed.[10]

This did not sit well. Mästlin did not respond immediately, but as years
passed, his relationship with Kepler diminished, partly because of time
and separation and partly because of old pique. Here was Kepler, his old
student made good, now a famous man, and he had turned on Mästlin,
his teacher. Over the years, Kepler tried to return to Swabia, to attain a
position teaching at his old university, but was rebuffed each time. Bit by
bit, Mästlin pulled away, until in the time of Kepler's greatest need, only a
few years away, he turned silent. His letters never came. He sent nothing,
no reply; there was only a dull, sullen silence.

Kepler saw none of this at first, because he was too often adept at not
noticing. His wedding plans marched on, and for a time he was happy. In
the midst of his happiness, however, dark clouds begin to form, a wisp
here, a puff there. Old Jobst was not the only one to have had doubts.
Kepler too fretted about the wedding. Though he loved Barbara, he was
overwhelmed by the expense of the wedding celebration, much of which
he had to pay himself. On April 9, only a few days before the wedding, he
wrote again to Mästlin: "The state of my affairs is continually such that, if
I should die within one year, no one could leave greater mayhem behind. I
have to pay large expenses out of my own pocket, for it is the local custom to celebrate a wedding most glamorously."

But money alone was not his greatest concern. What will happen to his
hopes to return to Swabia and take up the ministry? "I shall surely be tied
and chained to this place, whatever happens to our school, for my bride
has property, friends and a rich father here."[11] Once he married Barbara,
Graz would be his home. Although Barbara's money would never be his
personally, part of his task in life would be to manage her money, lands,
and estates along with his own. If Kepler was the irresistible force,
Barbara was the immovable object. The Counter-Reformation was coming in force—everyone knew this. It was already there in the company of
the Jesuits, in the person of the archduke. Kepler knew as well as anyone

that life in Graz could get warm indeed for a staunch Lutheran. If he married Barbara, with all her lands, his future could, indeed would, include a difficult choice between Barbara's desire to remain in her home and his own Lutheran faith. Kepler could not quit the land just to return to Swabia, to his own homeland, or to finish his studies for the ministry. If he married Barbara, he would be a family man, with responsibilities; he no longer had the freedom of his single youth.

Meanwhile, the forces of the Counter-Reformation were growing all across the Habsburg lands, including Austria, including little Graz. The Lutheran community, always at a disadvantage with the archduke, was beginning to watch the horizon for the coming storm. It would begin with an announcement, a tax on Protestant churches or a banishment of Lutheran preachers from the pulpit. Then it would escalate. The archduke could forcibly convert all Protestants in his territory if he wished, or he could exile them from the region. Either way, life for a good Lutheran would become increasingly difficult.

The Turks were gathering as well. Even as Kepler married Barbara that April 1597, "under pernicious skies," an army of over six hundred thousand Turks pressed the frontier. Would they attack? In spite of all, Johannes Kepler promised himself on April 12, 1597, "under the hand of the priest" to marry the "honorable, virtuous Frau Barbara." The wedding ceremony took place a few weeks later, on April 27, in the house of Herr Hartmann von Stubenberg, on Stempelgasse. The *Stände,* or church authorities, in Steiermark sent a silver cup worth 27 gulden. Kepler made sure they were all invited.

೧ꮗꮗꮩ

Presently, our Kaiser's (Rudolf II's) reign of power is
increasing, notwithstanding all arguments, regardless of the
Spanish declaration. Indeed, those from our country would do
well do be on the alert. It appears that our Kaiser possesses an
Archimedean calculation of motion, which is so slow that the
eye barely sees it, but as time goes on a tremendous mass is
put into motion. He sits in Prague, without the slightest
experience in the art of war, without total power, as was
believed before, and still he is accomplishing miracles; he
holds the dukes in obligation, has them compliant, obedient.
Candidly, he takes royal power upon himself, powerful
throughout the centuries. He wearies all through drawn-out
wars, so their only greater complaint would be against the
enemy. He uses those tools as a basis for undivided power, so
that only the subjugation of the Turkish Empire seems to be
missing. By what fortunate circumstances he gave
Siebenbürgen back to the Austrians, an act admired by all.
Yes, we are generally assured that our Kaiser is nominated as
third among the heirs of Moscow. In this matter a most
brilliant legation has been prepared to achieve that only our
heir will be nominated. Only the death of the Moscowitch
thwarted the legation. In any event, everybody has to
recognize that God blesses the cause of the Kaiser, and still,
everywhere in Austria, everything tragic becomes connected to
his name.

LETTER FROM KEPLER TO HERWART VON HOHENBERG
DECEMBER 9, 1598

In the month of August, the prelude to the tragedy took place.
The town pastor officially prohibited our preachers from all
practices of religion, the administration of the sacraments, and
the consecration of marriages. He gave as a reason a privilege
that all town pastors have had for a long time, and whose fees
are reduced when other clergy do the work. The head of the
church took our side. The town pastor replied once and again
and finally referred to the secular branch. The archduke
claimed he has to not only protect us but also his believers, so
now he does as requested what before he did out of his own
devotion. He decreed that all ordinations be stopped, and that
all servants of the church and school in Graz and Judenburg
be banished within fourteen days, ordering them to stay away
forever from his lands under penalty of death. Such happened
on September 20. The church authorities answered that their
sole desire is to protect people from violence; the power to let
people go is not in their hands, but is a matter of the whole
assembly.

Meanwhile, Spaniards were here to accompany the bride of
the king. On September 24, the archduke himself ordered us,
under penalty of death, to leave his provinces within eight
days. For a long time an argumentative correspondence took
place between the archduke and the ordained ministers; the
latter had called upon the "Assembly," members from regions
around Graz, but because of the unusually severe floods,
which occurred in August, not more than thirty came together.
In the end, the archduke ordered us in a stern directive to
vacate the city by sundown and leave the provinces within
seven days under penalty of death. Thereafter we left,
scattered toward Hungarian and Croatian areas where the

Kaiser rules, leaving behind the wives, in accordance with the advice and order of the "Assembly." Despite all this we received our salary, in addition to travel money, and were told to bear our fate until an assembly had taken place. We are hoping for this to this day. As for me, I returned after a period of one month, called upon by officials of the archduke, who described me as exempt. Despite this, I requested the archduke declare my servitude as neutral and exempted, for the decree statement was general, so I shall not be in danger while continuing to live in the country. The answer was worded as follows: "His Highness is herewith granting, out of special favor, that the petitioner, notwithstanding the General Dismissal, shall be allowed to remain here. But he shall maintain appropriate modesty everywhere, so that this exception will not be subject to cancellation."

People say the archduke likes my discoveries and Manecchio, his adviser, is in the habit of writing to me, so I enjoy a favorable court. What now, shall I stay? The preachers are expelled from all three countries. Some are hiding privately in castles. If one, however, administers the sacrament to one of the archduke's subjects, he will also be banished. On November 26 the citizens were advised to visit the town church for baptisms, marriages, and to receive other sacraments and to attend services there. That is why I shall leave. Meanwhile, my wife is attached to her property and the hope for her father's assets. It appears as if she would have to leave everything behind, even my stepdaughter of eight years, who would soon have to join the papists, after losing her mother. Confusion is everywhere. Therefore, I shall act as before. Also, I will not ask the Duke of Württemberg, before I am not banished entirely. If, on the other hand, there is an appointment in Stuttgart in my future, I will not decline it, nor wish to have it declined. If I should receive an appointment for my *"Model of Mysterium Cosmographicum as a Clockwork,"* I shall be at liberty to contemplate whether to accept or to decline. It is better perhaps if I act as the mathematician here, as Tycho has done in Denmark, as an individual, work with the written word only or, if the opportunity shall arrive, give lectures, perhaps at a university.

Archduke Ferdinand, later Emperor Ferdinand II

VII

An Archimedean
Calculation of Motion

*Where the Lutheran community of Graz is
persecuted, then banished, and where Kepler, who
must choose between his faith and his position,
is finally banished with them.*

THE RENAISSANCE ENDED BADLY for Rome. Pope Leo X excommunicated Martin Luther and called upon Habsburg Charles V, Holy Roman
Emperor, King of Spain and Naples, Overlord of the Netherlands, and a
good Catholic, to round up Luther and his followers, try them, and execute them as heretics. Charles agreed, but demanded a quid pro quo from
the pope.[1] The Reformation had already won too many converts, so trying to squash it in Germany might have ignited a civil war. Charles was
willing to risk this, but if the pope wanted the emperor's support, the
emperor also wanted the pope's. Charles planned to attack the French
holdings in Italy, especially in Milan, and it would have been convenient,
though not strictly necessary, to have had the pope's blessing. Leo agreed,
and Charles attacked Milan, chasing the French back toward the Alps.
Then Leo inconveniently died, and his successor, the Medici pope
Clement VII, a suspicious and uncertain man, was not quite the decisive

leader his predecessor had been. He vacillated between France and the emperor, landing finally on the side of the French, who wanted to form an anti-imperial league.

The Emperor was livid. The pope had wanted him to punish the Lutherans—so let the Lutherans punish the pope! The emperor's brother, Ferdinand of Austria, gathered a vast army, an easy twenty-thousand German *Landsknechte,* low-level knights and warriors, very much like Kepler's father, Heinrich, and marched them across the Alps into Lombardy. They were not so much an army as a horde of locusts—undisciplined, sadistic, and bent on vengeance against the popish Antichrist. Not to mention the rich booty they hoped to get. Georg von Frundsberg, an old man, red-faced, corpulent, and given to rages, led them down the length of Italy. Rain didn't stop them. Blizzards didn't stop them. Nothing stopped them, for their fury against the Antichrist was enormous.

They destroyed one army on the way down, the army of Giovanni delle Bande Nere of the house of Medici, and then met, greeted, and joined up with the imperial army of Habsburg Spain, which had been marching with a force of Italians and even a few Frenchmen, under the rule of the Duke of Bourbon, the traitorous constable of France. The few professional soldiers like the duke quickly learned that they could not control their own men, that the emperor had punched a hole in a dam of religious hatred, and the army they commanded had poured through. Their men, in rags, nearly starved, with half them unable to communicate with the other half, marched on. Clement sent word to the generals offering payment in return for Rome's safety, but when the men heard about it, they turned on their own leaders and shouted them down—they would not go back without raping and thieving and murdering until they were sated. Raging at their cheek, Georg von Frundsberg had to be carried off on a stretcher after a fit of apoplexy. Now in charge of the army, the Duke of Bourbon looked around himself, a tiny boat caught in a roiling tide, and marched his men on toward the city.

Rome's defenses were a shambles, its army almost nonexistent. There were palaces with unimaginable wealth, but the walls of the city itself had crumbled. Too late, the pope tried to gather the money to build an army, appealing to the members of the Commune of Rome, who agreed but fretted over their own business interests. The commander of the defenses, Renzo da Ceri, reinforced the Leonine Wall, but was able to gather only eight thousand men, two thousand Swiss Guards, and two thousand survivors of the defeated army of Giovanni de' Medici Bande Nere to man the battlements. There they waited through the night and the next day for the monster lurching toward them.

The army appeared over the Italian countryside and encamped north of the city. The Duke of Bourbon sent heralds to demand that the city surrender, but this was only a formality and nothing more. Before the attack, the duke spoke to his men to rouse up their battle fire, but a murmur of excitement and lust ran through the camp before he could finish. His men needed little from their generals. They wanted only to kill and gnash and burn. At four in the morning on May 6, 1527, an exchange of harquebus fire from both sides started the battle. The imperial troops attacked, but the Roman artillery slaughtered them by the hundreds and sent them running out of range to regroup. Suddenly, a fog gathered over the Tiber, making the Roman artillery useless, and the imperial troops attacked again. This time the Romans threw rocks and shouted insults: "Jews and infidels, half-castes and Lutherans!"

A stray shot from a harquebus hit the Duke of Bourbon. The prince of Orange ordered his body carried off to a nearby chapel, where he died. The imperial troops were downcast for a short time, but they didn't really need their generals and pressed the attack once more. The Swiss Guards fought back courageously, as did a good portion of the Roman militia, the surviving soldiers of the Bande Nere, and the students of the Collegio Capranicense, who fought side by side until they were all butchered. Blood pooled ankle-deep in the streets and ran in little streams into the Roman sewers and then on into the Tiber. Many of the papal troops deserted. Many joined the refugees fleeing the city across the bridges over the river. The panic of the crowds crushed some to death, while others fell over the sides into the water.

The city was now open, defenseless. The Spanish commander Gian d'Urbina led his men through the Borgo, butchering everyone, armed and unarmed alike. They broke into the Hospitale de Santo Spirito and threw the patients into the Tiber while they were still alive. Then they murdered all the orphans. Once across the Ponte Sisto, the plunder started in earnest. Palaces, monasteries, churches, convents, workshops—they attacked them all, broke down the doors, and scattered everything they could find into the streets. Money, booty, plunder was everything to them. They dragged citizens, even those who supported the emperor, even their own countrymen living in Rome, and tortured them until they handed over whatever money they had. They assumed that everyone in the city was hiding some secret treasure and pulled them, beat them, burned them until they handed it over. Those who suffered the most were those who had nothing to give.

When they found a priest, they cut him open until his guts ran out onto the street. Some they stripped and at sword point commanded to blaspheme the name of God. They held satiric masses and forced what priests they could find to participate. The Lutherans killed one priest who refused to give Communion to a donkey. The marauders then shot at holy relics, spat on them, and played football with the severed head of St. John. They tortured on, grabbing any man or woman they could find, still searching for hidden riches. Some they branded with red-hot irons. Others they tied by their genitals. Some they hung up by their arms for hours, some for days. Others they forced to eat their own severed ears, noses, or penises. They raped every woman they could find, young and old, married and single, including nuns. Especially nuns. Many they sold at auction or as prizes in games of chance. They forced mothers and fathers to watch the rape of their daughters. Some they forced to assist in it. The city had collapsed; it was violated, alone, without hope.

Cardinal Pompeo Colonna, an enemy of the pope, rode in with two thousand men to join the attack. He wept when he saw the state of affairs in the city, but his men soon joined the revelers. Most of the soldiers were half-starved peasants who robbed people as poor as they were themselves. Soon the city was empty, and the invaders weary. The noise died down; the exhausted revelers gathered in stupefied clumps, heavy with wine, hung over, bleeding in places from wounds. No one seemed to notice anything. Grandmothers wandered through the streets looking for their children. Babies cried for their mothers. Children stood in the middle of the streets, stunned. Spot fires burned here and there in the city. Over all was the buzzing of flies and the stench of the dead. Here and there dogs gnawed upon the corpses.

The pope eventually fled the city disguised as a servant; he was dressed in a cloak and hood and had a basket over one arm and a sack over his shoulder. He stayed at the episcopal palace at Orvieto, where he waited out the storm, shrunken, jaundiced from a diseased liver, one eye nearly gone, like Dante in hell crossing a sea of shit. Incongruously, a delegation from Henry VIII appeared, seeking the king's divorce from Catherine of Aragon. Finally, on February 11, 1528, the imperial army received its back pay and set out for home. Soon after, the pope returned to the city, or what was left of it.

"Rome is finished," said Farrante Gonzaga, a nobleman of the city. The city was nearly empty, the citizens dead or dying. All trade had stopped—the shops were closed, the people gone, while the streets stank with the putrefaction of the unburied dead. The glory of Renaissance Rome had been plundered. In 1534, Clement VII died from a fever, and the people of Rome danced in the streets. They stuck a sword into his tomb, smeared dirt and filth on it, crossed out the words *Clemens Pontifex Maximus,* and wrote *Inclemens Pontifex Minimus* in its place. They would have dragged his body through the streets on meat hooks, if they had had the chance.[2]

Little by little, the city raised itself from the dead. The next pope, Paul III (1534–49), gathered what money he could and what artists he could find who were still alive to rebuild the fountains and the palazzos. Michelangelo Buonarroti returned to the Sistine Chapel and painted his greatest fresco, *The Last Judgment.* When he had finished, the pope fell to his knees in front of it and trembled for the state of his own soul. The pope then commissioned him to rebuild the Piazza del Campidoglio on the Capitol and to design the steps, the Coronata, leading to the top. Ten years later, Emperor Charles V, the man behind the rape of the city, would visit Rome. Paul III would order houses, even churches torn down to make a straight path for the emperor and his retinue of four thousand knights. Rome would do its best to hide its wounds as Charles rode in under the Arch of Titus, down the Via di Marforio to the Piazza di San Marco, and over the river to St. Peter's.

Paul III tried hard to recover the festival spirit of the Roman Renaissance, but something had changed in the city. As Rome rose from the dead, a spirit of reformation rose with it. Paul III himself approved the founding of the Society of Jesus, the Jesuits. He called the Council of Trent and set the Catholic church on the path of Counter-Reformation. This new reformation was both in the spirit of Lutheranism and against it. It cleaned out the old evils, the simony of the past, the selling of indulgences and of spiritual privileges. Though these practices returned from time to time, they were never the same and soon died out. The Counter-Reformation also called priests to a higher standard of education and a higher standard of behavior. Most of all, however, it shifted the church's attention away from a reconstruction of the glories of the classical age to a reconstruction of the roots of Christianity. Catholicism once more became interested in Scripture and in the writings of the church fathers. In the end, Luther finally got most of what he originally wanted from the church, though by this time, it was far too late.

Paul III's successor, Julius III (1550–55), continued with the reconstruction of the city and with the Council of Trent, but after Julius came Paul IV (1555–59), a sour fellow who cared little for the glories of Rome. A devotee of the Inquisition, he refocused the Counter-Reformation along lines that fit his own rigid orthodoxy. Virtue, and not beauty, he said, was the concern of popes. He hated the nudity of Michelangelo's frescos and threatened to destroy *The Last Judgment.* His rules for the city were equally harsh, ferociously punishing sexual misconduct. Homosexuals he ordered burned. His successor, after Pius IV (1559–65), was Pius V (1566–72), a former Dominican friar like Paul IV, who continued with the new strictures, adding new and more austere rules for religious, secular priests, and even bishops. He suppressed nepotism and restricted the granting of indulgences and the giving of dispensations. He also drove the prostitutes out of the city, expelled the Jews from the Papal States, and ordered the Congregation of the Index to draw up a list of forbidden books. The printers quickly followed the prostitutes and Jews out of the city.

The Counter-Reformation was as strict as the Renaissance was loose. Pius V's successor, Gregory XIII (1572–85), instituted the Gregorian reforms of both the church and the calendar and founded the Collegio Romano, the great Jesuit college from which men of learning would spread out into the world, to Bavaria, to Austria, to little Graz, even to China. Galileo would visit the Collegio from time to time, have friends there, and also enemies.

In 1596, the Counter-Reformation arrived in Graz, two years after Kepler. It appeared in the person of young Ferdinand, the fervent Catholic archduke, who received the oath of allegiance of all the representatives on December 16, 1596.[3] He was eighteen years old at the time, seven years younger than Kepler. Up until then, the Protestants had dominated the region, which infuriated the archduchess Maria, Ferdinand's mother. She had no room for tolerance and raised her son accordingly. As Ferdinand ascended the throne, those who were paying attention to the shifting political winds sniffed the air. They knew that the young archduke had just returned from study with the Jesuits, the one order in the Catholic church that the Lutherans feared the most, for they were educated men, able to hold their own against the best Lutheran preachers. Many of them knew the Scriptures as well as any Protestant, and many were as dedicated to reform. But their reform always led back to Rome and to the hated pope, not away from him, so they could never be anything but enemies.

By this time, the Reformation had become an entity unto itself, no longer a movement within Christianity, but a new church, a new way, with its own structures, its own laws and traditions, its own theology. Sadly, if all the bile of the previous eighty years could have been set aside for just one hour, people would have seen that the differences between them were not all that great, but such a thing was impossible by that time. History had rolled on, its momentum unstoppable, and the religious world of Europe would never be the same.

Kepler, alert for trouble, was one of those who sniffed the air. It worried him greatly to watch his fellow Lutherans taunt the religion of their new ruler. One does not ridicule a Habsburg lightly any more than one teases a leopard. From April 22 to June 28, 1598, young Ferdinand traveled to Rome to meet with the pope and to pray at the shrine of Loreto. There, some say, in the midst of his prayers, he vowed to God to lead his divided people back to the arms of the true faith. Back in Austria meanwhile, while Ferdinand was still away, Graz and all the countryside prickled as on a hot day before a summer thunderstorm. Those like Kepler who paid attention heard stories of events along the young prince's journey, rumors and whispers, and found evil omens in them. "Everything trembles," Kepler wrote to Mästlin, "in anticipation of the return of the prince. One says that he is at the head of Italian auxiliary troops. The city magistrate of our creed was dismissed. The task of watching the gates and arsenal was transferred to followers of the pope. Everywhere one hears threats."[4]

When he returned, Ferdinand was even more ardent than when he left. The rumors, at least the ones about the prince's new determination, were all too true, and the Lutheran church argued over what to do about him. Some of the more rash preachers circulated profane caricatures of the pope, infuriating Ferdinand. One, a certain Balthasar Fischer, punctuated his sermon on the cult of Mary with an obscene gesture, opening his robe and asking whether it was proper for women to crawl inside it. Ferdinand then called in the Lutheran chairman for church ministry, accusing the Lutherans of bad faith. "You would spurn peace even if I would give it to you," he told him.[5] Things spun out of control from there. Ferdinand ordered arrests and levied a new, burdensome tax on Lutherans trying to bury their dead. Poor Protestants languished in hospitals untreated. Murmurs and angry whispers flowed around town—perhaps riot, some said, perhaps rebellion.

In the midst of this—joy. Kepler had married Barbara Müller on April 27, 1597, as the flowers bloomed and the farmers gathered for the planting. Copies of his first book, the *Mysterium Cosmographicum,* had just arrived, and everyone fussed over the young author. Marriage and publication both, in just two short months! Kepler sent copies of the book to everyone who mattered. The archduke, of course, and the emperor, of course. Tycho Brahe in Denmark, certainly. And one obscure mathematics professor in Padua, Galileo Galilei. Galileo wrote back, ecstatic to find another Copernican. He too was a Copernican, he wrote Kepler in a letter, but he was afraid to tell anyone after the treatment that Copernicus himself had received. Kepler wrote back, encouraging him to come forward, but Galileo did not acknowledge his letter, and Kepler did not write again for another thirteen years.

Then on February 2, 1598, more joy, but joy short-lived. Just before young Ferdinand left on his pilgrimage to Rome, Barbara gave birth to a son, little Heinrich, named, perhaps unpropitiously, for Kepler's misbegotten father and misshapen brother. Kepler, typically, cast a horoscope. In his journal for that year, he wrote: "A son! Heinrich Kepler was born on February 2. The stars' constellation promises a noble disposition, a strong body, strong fingers, agile hands, with a capacity for the mathematical and mechanical arts." Then he went on to explain why: "The moon in quadrant to Saturn promises a vivid imagination, diligence, though some mistrust and some stinginess. Both indicate compassion, deep thinking, piety, empathy, sadness, and grace. The ascendant in quadrant to the sun supports those attributes and with Libra full of light south in ascension indicates one who is stubborn, unruly, and also one who admires greatness."[6] A proud father, certainly. He could have been describing himself.

With his stepdaughter, Regina, brought into the marriage by Barbara, and now little Heinrich, Kepler had become a family man, the one thing that he had feared most before his marriage—that he would be tied to Graz not only by Barbara's estates, but also by family obligations. By now he knew that it was less likely that he would return to Tübingen to finish his studies for the ministry. He had become the comfortable burgher everyone had wanted him to be.

Perhaps it was an omen, or would have seemed so to Kepler, but little Heinrich died, however, after only two months from what Kepler called *apostema capitis,* possibly meningitis. Kepler's joy was crushed. Unfairly, he blamed the boy's ill health on Barbara's diet, writing to Mästlin that the boy's deformed testicles looked like a cooked tortoise, which was one of Barbara's favorite foods. Even so, Kepler and Barbara tried again, and this time they had a little girl, Susanna, who also died after little more than a month of the same disease as her brother.

After Susanna's death, Kepler fell into a black depression, seeing omens of death everywhere. "No day can soothe my wife's yearning and the scripture is close to my heart: O vanity of vanities, and all is vanity."[7] He reported that bloody crosses were appearing on the bodies of people all over Hungary, a sure omen of the pestilence, and that he himself had found a tiny cross the color of blood, which had then turned yellow, on his left foot. He was the first to have seen the omen in all of Graz, he said. His depression took him into dark places, as if death had entered somewhere deep in his soul, and for some months all he could see about him was death. Between Kepler's sighs and Barbara's epic weeping, the Kepler household teetered on the brink of despair.

Then the storm struck.

The first time it hit him personally was when the Counter-Reformation officials charged him ten taler to bury little Susanna, simply because he was a Lutheran. The fact that they reduced the tax in his case only halved the anger, but did not remove it. Still, Kepler tried to stay out of the fight. No one liked a good intellectual dustup as much as he did, and he would defend his position vigorously if put to it, but in his deepest heart he disliked war, and he hated wars of religion most of all. Sacred things, he believed, ought to be sacred—the things of God should be protected by godly means, by theological debate, by ministers and priests, not by soldiers. Kepler knew that one's conscience should be free to act, free to believe, and should never be subject to outside pressure, whether from the prince or even from the church. No one, he believed, should abuse the faith by slandering others for their beliefs.

In fact, for years he blamed his fellow Lutherans even more than Ferdinand for the troubles in Graz, especially the radical preachers such as Fischer and Kellin, who ridiculed the images of the archduke's religion. Ten years later, he wrote to Margrave Georg Friedrich von Baden: "Some of the appointed teachers confuse the positions of teaching and ruling, want to be bishops and have an ill-timed zeal with which they tear everything down, defiantly relying on their prince's protection and power, which they often lead to dangerous precipices. This has long ago ruined us in Styria."[8] In saying this, Kepler took a stab at his old teachers at Tübingen, who led the pope-baiting chorus. "Werewolf, whore of Babylon, Antichrist," they called him, and Kepler, who was a peaceful man in his soul, could not abide it.

Then the hammer fell once more. The archpresbyter of Graz, Lorenz Sonnabenter, dug up an old medieval tradition, one that had been practiced in theory only, and used it against the Lutherans, effectively shutting them down. The tradition had been designed to protect the livelihood of the archpresbyter of a place, so that the fees he charged for his ministry could remain constant. He could, quite legally, forbid the practice of any competing ministry that would undercut his own, and he used this right to prohibit any kind of ministerial practice by Protestants.

In a letter to Herwart von Hohenberg dated December 9, 1598, Kepler wrote: "In the month of August the prelude to the tragedy took place. The archpriest officially prohibited our preachers from all practices of religion, the administration of the sacraments, and the consecration of marriages." The Lutherans complained to the archduke, but he shrugged, saying that he had to protect his own Catholics as well as the Lutherans, and so, where his own devotion had led him before, now he would act because his people had requested it of him. On September 13, Ferdinand promulgated an order to dismiss the Lutheran preachers and to break the collegiate faculty, and that this had to be done in two weeks' time. On September 20, the archduke "decreed that all ordinations be canceled, and then let go all the servants of the church and school in Graz and Judenburg, and that within fourteen days they must leave his territories and stay away forever."[9] On pain of death. The councilors of the city begged Ferdinand for a repeal of the banishment, but he refused, and on September 23 gave them only eight days to get out and never come back.

The town tried to call for the *Stände,* the assembly of representatives, but there were floods that year, and only a few of them could get to Graz in time. Ferdinand then called out the troops. The city roiled, near riot, near rebellion, while the troops stood by silently, waiting, ready for slaughter. There was nothing else the Lutheran teachers and preachers could do, so they packed their bags and on the appointed day trickled out of the city, some to Hungary, some to Croatia. They left their wives and families behind, an act of vain hope, perhaps, trusting that because the archduke was young, he might be swayed by other voices and allow them to return. Kepler was among the banished and left with the others. In his letter to von Hohenberg, he wrote that somehow, in spite of all the persecution, he was still paid. As for the situation, there was not much anyone could do. The faculty at the school was told to hope for some change after the *Stände*, the council of representatives, met.[10]

Only Kepler was allowed to return finally, which he did in October. "As for me," he wrote, "I returned after a period of one month, called upon by officials of the archduke, who described me as exempt." Apparently, Ferdinand liked him and was proud to have a scholar of Kepler's growing reputation working in his duchy. He could return, not as a teacher in the Lutheran school, but as the district mathematician. After all, the great Tycho Brahe of Denmark had read the *Mysterium Cosmographicum* and liked it. He had even written to Kepler, inviting him to come to visit him in Prague. That fellow Galileo, down in Padua, had also written a letter praising it, one Copernican to another. But as nice as Kepler's growing reputation was, it was more to the point that Kepler was a modest man and was known to be a modest man, a man who disapproved of the antics of his fellow Lutherans. "We see," wrote Mästlin to Kepler, "with what raging fury the devil incites the enemies of the church of God, as though he wanted to devour it completely."[11] Kepler kept his own counsel on that point, writing in his diary: "I am just and fair toward the followers of the pope and recommend this fairness to everyone."[12] Kepler then petitioned the archduke for his exemption and received it. "His Highness is herewith granting, out of special favor, that the petitioner, notwithstanding the General Dismissal, shall be allowed to remain here. But he shall maintain appropriate modesty everywhere, so that this exemption will not be subject to cancellation."[13]

Oddly enough, the Jesuits may have had something to do with this. Although Kepler was a committed Lutheran and could never have accepted the Roman Catholic faith, he was on good terms with some of the fathers at the Jesuit college, for Kepler knew that Father Clavius and a handful of other Jesuits were some of the best astronomers in Europe. In September 1597, a few months after Kepler's wedding, one of the fathers in Graz, Father Grienberger, came to Kepler at the behest of the Bavarian chancellor, Hans Georg Herwart von Hohenberg, an important man and one of the first of Kepler's important patrons. No doubt, there was some hope on the part of the good father, and on the part of the chancellor, of converting Kepler to Catholicism, but it was a vain hope. Not an irrational hope, though, becauce one of the archduke's teachers at Ingolstadt had been Johann Baptiste Fickler, a distant relative of Johannes Kepler's.

Fickler's mother, Benigna, born on the last breath of the Renaissance, was one of the great women figures of the time. Like Meg More, the daughter of St. Thomas More in England, Benigna was a well-educated woman. Her father, a wealthy merchant in Ulm, had insisted she read law and theology, so she could write contracts and dispute theological questions. She could read and write Latin, and some said she was the best lutenist in town. Wanting children, she married very young, buried her first two husbands, and finally married Michael Fickler, Johann Baptiste's father. Had she lived at a later time, she would most certainly have been accused of witchcraft, because one of her many accomplishments, like Katharina Kepler's, was the knowledge of herbs and the making of potions.

In his first letter to Herwart von Hohenberg, Kepler made a point of sending his greetings to his relative, which were passed on and returned with thanks. These letters were all sent by personal courier from Bavaria to the emperor's court in Prague, and from there back to Graz by way of the archduke's secretary, a Capuchin father named Peter Casal, and then back again along the same route. This set of events raised Kepler above the rest of the Lutherans quietly leaving the city and set up his later exemption and return.[14]

But was Kepler right in assuming that if the Lutherans had only been less provocative, the archduke would have left them alone? Probably not.

The land was itching for war, and young Ferdinand itched along with it. He had determined to rid the land of heretics, and that determination would embrace an astronomer or two. Besides, he knew that the Lutherans weren't much better. Lutheran dukes and princes had merrily expelled Catholics from their territories, so why shouldn't he? People of all stations, all classes in Germany and Austria, from kings to peasants, were tripping over their own piety. In some secret, unlit room in the human soul, they all wanted war. And eventually, they got it.

Kepler's own openness did not sit well with his fellow Lutherans any more than his stubborn Lutheranism sat well with the Catholics. The other teachers at the Lutheran school resented his exemption, wondering if he had compromised his faith along the way just to please the papists. He had certainly broken ranks with the other teachers by petitioning, but there was more to it than that. His beliefs, like his thoughts, were subtler than theirs. Some later scholars accused Kepler of indifference to dogma, but this too was unfair, for he did believe that there is but one truth, and he pursued that truth, which he identified with God, with all his heart all his life. Kepler had to go his own way, for the truth he pursued was inside his own conscience and did not float outside him among the dogmas of the churches. Sometimes he accepted the teachings of his church, and sometimes he did not, for he had to find the authentic truth for himself, and no one could find it for him. This meant that he could not accept the truth of others just because they told him to, no matter how authoritative they were. Neither armies nor pulpits could sway him by themselves, but only reason and faith.

Kepler's doubts about the ubiquity doctrine and about the sacrament had been paddling about in his head for years, since the days he had first studied at Maulbronn. Now they came to the surface. Perhaps unwisely, he confessed his doubts to the exiled church members, who were shocked and suddenly suspicious. Those who had envied his return and suspected him of compromise suddenly had new ammunition. The others merely shook their heads, wondering. None of them really understood him, though. His theological meanderings were well within the bounds of the

Lutheran faith. Or so he thought. But differences over what constitutes the heart of the faith are often the source of schism and of heresy. Kepler never quite understood this for himself, but others did, and from that day on they watched him carefully.

Once back in Graz, the hardest thing for Kepler to endure was the sudden loss of the church life that had kept him close to God. He felt the emptiness of the place—he missed the sermons; he missed the sacraments. There were preachers about still, tucked away in the castles of the various Protestant estates, but if any subject of the archduke were to ask for the sacraments, and if the preacher agreed, then that preacher could find himself on the road out of the archduke's territories, which included all of Styria, or Upper Austria. Surprisingly, though the Lutheran school had closed its doors, the school officials had kept Kepler on, partly because they felt sorry for him. They even paid his meager salary, but offered him no raise. Like Jobst Müller, the school councilors were practical men who enjoyed Kepler's growing fame, but did not understand his work and could see little value in it.

Kepler thought about leaving Graz, about finding another post somewhere else, at some university perhaps, maybe even at Tübingen, though he couldn't have known that the faculty there was set against him. But how could he leave? Barbara was attached to her inheritance, though Jobst balked at giving it to her. She had not yet seen a taler of her money from dear Papa. And the way things worked in Austria at that time, old Jobst, and not Barbara, was little Regina's guardian. Regina had inherited some 10,000 gulden from her own deceased father, and Jobst, as head of Barbara's household before she married Kepler, took control of it. After the wedding, Kepler received 70 gulden annually for Regina's upkeep, as well as the proceeds from a vineyard and the use of a house. Old Jobst had been trying to estrange the girl from Kepler in every way he could, though everyone could see that her stepfather was devoted to her. For Kepler's part, he would have missed his dear little stepdaughter if he had left town without her, because old Jobst, as her guardian, would have kept her with him. Most of all, however, he worried that she might fall under the influence of the Catholics. Jobst had always been a Lutheran, but the depth of his faith was showing under pressure from the Counter-Reformation. When the time came, he scurried to Catholics, as did many of his relatives. No need to upset the archduke. Bad for trade.

So Kepler stayed on.

In the middle of all this, he naively found himself caught in a feud between two prodigious scientific egos. On one side was Tycho Brahe, the famous Danish nobleman turned astronomer. On the other was Nicholas Reimarus Ursus, "the Bear," and Tycho's predecessor as imperial mathematician. Ursus had once been a swineherd, and his manners hadn't changed much since then. He was crude, abusive, and crafty, and his position at court was as much the result of theft as it was of accomplishment. When the *Mysterium Cosmographicum* came out, Kepler sent a copy to Ursus, asking his opinion. He wasn't above a little flattery, and heaped it on, calling Ursus the greatest mathematician of the age, and so on and so forth. Ursus never wrote back, but he kept the letter tucked away and dropped it into his next book, whole and entire. That was the book in which he stole Tycho Brahe's idea about the solar system and claimed it for his own, and then used Kepler to verify his brilliance. Tycho couldn't stomach Copernicus's idea of a moving earth, so he had cooked up a new system for himself, one that set all the other planets together orbiting around the sun, as Copernicus did, but then set the sun and the planets orbiting the earth, as Ptolemy did. Though it was more complex than it needed to be, it was a nice little bridge between the two systems.

When Tycho read Ursus's book, he was furious. Who was this Bear, this swineherd, to steal his work? And who was this Kepler to heap such praises on him? The first Kepler heard about it was in a letter from Mästlin. Kepler had sent two copies of *Mysterium Cosmographicum* along with a letter to Tycho at Wandsburg, and Tycho complained to Kepler by letter, protesting the theft and Kepler's part in it. That letter never arrived, but Tycho had sent a copy of his letter to Mästlin, and Mästlin sent a frantic letter off to Kepler, which did get through in late November 1598. Mästlin warned Kepler that he had made a fool of himself, that he had made a terrible mistake, and that his career was on the line.[15] Mästlin said that Kepler should never have praised Ursus and re-minded him that he had once told Kepler that the man's work was trash. Kepler didn't understand a thing, because he had never received Tycho's original letter, so he asked Mästlin to send a copy of the letter he had received from Tycho.

When the letter arrived, it wasn't as bad as Mästlin had made out.[16] Tycho had a few doubts about Kepler's polyhedral theory, but he found it ingenious and asked Kepler to try it on the Tychonic system, as he had done on the Copernican. Then he remarked that Copernicus's measurements of the planetary distances were not accurate enough to prove Kepler's theory, and that Kepler might find Tycho's measurements more serviceable. Then in a long postscript he complained to Kepler about his letter of praise to Ursus. He had no doubt that Kepler never realized that Ursus would include his letter in that "defamatory and criminal publication," and he hoped that Kepler would supply Tycho with a short statement of the facts that would help Tycho in a court action he was planning against Ursus. The letter was velvet and iron, praise and complaint, and seemed to be an act of kindness. Still, Mästlin had informed Kepler that Tycho had written to him in a separate letter, complaining much more directly about Kepler's praise and criticizing the *Mysterium Cosmographicum* much more sharply.

Kepler was quite chagrined, for he did not want to burn any bridges with Tycho, a great man who looked kindly on his work, and so he wrote back at once, apologizing and defending himself, explaining that he had only meant to praise Ursus in a general way and had no idea that the Bear would publish his letter, implying that Kepler was a party to the scheme. "That nobody that I was then searched for a famous man who would praise my new discovery. I begged him for a gift and, behold, and it was he who extorted a gift from the beggar."[17] Tycho was eventually mollified, but it took more than a few letters to do so. The incident jarred Kepler from his näiveté and forced him to enter the politics of the age.

As it turned out, Kepler certainly needed Tycho more than he needed Ursus to keep his own work going. What Tycho had said in his letter was correct. Tycho had the observations that he needed. While cooling his heels in Graz, Kepler decided to push on with his astronomical studies and realized from Tycho's letters that Copernicus's astronomical data was not accurate enough. Kepler could see that his regular solids did not quite fit the data and that he would need Tycho's more accurate observations to verify the *Mysterium Cosmographicum*. But Tycho kept his observations to himself and would not pass them around, so Kepler would need to wheedle them out of him. Writing to Mästlin on February 26, 1599, Kepler said: "This is my opinion of Tycho Brahe: He has riches in abundance, which he does not quite use the way most talented people do. One has to take pains, therefore, which I did for my part, with all appropriate modesty, to wrest those riches from him, to beg him to disclose all of his observations, unchanged."[18]

The next few years would draw the two men closer together and into a collaboration that would change Kepler's life. After Tycho's death, Kepler would publish the data in Tycho's name and title them the *Rudolphine Tables* after Emperor Rudolf II. The one major disagreement between them, however, was over Copernicus. Tycho rejected Copernicus on theological grounds, while Kepler embraced him, also on theological grounds. Tycho believed that because humans were created in the image of God, then their little place in the universe had to be at the center. God placed them there so they could get a good view of the great cosmic show and thus come to praise and honor God even more. There was no greater theological master than the universe itself. Kepler agreed in principle, but Tycho's system still seemed overly complex and unwieldy to him, a patchwork compromise that was unmindful of the elegance of God's plan. God's plan must be rational, and rational meant geometrical, and geometrical meant elegantly simple. And so, while Tycho was thinking like an astronomer, Kepler was thinking like a mathematician.

However, there was still one worry, one tiny dark cloud, that bothered him. If Copernicus was right and the earth moved around the sun, then there should be some visible parallax, a shifting back and forth, of the stars from one season to the next, as an astronomer would first observe them from one side of the orbit and then from the other. However, no one could find any parallax. Tycho couldn't find it and used that fact as evidence for the centrality of earth. But there were other possibilities. One was that the universe was infinite, as Nicholas of Cusa said, and then because the stars were infinitely distant, there wouldn't be any parallax visible. Still, Kepler couldn't accept that on general principle. The fact that the stars might not be at an infinite distance, but might be very, very far away was also possible, but Kepler had no way to prove that. Even Tycho's fine instruments couldn't have measured that. Such a discovery had to wait until 1838, when Friedrich Wilhelm Bessel observed the stars using instruments more refined than anything available in Kepler's day.

So Kepler went on to other problems. Why was the period of the moon's motion apparently longer in winter than in summer? Kepler explained that by mechanical forces, his *vis motoria* of the sun, coupled with another *vis motoria* of the earth itself. He also asked why there was a reddish color to the moon during lunar eclipses. This led him to raise questions of optics, which became important to him later. Finally, he burrowed into the question of chronology, trying to pinpoint biblical events through astronomical means. Herwart von Hohenburg involved him in trying to find the year of the Roman civil war between Caesar and Pompey, which Kepler placed at 51 B.C.

That summer of 1599, in the midst of all his astronomical work, was when his little daughter Susanna suddenly died and Kepler nearly fell into despair. Oddly enough, in the midst of his sadness, perhaps because of his sadness, he began researching a new book, a book that would eventually lead to his great work on world harmony, the *Harmonice Mundi*. This was perhaps more revealing of his personality than anything else. His world was collapsing about him. He was about to be chased out of his home and lose his career. His little son had died and now his daughter, of the same illness. His wife's family gnawed at him over everything, and there he was in his study thinking about a book on the harmony of the universe.

His faith told him that God does nothing without a plan and that God's wisdom and goodness made the whole world beautiful. Buried in the heart of the world is the image of the Creator, a copy of the divine mind. God gave human beings the power to reason, the most godlike characteristic of all, and out of that rational power would come the ability to see God's own plan in action. Copernicus had seen it, that marvelous harmony, that order to things, like the order of music that heals the soul and harrows sin from the world. The sensation of joy that different tones played in orderly succession or played together in chords brings to the human soul speaks of a perfect world, higher than this earth where, like little Susanna, things are born doomed to die. But harmony speaks to the soul of the transcendent world where God sits enthroned above the stars and listens to the music they make. The pleasure people feel in music partakes in this and encompasses each whole person in the place where the intellect joins with the senses, just as mathematics joins with sound. Major thirds, minor thirds, major fourths, major fifths, each with its own unique feel. So what is it about these proportions that stirs the human soul so? Where does the joy come from?

Kepler, following Plato, believed that music, and the simple mathematical relations that music embodies, reveals a cosmic harmony. This was not a rough Pythagorean number mysticism, by any means, but an icon of the mind of God embedded in the human soul. "God wanted to make us recognize him, when He created us after His image, so that we should share in His own thoughts. For what is implanted in the mind of man other than numbers and magnitudes?"[19] As he later wrote in his defense of Galileo, his *Dissertatio cum Nuncio Sidereo:* "Geometry is one and eternal," a reflection out of the mind of God. That humankind shares in it is one of the reasons to call the human being an image of God.[20]

For Kepler, these world-forming harmonies are ubiquitous, for they make up the first pattern of the universe. He found them equally in the sound of music and in the velocities of the planets, in the pleasure of the human senses and the movements of the heavens—the "music of the spheres" in a new package. "Give air to the heavens, and truly and really there will be music," he wrote to Herwart von Hohenberg.[21] And he was certain of it. The harmonies, like the Platonic solids, were the key to the planetary orbits. They had to be. The idea was too beautiful for them not to be. Still, he needed more accurate data to prove it, and for that he needed Tycho Brahe.

In Kepler's study, the candle burned long into the night while he scribbled out calculation after calculation, looking for the harmony, the perfection of the universe. Outside his study, however, the world was growing more dangerous by the day. The Counter-Reformation was heating up in little Graz. The archduke was unsatisfied with merely banishing the preachers and teachers of the Lutheran community, for he had sworn to bring his part of Austria back to the true faith, under obedience to Rome, and that meant everybody. He announced a new set of regulations, one after the other, this time aimed at the people themselves. Many continued to attend services and to receive the sacraments at the castles of the nearby Protestant estates, and for a time the archduke did not pursue them. Only the minister was liable to banishment if caught. But then, suddenly, the rules changed, and the archduke promised to punish anyone caught attending Protestant services. Moreover, Ferdinand demanded that all children in the region be baptized in the Catholic faith and that all engaged couples be married before a Catholic priest. Soon the archduke ordered all Protestant clergymen—Lutherans, Calvinists, and Anabaptists—out of Styria. No exceptions. No excuses. He made it a crime, punishable by confiscation of property and even prison, to offer them asylum in any way. Jesuit schools alone remained open, and anyone who wanted an education went there.

Anyone who sang Protestant hymns or read Protestant homilies, read Luther's Bible or recited Luther's prayers could be fined or banished, for these were now works of heresy. The archduke set guards around the city to check wagons and mules and bags of personal belongings and to watch for forbidden books. Lutherans complained to him, but he pointed out how Protestants had forced Catholics to convert in other parts of Germany, in Saxony, in Württemberg, and in the Palatinate, and that he was merely exercising his rights under the Peace of Augsburg. Little Graz buzzed and hummed with rebellion. Riots popped up in parts of the city, and there were finger-pointing arguments on every street corner. Rumors flowed from street to street; soon, they said, no Lutheran could live in the city, and those who tried to emigrate would no longer be able to sell their possessions and would have to leave penniless. And they were right, or nearly so.

Kepler felt this oppression as much as anyone, as his position at the school became increasingly tenuous. The school councilors were growing impatient with him. What was the point of a speculator like Kepler when the Lutherans were at war? Kepler wrote to Mästlin in distress to ask for advice and for some "little professorship," but his old teacher had nothing for him. Tübingen would not have him, even then, even with his growing fame, even in his great distress. The faculty was all too chummy, all too connected by blood or marriage. Their world was too insular, too narrow, too orthodox for them to accept someone like Kepler. Their little ship was at war, and Kepler was a loose cannon.

Kepler, however, had one last card to play—Tycho Brahe. The Danish astronomer had come to Prague in June 1599. After Tycho had lost favor with his own king in Denmark, the emperor Rudolf replaced Reimarus Ursus with Tycho, at a salary of 3,000 gulden a year, bringing him first to Prague and then later, because the city had become too noisy, to Benatky Castle, about twenty-two miles northeast of the city. He and Kepler had patched up their differences over Ursus, more or less, and Tycho had in a friendly letter invited Kepler to visit him, so Kepler decided to make the journey, even though he had no money and no way to get there.

Then, in Kepler's mind, God opened the way. Right after the first of the year in 1600, a friend of Kepler's, Johann Friedrich Hoffmann, Baron of Grünbüchel and Strechau, was returning to Prague after meeting with the Styrian *Stände.* As wealthy men do, he had his own coach and four and the best of everything. He understood Kepler's situation in Graz and, though the two men were in vastly different stations in life, he respected Kepler's talent. The two men shared a love of astronomy, and the baron had taken Kepler's case to heart, offering him advice and help along with a personal introduction to Tycho Brahe and a ride to Prague. That December, Tycho had repeated his invitation to Kepler, saying that he hoped Kepler would come to Prague to pursue joint astronomical study and not just because he had nowhere else to go. Tycho stood ready, as he said in his letter, to aid and advise Kepler in whatever way he could. Oddly enough, the letter did not arrive in Graz before Kepler had already left for Prague, traveling with the baron. He had left trusting in the baron's assistance and in the generosity of Tycho's earlier invitation. The journey was not an easy one in the middle of winter, even in the best of times, even with the best of equipment. By modern standards, and even by ancient Roman standards, the roads were rough and unpaved, sometimes narrowing to a few ruts.

Eventually, they arrived in Prague, and Kepler stayed with Baron Hoffmann for a few days until Kepler could send word to Tycho out at Benatky Castle, who was thrilled at Kepler's arrival. He sent a return note, calling him not so much a guest, but a friend and companion in the observation of the heavens. A few days later, on January 4, Kepler rode out to the castle with one of Tycho's sons and one of his assistants. The weather was often cold around Prague in January, with mists rising from the river and a cold drizzling rain falling from a low, gray sky.

No two men could have been more different. Tycho was noble, used to wealth and privilege, often suspicious of others, and fearful that someone would steal his discoveries from him, cadge his fame in the night. Also a Lutheran, he had not received Communion for eighteen years, though he believed in the "new way" and often let theology guide his mind. Kepler on the other hand was open, sometimes to the point of naïveté. He could be suspicious, stingy, and irascible, but he could not imagine brooding over his ideas and discoveries as Tycho had done. His Lutheranism was his heart; his belief in God guided his every action. Moreover, Tycho was the consummate extrovert, constantly surrounded by family, friends, and servants. He lived in a great Danish-style hall full of noise and enjoyed endless meals spiced with conversations on science, philosophy, religion, and the news of the day. Kepler, on the other hand, preferred his own company and the quiet of his study, where he recalculated his figures and worked out his ideas.

Of the assistants who worked and lived in Tycho's household, two would affect Kepler's later life. The first was Christen Sörensen Longberg, who had latinized his name to Longomontanus. Another suspicious man, he had been with Tycho for eight years and had worked with him in Denmark, at Uraniborg on the island of Hveen, and he jealously guarded his own position with the master. The other was Franz Gansneb Tengnagel von Camp, a Westphalian nobleman who was far more interested in Tycho's daughter than in Tycho's astronomy. Because he was an aristocrat and because he was about to join Tycho's family, he was arrogant and treated the other assistants as if they were servants. Years later, he would give Kepler no end of trouble over control of the *Rudolphine Tables*. Kepler, for his own part, must have seemed threatening to the other assistants because he had brought to Prague a reputation of his own, separate from Tycho's, and although his fortunes had fallen because of the Counter-Reformation, he had more independent standing as an astronomer than anyone else in the household except Tycho.

What started as a visit soon became a collaboration. Most of all, Kepler wanted to examine Tycho's observational data so he could prove his theory in the *Mysterium Cosmographicum,* but Tycho wasn't about to hand over his observations to a relative stranger, especially anyone with Kepler's talent, talent enough to misuse them for his own purposes. Kepler, therefore, was kept far from Tycho's observations. Unfortunately, observing was something Kepler had a hard time doing for himself, because, like Beethoven two centuries later, whose deafness kept him from hearing his own music, Kepler's childhood smallpox had clouded his eyesight.

But his mind was active, and he could easily see the richness of the scientific meal available at Tycho's house from the crumbs that Tycho let fall at the dinner table. At one meal, Tycho happened to mention the apogee of one of the planets, the next day he mentioned the nodes of another, and Kepler began putting things together. Within a short time, Kepler had already fashioned a bold new theory of the orbit of Mars. Seeing this, Tycho was impressed, and he opened some more of his data to Kepler, who saw at once that Tycho had the best data around. He also had the best setup, with the most assistants and the best equipment. All he lacked was someone with the ability to take that data and construct a general theory out of the parts. "Now old age steals upon him, weakening his intellect and other faculties or, after a few years will so weaken them that it will be difficult for him to accomplish everything alone."[22]

After a short time, Kepler took on a new project. Longomontanus had been working on the orbit of Mars, and though he had been able to represent the orbit in terms of latitude, he was unable to do so in terms of longitude. Kepler asked if he could try, and Tycho agreed, assigning Longomontanus the problem of the moon. Longomontanus was not happy, and said so, but he admitted that he could go no further and surrendered. Quickly the new work absorbed Kepler, and he lost himself in it for a time, making one calculation after another, eating up the days and nights, until he suddenly noticed that he was staying in Prague longer than he had intended. He knew that if he wanted to make use of Tycho's observations, he would have to stay on as long as one or two years. But to do this he would have to resign his position in Graz, and though things were shaky there, he still had a job and a family to attend to. To keep his position, he would have to receive the school councilors' permission to stay in Prague for a while, and under the circumstances this would be unlikely, because they wanted to get rid of him anyway. He wondered if he should get Tycho's support, or if Tycho could ask the emperor for his intervention.

Whatever happened, Kepler did not want to be seen as driven out of Graz by poverty and misfortune—the shame of that would have been too great to bear. Besides, even if he immigrated to Prague, how could he survive there? How could he support his family there? Everything was far more expensive in the imperial capital than it was in little Graz, and Barbara's lifestyle would be seriously curtailed. Moreover, what would be the nature of his new position with Tycho Brahe? Would he be an assistant like all the others? But surely he was more than that, for he brought to the enterprise a reputation and a work of his own. He brought the ability to build a general theory and support it by mathematics. Not even Tycho could do that as well. In many ways, Tycho needed him as much as he needed Tycho. Tycho was nearly fifty-seven, and his time was passing quickly. He did not possess Kepler's theoretical talent, and he could not make use of his own data to the degree that Kepler could. Tycho must have seen this himself, at least in some unconscious way, or he would not have welcomed Kepler with such eagerness. On the other hand, Tycho was still very suspicious of strangers, and, after all, this young man had given praise to that pig herder, Ursus.

Meanwhile, Kepler sat down with a few of his friends and wrote out a long memorandum detailing all of the things he would need if he were to stay and collaborate with Tycho. This habit was a byproduct of his tidy mathematical mind, and whenever he had a serious question to ponder, he would write it all out, pros and cons, so he could see it in front of him. This habit got him into an awful bit of trouble, and more than once. Kepler wrote that he would need a place to stay, one that fit not only his needs, but his wife's needs. He would need 50 gulden four times a year if he were to survive and raise his family. Moreover, he would need Tycho's help in obtaining a salary from the emperor. But most important, he would need independence, a chance to work without Tycho looking over his shoulder, as if Tycho were the master and Kepler were the apprentice. Undoubtedly, for a young man with only one book and a career that was just beginning, as talented as he was, this memorandum would have seemed cheeky, though it was never meant for anyone to see. Kepler wrote these things down to clear his mind and as an aid for discussion.

At Longomontanus's suggestion, however, Kepler showed the memorandum to Johannes Jessenius, a doctor of medicine teaching at Wittenberg University and a longtime friend of Tycho's. Jessenius was friendly with Kepler as well and agreed to act as a go-between, further agreeing to keep Kepler's draft memorandum a secret and not to show it to Tycho. Somehow, however, whether by betrayal or by accident, the draft ended up in Tycho's hands. One can only imagine how hurt the old Dane must have been. Here was this young man coming to him on the bare edge of poverty, about to be thrown out of Graz by the Counter-Reformation, and he was dictating terms. Tycho was a proud man, and though he was generally fair with others, he expected them to be fair with him, that and the deference owed to the nobility.

On April 5, Wednesday of Easter week, Tycho and Kepler's relationship began to fall apart. Jessenius was quietly presiding over a negotiation between them about the nature of Kepler's position within Tycho's organization, when somehow the conversation took a nasty turn. Hurt feelings bubbled up like hot mud, and Kepler, always quick to anger, at some point stopped listening. Tycho tried to take some time so that he could write down his own remarks and then have Jessenius certify them, but Kepler brusquely announced that he wanted to return to Prague the next day. Tycho tried to calm him and asked him to stay a few more days at Benatky Castle, explaining that he had hopes for Kepler's employment by the emperor and that he wanted them to wait for word from the imperial court. But Kepler, his nose now thoroughly out of joint, wouldn't have it and decided to leave the next day, April 6, with Jessenius. From there, things just went from bad to worse, as two oversized egos batted against one another like mythical frost giants throwing stones.

As he was leaving for Prague, however, Kepler calmed himself and showed some remorse for his earlier comments. He apologized to Tycho. He had always had difficulty controlling his feelings in the middle of an argument, as he well knew, and at times he behaved just like a mad dog. This temper of his had plagued him all his life, in Maulbronn and in Tübingen, and now even with Tycho. But Kepler's apology wasn't quite enough for the old Dane. His noble vanity had been pricked, and he whispered into Jessenius's ear that he wanted Kepler to give him an apology in writing. He asked Jessenius to speak to Kepler on the way back to Prague and set the young man straight about his temper. Jessenius must have thought that Kepler was largely at fault, for he agreed, and on the way back to the city he remonstrated with the young man about his bad behavior, telling him that a learned man should not be so quick to anger and that a gentleman should always control his feelings. This didn't help—Kepler took it all the wrong way. Perhaps he saw more deeply into Jessenius's words than Jessenius wanted him to. Perhaps he saw behind those words their true meaning—that a servant, no matter how brilliant, should always be docile before his master.

The dam of Kepler's emotions finally burst, and after his return to Prague he wrote Tycho a letter, unloading every spoonful of his anger. Ego upon ego upon ego—Tycho responded in kind, and on April 8 he wrote a letter to Jessenius complaining about Kepler's further bad behavior. He told Jessenius that he no longer wanted to deal with a man such as Kepler, who talked so impudently and with such little balance. In the future, he wanted nothing to do with Kepler and wished he had never met him. He said that Kepler could not use the excuse that he was drunk on Tycho's wine, for Kepler rarely drank to excess; nor could he use the excuse that he had been ignored by Tycho, for he had not been, so he must have been angered by something that was said at the dinner table.

Then suddenly, like a passing thunderstorm, the argument ended. Kepler, who could always see the other side of every argument in time, suddenly saw Tycho's. One minute he was the victim, and the next he was the victimizer. He beat his breast and was thoroughly ashamed of himself. Like a deflating balloon, his anger collapsed inward, and all he could see were the blessings that had come his way since he met Tycho Brahe, and as he counted them, with each new blessing he reproached himself once more. Kepler fell into a depression knowing full well that his anger had gotten the best of him. He felt as if God had abandoned him and prayed fervently for forgiveness and reconciliation. When Tycho came to Prague sometime later, Kepler wrote him a letter begging his forgiveness and accusing himself of gross impropriety. In his sudden burst of humility, he offered his service to Tycho in whatever manner would satisfy him, and he promised to prove to Tycho that he had become a new man. Suddenly Tycho was satisfied, because he could see that at least to some degree Kepler had been right in his complaints. Moreover, he needed Kepler to finish the work. Not that he would have admitted it, for his pride was too massive a thing, but he was secretly glad that Kepler had apologized because it gave him an opportunity to be generous.

Finally, Tycho and Kepler settled the terms of their relationship. Tycho promised to ask the emperor to summon Kepler to Bohemia for two years to help him with the publication of his work. They would have to get permission from the representatives in Graz for this, for they needed the school councilors' agreement that they would pay Kepler his salary during this time. Tycho, or the emperor, would add 100 gulden to Kepler's salary.

On his part, Kepler promised that he would focus on the glory of God first, and then on the glory of Tycho Brahe, and lastly on his own.

Soon after, Tycho spoke with Corraducius, the emperor's vice counselor and asked him to request that the emperor intercede with his cousin Archduke Ferdinand and the representatives in Styria. Corraducius agreed. Moreover, whatever moving expenses Kepler incurred bringing his family from Graz Tycho promised to pay. Suddenly, everything was moving along swimmingly. Kepler was happy with Tycho, and Tycho with Kepler. Only one dark cloud remained in their relationship, though Kepler likely knew nothing about it. If all worked out as planned and Kepler returned with his family to Prague, Tycho decided to keep Kepler close by, either at Benatky Castle or in one of the surrounding towns. He didn't want him living in Prague close to the imperial court, where he could make other contacts and build a political base for himself. Most specifically, he wanted to keep Kepler away from Ursus, for Tycho was even then bringing a lawsuit against him. That summer, however, the problem evaporated. At the request of the emperor, Tycho moved from Benatky Castle into Prague, for the emperor too liked to keep his mathematician close by. Then in August of that year, Ursus died.

Kepler set out for Graz on June 1 with a solid letter of recommendation from Tycho in his pocket, accompanied by a relative of Tycho Brahe's, Friedrich Rosenkrantz, whom Shakespeare had memorialized in *Hamlet*. The letter of recommendation, filled with golden praise, was meant to sway the school councilors, to impress them with Tycho's fame and daunt them with his imperial connections. It had little effect, though. The entire city of Graz was in turmoil over the measures of the Counter-Reformation, and the councilors at the Lutheran school had no time for a famous mathematician.

Instead of granting their permission for him to return to Prague, they ordered him to abandon theoretical astronomy altogether and to take up the study of medicine. Who cared about the heavens, when events on earth were so dire? Kepler, they said, needed to compose his mind and to dedicate himself to the common good. They ordered him to leave for Italy in the fall and to prepare himself for a new profession as a physician. His family could do without him for a time—he proved that by his long stay in Prague. Better for them and for everyone if Kepler made better use of his skills.

What was Kepler to do? His entire deal with Tycho depended upon the agreement of the school councilors. He needed them to continue to pay his salary, but now that seemed impossible. Should he go to Italy as the councilors wanted? He could not do without his salary from the Lutheran school, for money promised by the emperor was more speculative than astronomy. Even Tycho had a hard time getting money out of the emperor. For a time Kepler considered taking their advice, or rather following their order, because as a practiced astrologer, his study was not that far away from medicine. Most good astrologers compounded their own cures, which required a good deal of understanding of the language of the heavens. A person's health depended upon the good favor of the stars, and the compounding of medicines was done under the aspects of specific planets. Both Copernicus and Tycho were physicians as well as astrologers and often prepared their own cures, a practice that finally killed Tycho from an overdose of mercury.

But Kepler was a purist, and although the healing of the sick was a good thing, stalking the mind of God was far better. To give up his speculations on the nature of the universe would be to deny his very self. Everything seemed to be in his grasp—Tycho's observations were available to him and a position was waiting for him, a position that included the approval and protection of the emperor. How could he give up the study he had spent his life working for? He contacted his friend Herwart von Hohenberg, the Bavarian chancellor, to inquire about the possibility of working for Archduke Ferdinand as his ducal mathematician, but the chancellor could see the complications in his proposal, given Ferdinand's character, and advised Kepler to trust in Tycho, regardless of their previous agreement.

Undaunted, Kepler did his best to curry favor of the archduke. Shortly after his return to Graz, he wrote an astronomical essay on the unexpected lunar eclipse that would occur on July 10 of that year. He calculated the eclipse and included some especially detailed arguments about the theory of the motion of the moon. Along the way, he developed some new ideas and insights that would serve him well in the writing of his *Astronomia Nova*. Up until Kepler's day, astronomers held to the Aristotelian notion that all the orbits of the planets had to be perfectly circular. This had to be so, according to Aristotle, because the heavens were perfect and all movements in it had likewise to be perfect. The circle was the perfect shape, and therefore the planets had to move in perfect circles.

Everyone accepted this, even Kepler, for it seemed as if it ought to be true. The problem was that the theory did not explain the data and, instead, spawned theories that were ever more complex. All of them were built out of perfect circles, but offered little or no explanation about why the planets moved in this way. To make sense of his own data, Tycho had to create ever newer and more complicated levels of circles in order to explain the motion of the moon and still keep Aristotle's assumptions about the heavens. Kepler had opposed Tycho in this regard, feeling that he was on the wrong track, that the system had to be simpler than Tycho envisioned it and that Tycho's systems were too complicated to defend. He believed this, because he believed that "simplicity is more in agreement with nature." To solve the problem, he at first sided with those who believed that the motion of the moon was nonuniform, that the secret of this nonuniform motion could be found in the earth itself, and that whatever it was that caused the motion faded with distance. This little insight was a second glimmer of the theory that would eventually become Newton's law of gravitation. Kepler believed that there was a force in the earth that caused the moon to move; as the moon grew farther from the earth, the force grew weaker, and as the moon grew closer to the earth, the force grew stronger. Although it lacked Newton's mathematical schema and his subtlety, it was a groundbreaking idea, one that Newton could build on to fashion the inverse-square law of gravitation.

The archduke received Kepler's essay kindly and in return presented him with a small gift, but he offered no permanent employment. He had other things on his mind. Meanwhile, Kepler happily went back to his observations and developed a new instrument to observe the lunar eclipse of

July 10. Standing in the marketplace in Graz with his new instrument, he watched the eclipse, making careful observations at each stage. By July 22, while in the midst of his calculations, he realized the solution to one of Tycho's nagging problems. He realized that the reason that the lunar disk shrank at the time of solar eclipses was optical rather than astronomical. Then a short time later, he formulated the laws that govern light as it goes into a pinhole camera, a camera obscura.

In the midst of this happy moment, however, the hammer fell once again. Everything that Kepler had feared, the final gathering of the Counter-Reformation, occurred on July 27, 1600. The archduke's final decree on the Protestant problem was proclaimed in the marketplace and on the street corners, then published on church doors throughout the city. All citizens, burghers, and all inhabitants of the city of Graz, including all doctors, preachers, and nobles—except those members of the old nobility—were commanded to gather in the church on July 31 at six o'clock in the morning to be examined about their faith. This was the day that Kepler had long feared, for there was no more hiding, no more dissembling, and no more escape. The archduke presented Kepler with a simple choice—either his home and his career or his faith. There was no doubt in Kepler's mind about which way he would choose—he was a Lutheran and no prince of this earth could take that from him. Jobst Müller had determined early that he had too much to lose in his investments and that his lands were too precious to him, so if the archduke demanded it, he would turn Catholic. So did many of the men and women of his family. His granddaughter, the oldest daughter of Jobst's son Michael, eventually became a Dominican nun.

Kepler gathered with all the others early in the morning on a hot summer day, knowing full well that his fate had already been decided. The archduke himself, with a full company of guards, nobles, and other retainers gathered around him, sat in state in the cathedral. Bishop Martin Brenner of Seckau preached a sermon, and then the Counter-Reformation commissioners took their places at a long table set up in the middle of the church. They pulled up a long list of the city rolls and called the names one by one. As each man stood, the commissioners questioned him about his faith. Was he a Catholic? If not, would he ever become a Catholic? Would he swear allegiance to Rome and to the pope? If so, he could remain in Graz. If not, if he stubbornly held to the Lutheran heresy, then he would have to leave and take with him the displeasure of his liege lord, Archduke Ferdinand. Those who would not go to confession and take Communion in the Catholic church would be banished and would have to leave the city almost immediately, after paying 10 percent of all their assets to the archduke.

On August 2, they called Kepler's name, and he stood, ready, in front of a thousand witnesses, and moved to his place in front of the long table with the archduke looking at him from his throne on the dais, while the commissioners posed the questions to him one by one. Are you Catholic? He was not. Are you willing to become a Catholic, to go to confession and to take Communion? He was not. He was the fifteenth name called that day; the register listed him as Hans Kepler with a terse note beside his name saying that he should quit the land inside of six weeks and three days. Ten days later, on August 12, the archduke discharged him from his position as the district mathematician for Upper Austria. Still, the archduke respected him in spite of his Lutheran ways. Kepler petitioned to Ferdinand and the *Stände* to give him a written letter of reference and to allow him a half-year's salary as severance pay. The *Stände* agreed and paid him on August 30. Then, on September 4, the councilors at the Lutheran school gave him a second letter of recommendation, stating that he was a highly praised professor, that they were sorry to see him go, and that they gave him their full recommendation most heartily.

In truth, the Catholics in Graz hoped that he would turn Catholic under pressure as so many others did. For them to convert to Catholicism a man with his growing reputation would have given great honor to the Catholic church and would have embarrassed the Lutherans no end. The Jesuits especially were interested in this, for they could see in Kepler a searching and sincere mind. They were in some ways kindred souls, separated only by a divided faith. The rumors seem to have originated with a Capuchin Father Ludwig, a man who had once spoken to Kepler soon after his refusal to convert. In that discussion, Kepler claimed to be a

Catholic, which Ludwig imagined was the first sign of conversion, but it wasn't. Kepler did not mean he was a follower of Rome, but that he was a child of God, like all children of God who are brought into the Christian faith through baptism. The Jesuits were disappointed in their expectations, for Kepler would not bend, no matter what they said or did.

But where to go? All that was left to him was a bare hope in Tycho Brahe. Kepler wrote to Mästlin once more, describing his misfortune and once more seeking some position at Tübingen. He planned on traveling to Linz, he told Mästlin, where he would leave his family and travel on to Prague alone to look for employment. He hoped something would come along. He wondered if he could return to Tübingen to study for his *"professiuncula"* in medicine, and would await Mästlin's return letter in Linz.

Suddenly free of Graz, he left the city on September 10, 1600, with all his goods stuffed into two small horse-drawn wagons. Barbara and little Regina sat with him in the lead wagon as they rolled along through the city gates, leaving behind their home, leaving behind Barbara's inherited lands, leaving behind the sad graves of their two babies, now gone to God. "I would not have thought," he wrote to Mästlin, "that it could be so sweet, in union with my brothers, to suffer injury for religion, to abandon house, fields, friends, and nation. If real martyrdom is like this, to lay down one's life, our exultation is so much the greater, the greater the loss, and it is an easy matter to die for faith."[23]

෧ᴍᴜᴏ

Just as in the city of Prague the grief is spreading,
To the north and the south, it is reaching all nations.
Denmark laments, as everyone knows, for it is Brahe's home
    and native land,
It did not want to lose its Atlas so soon.
For the proud blood of Brahe is the pillar of the land,
When the sun sets, it is robbed of its light.
In sorrow, it tells us to mourn the others—Rantzau,
Billau, Rudro—and what grew from the seed of the rose.
Such fame wants a realm all its own,
Appreciation of his teachings, meanwhile, spans all nations.
News crossed the ocean to faraway Scotland,
King James himself was told of the sad fate.
He, friend of the muses, with whom many in the spiritual
    realm
are prepared to share the holy mysteries,
Prince of oratory power or valued priest—
May it be spoken by word of mouth, or written in his own
    book,
Or may it be found in other disciplines of expertise in the
    science of the stars,
Where Tycho pledged his passion.
Do not the Delphic prophets walk among their crowded flock,

With attentive ear perceiving mystifying sounds,
When in heaven the flock of secret movers
Were delivered by Tycho Brahe's written work?
The prophets are silent now; oracles stay away from Earth,
Go! Delphic flock, look for gods in another place!

Tycho Brahe

VIII

When in Heaven the Flock of Secret Movers

Where Kepler takes employment with
Tycho Brahe and moves his entire family to Prague.

PRAGUE, THE GOLDEN CITY, was founded by a witch. A seer perhaps, a prophetess, but in those days, it was all the same. According to legend, her name was Libuše, and she was a princess in her own right, a princess of the Czech people, who had lived in the land for generations and who were flourishing under her rule, turning old forest to new farmland to feed the scores of new babies born to them every year. Knowing she was a wise woman, her people came to her for advice. "Where shall we settle?" they asked. "Where shall we build our farms and have our babies?"

"Look for the place where the four elements dwell in abundance and settle there," she said. "Look for fertile soil that gives life. Look for water, clean and pure. Look for fresh and healthy air. Look for fuel to give you warmth, where the trees grow down to the waterside to give you both wood and shade. If harmony among the four elements rules a place, then you will want for nothing." The people followed her advice and prospered.

When the time came for her to marry, Libuše took her own advice and chose a man to balance her gifts. She was undoubtedly beautiful and had suitors from all over Europe seeking her hand, but out of them all she chose a simple plowman, a man of the people named Přemysl. Where she was air and light, he was water and earth. They lived together in his castle, which stood on a cliff above the Vltava River, the Moldau of legend. Přemysl had named the castle Vyšehrad, which meant, prosaically, High Castle, because presumably that's what it was. All around them, the people flourished.

As royals do, the couple often moved between castles, sometimes to his, sometimes to hers. One day, when they were living in her castle, which they had been doing for some time, the two of them wanted to see the wide lands unrolling all around them, so they climbed the battlements up into the highest tower, to the tallest spire, where there was a room made of stone with windows all around and firing slits in the walls for archers. Behind them, the whole court huddled on the steps, breathing hard from the climb, gossiping here and there and speculating about whatever business the prince and princess had up in the tower. Outside, the evening gathered, rose-colored, with green and brown fields all around, bluing slowly in the distance, out to the edge of the forest, and then deepening to purple near the horizon. Swallows wheeled and hummed, snatching mosquitoes from the air. It was an evening ready for prophecy.

Suddenly, the world hushed. Princess Libuše stretched out her hand to the evening and pointed into the distance. A light gathered around her. The court quieted, fearing to breathe and break the spell—even the birds fell silent.

"I see a great castle and its glory reaches to the stars," the princess said. "The place lies deep in the forest—the Brusnice river valley guards it from the north, while a rocky hill protects it from the south. The Vltava winds between, at the bottom of the great hill. Go there," she said, "and on the left bank of the river, you will find a man carving the doorway of a new house. You will build your castle there and name it 'Threshold,' *Praha*, the sacred doorway leading into hearth and home, to warm fires and to meat on the spit, and finally leading into the place of the dead. You know that tall men must bend their heads to enter any doorway. So will all, tall and short, men and women, bend before this castle, and one day it shall be a great, golden city."

The prince and the men of the court tried to follow where Libuše pointed, but could see nothing but the night. Then, after a time, the princess fell back, the light of prophecy fading from her. The birds sang once again, and the court buzzed with gossip. The next day, the men packed their horses and set out to find the place. Quite soon they found the valley, just as the princess had described it—the narrow river valley to the north, the great hill to the south, the winding Vltava in between. And there, near the river, they came upon a man carving the lintels of a doorway. The men of the court set to work building the castle. They felled trees and opened up new lands; they built a rampart and a great hall, just like the one at Vyšehrad, at the court of Prince Pøemysl. This new castle was bigger, however, greater and taller than any other, built as Libuše had said, on the left bank of the Vltava. Its fame grew and has lasted even to today.[1]

◠◡◠

IT IS LIKELY THAT KEPLER, on his way to Prague via Linz, never realized his good fortune, never realized that his life was about to change for the better. It is likely that all he could see was a future full of darkness. Tübingen had abandoned him, though. Hoping against all evidence to the contrary, he still waited for word, waited for some letter from his old school that would change his life. Other prospects and other universities were vague chances at best. And attaching himself to Tycho was risky. Kepler, his wife, Barbara, and his daughter, Regina, with all their worldly possessions stuffed into two little wagons, rode behind a plodding horse on their way to Linz. Kepler, his face drawn with fear for the future, wondered how they would all survive. The trip to Prague was no great adventure, no wonderful new thing—it was an act of desperation. Kepler was a refugee—the greatest tragedy that could befall a man, finding himself on the highway, banished from his home, with no way to support his family and with

147

nothing but poverty following behind. He was fearfully, agonizingly homeless.

When he and his family arrived at Linz, Kepler asked about the letter he had hoped would be waiting for him there. He was praying, irrationally perhaps, for some magic word from Mästlin, a change of heart from the faculty that would, in spite of everything, call him back to Swabia, where so much of what he knew and loved resided. But no such letter had come, and Kepler's last hope of returning home died. What should he do? What could he do? He decided that he could not leave his family behind in Linz among strangers, for fear that one of them might sicken and die, far away from her friends and family. So he left his household goods in the care of someone in Linz and, taking Barbara and Regina with him, he set off for Prague.

On the road between Linz and Prague, he took sick. The distress, the exhaustion, the desperation encircled him, and he fell into an intermittent fever, a *febris quartana* that would level him for the rest of the year. The fever swelled for about four days, sucking the life out of him, and then it would fade only to return a few days later. Was this malaria? Or typhus? While on the road, weakened and shivering, he sent a letter to Tycho Brahe announcing his arrival.

Tycho had hoped to establish a new Uraniborg at Benatky and had sent his son with Longomontanus to fetch the four great instruments he had abandoned on Hveen when he left the country. Because of bad rains and swollen rivers, because of one bureaucratic tangle after another, the instruments did not arrive in time. A recent outburst of plague had sent the imperial court scuttling from Prague, but then after the plague subsided, the emperor returned. Fretting over the future, Rudolf suddenly needed the astrological advice of his imperial mathematician and on June 10, 1600, sent for Tycho. Tycho took residence near the palace, because the emperor often commanded his attendance twice a day.[2] He set up his household at an inn, the Sign of the Golden Griffin, built on the slopes of the Prague Castle, the emperor's palace.

The letter from Kepler was businesslike, full of hidden bravado.

Though I was not able until now to write to you about my business with you, I am writing now partly because of my desire to communicate to you my reasons for travel to Bohemia, which I have been best able to do while on the road. Your letter containing my appointment only reached me after I had left the Steiermark and had come to Bohemia. I have several points. The first arises from the contract that we signed last May and have already closed. In it, I promised, with the consent of the emperor, to work at your disposal on some area of astronomy for two years, in which time you would help and support me. The second point refers to my present state of distress, because it is the will of the local authorities that I emigrate, and I request that you give insurance, either by recommending me to the emperor or by helping in my state of emergency to relocate to Prague.

Bereft of everything, Kepler's pride was touchy. He certainly did not want to appear like a supplicant before Tycho; and he would not allow himself to become one more in Tycho's battalion of servants or one more mouthpiece for Tycho's theories. He had a reputation of his own to defend, a theory to prove, and, after all, he did not completely agree with Tycho. Perhaps he had overstated his case in his letter, pretending that he had more prospects than he had. Perhaps he had hidden Tübingen's rejection, for he didn't want to admit that his own people had abandoned him. It was true that Tycho had invited him even without his salary from Graz, but he was a destitute man, and he did not want to look destitute. In the letter, he told Tycho that as a scholarship student, he needed to travel to Württemberg to ask the duke for his support and for the support of the duke's ambassador at the imperial court. He was looking into a position at a university, perhaps Wittenberg or Jena or Leipzig. He made it sound as if the professors at Tübingen had also offered him a position and that he was trying to make a choice between several lucrative opportunities. However, he said, if Tycho had a position for him, then he would certainly give him first choice.

Kepler arrived in Prague on October 19, 1600. He was in a bad way—depressed, nearly broke, shaking with fever, with his wife and stepdaughter on the verge of illness themselves. He sought refuge in the house of Baron Hoffmann, where he was warmly received. The fever carried on, however. He began to cough and feared tuberculosis. Added to that were money problems. The cost of living in Prague was so much higher than in Graz that what little money he had was rapidly disappearing. In Graz, he had made only 200 gulden a year, and it had cost him 120 to move his family to Prague. Barbara complained constantly. Because of her husband's religious scruples, she had been torn from everything she ever knew and loved. Her father's conversion, on the one hand, and her husband's refusal to convert, on the other, must have torn the poor thing apart. Besides, as the daughter of a rich man, she was used to a much higher standard of living. Everything was so much more expensive in Prague that Barbara had to live like the poor. She soon fell into a terrible depression, missing her family, missing her city and all that was familiar to her, and then she too fell sick.

Tycho, meanwhile, received Kepler warmly. Tycho was a builder and an observer, a gatherer of information, and although he was a master of astronomy and calculation, he lacked Kepler's sublime fire, his lightning insight into how things fit. If he wanted a true victory, he needed someone who could take his observations and fashion a proof out of them. The bare observations alone wouldn't be enough. They had to be sorted, calculated, and woven into the right patterns to prove the Tychonic system, and that required Kepler. Besides, it was hard to hold on to good assistants. He had gathered good astronomical minds from all over Europe, but after a time, one by one, they found other positions, returned home, or set out on their own. Tycho wasn't all that easy a master. Therefore, he worked hard to secure a position for Kepler, not only with his own staff, which he was willing to do in spite of their often prickly relationships, but also with the imperial court. The emperor, for his part, was well disposed toward Tycho, partly because he needed him for his astrological advice and partly because he was an expert in all the little machines the emperor loved so well. He was therefore well disposed to any suggestion that Tycho might have had for Kepler.

But the emperor was not completely his own man. He was surrounded by an imperial bureaucracy that seemed to go on in Byzantine fashion for miles and miles. To secure a position for Kepler would not only require the emperor's approval, but the approval of every little bureaucratic head in every little bureaucratic office down every little bureaucratic corridor in the palace. One of their functions was to make sure that things did not happen, because when things happened, it cost money. They excelled at this job. As with all governments, without a strong hand from the emperor, the imperial bureaucracy was glacial, so that no matter what Rudolf wanted for Kepler's salary, the money somehow never showed up, and Kepler quickly became dependent upon Tycho.

In December, the last nail in the Tübingen coffin was hammered in; the last of his hopes for that university was crushed. The letter he had looked for arrived, but his former teacher offered him no advice, no comfort except for his prayers. Kepler wrote back at once: "I cannot describe what paroxysm of melancholy your letter has occasioned me, because it destroys all hope of going to your university. So I must stay put until I either get well or die."[3] Several weeks later, he wrote once again, this time begging:

I long for consolation, for I am still suffering from the *Wechselfieber*, the intermittent fever, and from a dangerous cough. I suspect consumption, which may take my life. My wife is also ill. Not four months have passed and I have spent one hundred taler in Prague. On top of that, little is left of the travel money. Tycho keeps promising me that if everything were up to him, no one would be happier than I. My impatience and the significance of Easter time are prophesying my impending death, and if so, I shall depart this world around the Easter holiday. The love of my homeland is tearing me away, whatever its future destiny shall hold. Once before, however, I was there when my world fell apart, so now I have a fearless spirit.

Tycho is very stingy with his observations, but I am allowed to copy them daily. If only the transcript were enough. Therefore, a selection is necessary. The illness leaves me gloomy. I have been burdened too much with the gift of darkness. I am unsatisfied with myself.

Longomontanus, homesick for Denmark, had already left Prague by the time Kepler had arrived, and so Kepler was able to take on the problem of the orbit of Mars once again. He worked on it diligently, as much as his health would allow, which would have been enough by itself, but Tycho assigned Kepler one more task, one that troubled him deeply. Returning to his old battle with Ursus, he set Kepler to work on a refutation of the dead man's ideas. Perhaps Tycho wanted Kepler to prove his loyalty, but more likely Tycho was just old-fashioned and an enemy was an enemy. Like Ahab after the white whale, he would not let go of Ursus and would chase the man into his grave and argue with his ghost. Kepler did as Tycho asked, though he found it a waste of time and believed that it was unseemly that men who studied the secrets of God would behave in this manner. As he wrote Tycho's defense, therefore, he did it with a remarkable sense of balance. He kept all personal rancor out of the text and, instead, refuted Ursus's hypotheses one at a time. In doing so, he defined what an astronomical hypothesis was, rejecting the idea that it was simply a matter of correct conclusions arrived at by calculation.

Then Kepler received word that old Jobst Müller had died, and in the spring of 1601 he traveled south to Graz once more to settle his wife's affairs. Surprisingly, in spite of all the threats that the archduke had fired at the Lutherans, Kepler wasn't arrested. No one in the Counter-Reformation seemed to care or even notice that he had returned. If he had stayed longer, perhaps they might have. The archduke's government in Graz looked the other way, and the Lutheran community, or what was left of it, warmly received him. The trip apparently did him a world of good. The fever abated, and he was able to travel in and around the city in some comfort. He had the energy to climb Mt. Schöckel, where he made measurements of the curvature of the earth and witnessed a strange thunderstorm.

In the midst of his joy, however, troubles exploded with Tycho once again. Tycho had promised Kepler that he would provide money to take care of Barbara and Regina while he was away, but then suddenly, while he was still in Graz, Barbara wrote Kepler a letter complaining that she was receiving too little money and that household goods were costing too much. She complained further that Tycho was too stingy and that life in Prague, for all its golden beauty, was not as gracious or as comfortable as it had been in Graz. Fired up, Kepler wrote to Tycho and passed on his wife's complaints is if they were his own.

Tycho angrily responded through his pupil Johannes Erickson that Kepler should consider what Tycho had already given to him and to his wife and that the Keplers should have more respect and be more moderate with their benefactor, who was helping them in every way possible. This angered Kepler in turn because he did not want to be a charity case; he wanted to be paid for his work, work central to everything Tycho was doing, and given the respect that one colleague would give to another. But as late middle age began to settle on Tycho, he was less and less willing to compromise. He had grown irascible, even with the powerful members of the emperor's court. He worried over his rejection by the Danish king and all the other events that had brought him to the emperor's court. For all his wealth and noble birth, Tycho had also come to Prague as a refugee. He was a man out of place in the city, and though he was well traveled, as he grew older he yearned increasingly for his homeland. As Kepler said, "He was not the man who could live with anyone without very severe conflicts, let alone with men in high position, the proud advisers of kings and princes."[4] Gradually, the loss of his homeland, the stress of his position as imperial mathematician, and the pull and tug of imperial court life began to show on the old man, and he acted more like a petulant child.

At bottom, Tycho was a good sort with a good heart; he tried to do the right thing, but he was also feisty, and his own pride subverted him. In that, he was a lot like Kepler. Tycho protected his observations like a wolf guarding her cubs, and he showed Kepler only those observations that were germane to the work he expected Kepler to do. Such miserliness was a symptom of the age. Kepler was more open-minded than most, and in some ways more modern. It infuriated him that other astronomers, including his teacher Michael Mästlin, refused to work with each other, guarded their own observations like dragons, and hoarded their data as if it were spun of gold. Throughout his career, Kepler encouraged other astronomers to share their work, to advance general knowledge, but got little response. Each astronomer was too worried about his reputation, his place in court, the opinions of his patrons, as if one man could own all knowledge. Tycho was typical of the times, more interested in advancing his own fame than the discipline. On Hveen, his island fiefdom in Denmark, he had set up a printing press to publish his works and observations. No other printer could touch the data, for Tycho demanded absolute control over his work.

After a few months, Kepler returned to Prague from Graz feeling well. The spat with Tycho had blown over, the fever that had plagued him had passed, and he had benefited from the kindness and hospitality of his old friends in Austria. Almost immediately, Tycho introduced him to the emperor, who received Kepler graciously and commissioned him to collaborate with Tycho in compiling new astronomical tables based on Tycho's observations. Tycho begged the emperor to allow him to name the tables after the emperor himself, which Rudolf agreed to, and so the new book would be titled the *Rudolphine Tables*. Kepler was included in the project as a collaborator, not an assistant or servant. Kepler's life had suddenly changed; his future was secure. Tycho also did something he had done with few others. Knowing that only Kepler could prove his Tychonic theory, Tycho entrusted him with his precious data, the keys to his heavenly kingdom. Kepler suddenly had everything he needed.

Then, however, his life changed once again. Kepler had been sick for nearly a year, and only through his trip to Graz had he been able to cleanse his body and soul. Now, suddenly, it was Tycho's turn. Things were going well between him and Kepler, and there was finally some peace in the house. Kepler had received the recognition he desired, and in the emperor's own words he was no longer a subordinate, but a collaborator. Then, suddenly, Tycho took ill, with an illness that would lead to his death. A few days after Tycho had presented his new collaborator to Emperor Rudolf, he went with a friend, Councilor Minckwicz from the imperial court, to a luxurious banquet given by the old patron of Edward Kelley, Peter Vok Ursinus Rozmberk, at his house near the gate to the Prague Castle. The rules of civilized behavior at the imperial court were decidedly medieval and required that all guests remain seated at the table until the host finished his meal and signaled the end of the banquet by rising and leaving the room. According to an account of the event that appeared at the end of Tycho's personal papers and observations, an account that scholars later decided was penned by Kepler: "Holding his urine longer than was his habit, Brahe remained seated. Although he drank a little overgenerously and experienced pressure on his bladder, he felt less concern for the state of his health than for etiquette. By the time he returned home, he could not urinate anymore."

Finally, with the most excruciating pain, he barely passed some urine. But it was still blocked. Uninterrupted insomnia followed, then intestinal fever, and little by little delirium. His poor condition was made worse by his way of eating, from which he could not be deterred. On October 24, when his delirium had subsided for a few hours, amid the prayers, tears, and efforts of his family to console him, his strength failed and he passed away very peacefully. With his death, his thirty-eight-year series of heavenly observations came to an end. During his last night, through the delirium in which everything was very pleasant, like a composer creating a song Brahe repeated these words over and over again: "Let me not seem to have lived in vain."[5]

Kepler stood at Tycho's bedside with the family, and just before Tycho died, at nine or ten in the morning, he begged Kepler to carry on with his work, asking him not to abandon the Tychonic model of the universe entirely in favor of the Copernican. If he had to turn to Copernicus's plan, then he should at least follow through with Tycho's as well. The family buried Tycho on November 4 at Tyn Church in Prague in an Utraquist service. The Utraquists were pre-Reformation reformers peculiar to Bohemia, the followers of the long-dead martyr Jan Hus. Kepler was part of the funeral procession as a collaborator and a colleague, not an assistant. He composed and recited a long elegy for the service, for he never forgot what he owed to his old master. Though he would never surrender his belief in the Copernican system, he did his best to show where the Tychonic system was still applicable to the motion of the planets.

Two days after Tycho died, the emperor declared Kepler his imperial mathematician. He sent an adviser named Barwitz to tell Kepler of his advancement and to inform him that the emperor had transferred the care of all of Tycho's instruments and works to him. Barwitz told Kepler that the emperor would bestow on him both a salary and a title, but of course he would have to apply for the salary. Given the state of the imperial bureaucracy, this alone signaled trouble for Kepler. Nevertheless, where he was once the assistant, now he was the master. He had control of all of Tycho's observations and could work freely for the first time in his life. His one great task was to prepare the *Rudolphine Tables,* and since little had been done on that project while Tycho was alive, Kepler needed to make mountains of calculations in order to finish it.

For the next few months, however, he stood in the hallways of the great Prague Castle, one of a small crowd of imperial employees who gathered in clumps about the castle. In a little over a year, he had risen from an outcast from Austria to the imperial mathematician. Surely, he thought, this was the hand of God. Congratulations flowed in from all over Europe. Kepler received a letter from Baron Hoffmann and another from Herwart von Hohenberg, both of whom were thrilled, and said so. Only Kepler, Herwart said, could replace Tycho Brahe. Knowing the Prague bureaucracy all too well, Herwart advised Kepler to ask for a salary commensurate with his mind and his importance and to seek an immediate down payment. Kepler listened to this, but sadly took the advice of other, less astute political minds and let the emperor fix the salary. Seeing a bargain, Rudolf offered him 500 gulden a year, one-sixth of what he offered to Tycho, a salary commensurate not with his mind, but rather with his social standing. The salary was expected to begin on October 1, 1601, but that didn't mean it would actually start then. Kepler, so unused to court life, spent the next few months shuttling back and forth from one office to the other in the imperial bureaucracy, looking for his pay.

Troubles also began about that time with Tycho's family. Although the

Brahe family legally owned Tycho's legacy, the emperor had given control over that legacy to Kepler since no one else could complete Tycho's great work. The family, led by the son-in-law, the same Tengnagel who had so abused Tycho's other assistants while courting his daughter, conspired to assure that the upstart Kepler would not profit from Tycho's work and that all fame, glory, and whatever money might come would go to the family—in Tycho's name, of course. Kepler tried to keep the peace and to respond with proper respect to Tycho's family, but the problems would only get worse as the *Rudolphine Tables* neared completion.

As a mathematician, however, Kepler had reached the heights. He was a member of the emperor's court, seemingly untouchable. But even there, the Counter-Reformation gathered around him. The old warfare between Protestants and Catholics was even more complicated in Prague than it had been in Graz. In Graz, there had been only two groups struggling for power. In Prague, there were three—the Catholics, the Utraquists, and the Bohemian Brethren. The Utraquists were founded when the more moderate followers of Jan Hus signed a treaty with the Catholic church on November 30, 1433. One of the Hussites' great complaints was the division between the clergy and the laity, symbolized by the fact that the clergy received Communion under both species, *sub utraque specie,* the bread and wine, while the people received Communion under only one, the bread. The Utraquists were therefore a form of liberal Catholicism, more schismatic than heretical. They held a middle ground, much as Anglicanism does in the modern world. The Hussite rebellion, however, was a violent one, with war and betrayal at the hands of church officials. In the fifteenth century, after a good deal of bloodshed, church officials invited Hus to a conference to discuss healing the rift between his followers and the church, and while he was on the road to the conference, the bishops sent soldiers to arrest him, try him, and finally burn him at the stake. Memory sears like a hot iron, and almost a hundred and fifty years later, the people of Bohemia could still smell the smoke of betrayal.

Into this mix of Catholics, Utraquists, and Bohemian Brethren, a more radical Hussite group, came Luther and his followers. Protestants began to split, and to split again. Some of the Bohemian Brethren formed a new group called the Bohemian Brethren in the Bohemian Confession, which followed the Augsburg Confession in most ways, with a pinch of Calvinism thrown in. In 1556, however, the Jesuits entered the scene, as they had done in Graz. Peter Canisius, later canonized a saint, lead them, and his zeal was aimed at one thing—the reconversion of Bohemia to the Catholic church.

The Jesuits were missionaries and educators, founding colleges all over Europe, and they admired reasonable men. Complete union with the Society of Jesus required that one take a fourth vow over and above the traditional three of poverty, chastity, and obedience, a vow of special obedience and dedication to the pope. The Jesuits were therefore the most Catholic of Catholic orders. Throughout his life, Kepler had a complex relationship with them. Some Jesuits befriended him and gave him sanctuary when members of his own church would not. They supported him, promoted his work, and prayed for his conversion. From time to time, they passed on hopeful rumors wafting around Prague, as they had done in Graz, about Kepler's imminent conversion, but these were pipe dreams.

Once in Prague, the Jesuits founded a new university, the Clementinum, which, like all Jesuit schools, grew quickly and attracted political influence. The already existing university, the Carolinum, founded in 1348 by the emperor Charles IV, had become the seat of Utraquist doctrine, and from there it had opened to all the other new theologies, and so the Catholics needed a university of their own. After the arrival of the Jesuits, however, the initiative that had once belonged to the Protestants shifted to the Catholics. Meanwhile, the Habsburgs battled one another over the question. Maximilian II, Rudolf's father, like Rudolf himself, was a Renaissance man. Theological niceties bored him, and what he wanted most was peace, so he oscillated between one stand and the next. When Rudolf II became emperor in 1576, he followed his father's example, but under pressure from the growing Catholic faction, led by his mother, he invited the Capuchins into the city to join the Jesuits, and so the Counter-Reformation began to grow. Suddenly, new antagonisms gestated between the emperor and his people, just as they had done in Graz while Kepler was there.

In 1602, two years after Kepler arrived, the emperor declared that only Catholics and Utraquists could live in the land of Bohemia. What was Kepler to do? Was his position once again in jeopardy? Would he be forced out, as had happened in Graz? None of these things happened, however. The hammer never fell. Partly, this was because Kepler had grown in stature. Where he was once a simple teacher in a Lutheran school, now he was the imperial mathematician. That, along with the personality of the emperor himself, made the difference. Although Rudolf's cousin Ferdinand and his younger brother Matthias were both devoted to the Catholic faith, Rudolf was much more open. Although Ferdinand was a zealous torch, ready to set the world on fire, dogmatic in his beliefs, and intolerant of anyone who was not in direct communion with Rome, Rudolf was too busy trying to solve the problem of himself. He was an eccentric and suffered from profound depression. Reclusive, he abandoned much of the business of the empire to his corrupt and treacherous court and hid himself in the labyrinth of his *Kunstkammer,* his personal imperial museum. Weighed down by sadness, he grew more introverted with each year and eventually suffered a breakdown, tormented by hallucinations. Fearful that he was bewitched, he gathered around himself astrologers and alchemists, mathematicians and astronomers, practitioners of magic high and low, and even a few mystical rabbis.

◠◡◠◡◠

RUDOLF II WAS AN INTERESTING MAN. King of Bohemia and Hungary, emperor of the Holy Roman Empire, he was a collector of oddities, antiquities, and wonders. His family, especially his ambitious younger brother Matthias, thought he was too soft on Protestants and should be replaced by, well, Matthias. Although Rudolf was a strong Catholic and promoted the works of the Jesuits and the Capuchins in his kingdom, he preferred peace and was at heart more concerned with the occult, with piercing the veil of the mystery of life. His grandmother, Juana "the Mad" of Castile, had died howling with insanity, and his mother, Maria of Spain, was a cold, difficult woman who had never wanted to leave Spain and wanted

159

everyone around her to be as Spanish as possible. She gave birth to Rudolf on a summer evening in July—July 18, 1552. His father, Maximilian II of Austria, was a different sort, an openhearted, friendly man, unlike his mother, who like her grandmother suffered from chronic melancholy. All her life, she remained cool and distanced from her sixteen children, as if she were living in a high, dark room.

Madness permeated both branches of the of the Habsburg line—the Spanish and the Austrian. Both of Rudolf's parents, who were cousins, were the grandchildren of Mad Juana of Castile, who lived from 1479 to 1555. According to the story, after her husband, Philip the Handsome, died on September 25, 1506, Juana kept his body beside her bed for the next nineteen years in the belief that on the anniversary of his death he would come back to life. Supposedly, she took his casket along whenever she traveled about Spain, and now and then, perhaps wondering if he were still dead or perhaps desiring to look one more time on his moldering face, she would open the lid and look inside. An odd connection: as a young man before his conversion, Ignatius of Loyola, the founder of the Jesuits, would accompany the royal treasurer to the royal villa at Tordesillas, where Juana lived with her youngest daughter, Catherine. For several years, Ignatius held a secret crush on the young Infanta, "the secret of his heart."

The Spanish king, Philip II of Spain, joined with his sister, Rudolf's mother, Maria, in lobbying for a Spanish education for all of Maximilian's children. Maria feared that her children would come under the sway of the Protestants, for Protestant ideas floated through Vienna like a morning mist, but not so in Madrid, which remained Catholic, austere, proud, and gloomy. Maximilian, who never liked the Spanish, put off his sons' departure year after year, until the pressure from the Spanish faction became overwhelming, and he gave his permission.

Rudolf arrived in Spain in 1564 with his younger brother Ernest and two friends, Wolfgang von Rumpf and Count Adam von Dietrichstein. Even by that time, his personality had begun to form, and he showed the first symptoms of his mother's melancholy. He was also probably bisexual. A serious boy, Rudolf was reserved but deeply intelligent and often given to flights of fancy and bouts of sadness. Eventually, he would prove to be a great linguist, able to speak, read, and write Latin, Spanish, German, and eventually a little Czech. He also developed a taste for art and loved mathematics and science. All his life, the natural world held him spellbound, and he wanted to know all he could about it.

Life in Spain was difficult for young Rudolf and his brother. Philip II's court was a cold place, mannerly, bigoted, and often ruthless. Spanish manners seemed to infect him, playing on his natural melancholy. While in Spain, he met Don Carlos, the insane son of the King Philip II. He stood by watching in January 1568 as the king ordered his own son locked into a room and then forbade everyone in the court to mention the boy's name, even in conversation and even in prayers. Don Carlos died in July that year, followed by the queen, his mother, in October.

A pall hung over the court after that; Rudolf and young Ernest hid behind their studies, the classics, Latin prose, fencing, and theology. On Sunday, they served as altar boys at Mass. In early 1570, Philip II married Rudolf's youngest sister, Anna, making her his fourth wife. Finally, the following spring, the king gave Rudolf and Ernest permission to return to Vienna. Rudolf later remarked that he had spent the next night so filled with joy that he couldn't sleep. Their years in Spain had been a difficult time for them, scarring them deeply. Once they were back in Vienna, their father, Emperor Maximilian, did not approve of the changes and commanded them to rid themselves of their "Spanish humors." He disapproved of the penitential gravity and prideful distance he found they had acquired. "Change your bearing!" Maximilian told them, but they could not; Spanish manners had become too solid a part of them. Soon after they returned, Maximilian's health began to fail. He suffered from numerous health problems—heart attacks, gout, and something called "kidney colic," all of which might have had their roots in syphilis.

Maximilian, knowing that his time on earth was coming to a close, arranged to have Rudolf, his oldest son, crowned king of Bohemia and king of Hungary. Soon after, the Imperial Diet met in Regensburg, so that Rudolf could be crowned king of the Romans, a necessary prerequisite to the imperial throne, as well. On Maximilian's way to Regensburg, his health collapsed. Though he seemed to rally for a short while, his health faded, and he grew weaker every day. Rudolf rushed to his bedside. His daughter Anna rode quickly from Bavaria in order to join the family. Both Anna and Queen Maria urged him to receive the last rites. The Spanish faction gathered, until finally the Spanish ambassador said: "I see from your condition, Your Majesty, that it would be time..." Maximilian, however, cut him off. "You are right, Mr. Marquis. I have not slept well and would like to rest a little." Finally, on October 12 Emperor Maximilian died, with his entire family gathered around.

Wasting little time, the German electors voted Rudolf the new emperor and crowned him on November 1. As with his mother and Mad Juana of Castile, Rudolf suffered from bouts of melancholy. These would only get worse in his life, as the troubles of his times encompassed him. Those were troubled times, indeed. Catholics and Protestants were at war all across his kingdom. An earthquake rocked Vienna, and local epidemics of plague flashed into and out of existence like brush fires. In 1577, the year that Johannes Kepler stood upon the hill outside of Leonberg holding his mother's hand watching the great comet, Rudolf II, king of Bohemia and Hungary, king of the Romans and emperor of the Holy Roman Empire, suffered an emotional breakdown. From that time on he almost never left the castle; he became so sick over the next four years and lost so much weight that people feared for his life. The responsibilities of power were too much for him. The threats of the Turks were constant. The struggles between Catholics and Protestants were endless.

Finally, he moved his court to Prague in Bohemia to get away from the crowds and pressures of Vienna. Almost immediately after moving into Prague Castle, he ordered the construction of a series of great cabinets and long shelves throughout the hallways of the castle, the beginnings of his *Kunstkammer,* his private museum. This museum was never open to the public, and only a few select guests, usually kings and important ambassadors, were ever given a tour. There, he began collecting exotic animals and gathering to himself mountains of art. He also gathered some of the great minds of Europe, not only painters and artists, philosophers and mystics, but scientists such as Tycho and Kepler.

Throughout the long galleries of the Prague Castle, which is actually a small city built on top of a high hill, he housed several thousand paintings (some by Arcimboldo, Breughel, and Correggio), sculptures, coins, gems, natural oddities, medicines, scientific instruments, and clocks as well as books on the occult and other curious matters. On the first floor were the Spanish Room and the New Room, where he placed his art collection. He was fascinated with little machines, which was one of the reasons he had appointed Tycho Brahe as his imperial mathematician.

In July 1599, just after Tycho first arrived in Prague, Barvitius, the emperor's private secretary, drove Tycho up the long hill in a magnificent carriage to a "splendid and magnificent palace in the Italian style, with beautiful private grounds." There, inside the palace, he waited for the summons, holding three of his books to present to the emperor. A court attendant appeared, calling him forward and leading him to the emperor's audience chamber. Reclusive Rudolf almost never received anyone alone. In a letter to Rosenkrantz, Tycho described the scene: "I saw [the emperor] sitting in the room on a bench with his back against a table, completely alone in the whole chamber without even an attending page. After the customary gestures of civility, he immediately called me over to him with a nod, and when I approached, graciously held out his hand to me. I then drew back a bit and gave a little speech in Latin." Afterward, Rudolf, trained in Spanish manners, responded graciously to Tycho, "saying, among other things, how agreeable my arrival was and that he promised to support me and my research, all the while smiling in the most kindly way so that his whole face beamed with benevolence. I could not take in everything he said because he by nature speaks very softly."

After the audience, the emperor called Barvitius into the audience chamber. When Barvitius emerged, he told Tycho that the emperor had been watching his arrival from his window and had seen a mechanical device on his carriage. He wondered what that was. Tycho told Barvitius that it was his odometer and sent his son to remove the device from the carriage and bring it back to give to the emperor. He explained its workings to Barvitius and showed him how it rang out the passing distances by "striking distinct sounds with two little bells."

Barvitius reentered the audience chamber and after a time emerged once again, saying that the emperor did not wish to accept Tycho's instrument, but wanted to have one built according to its specifications. Tycho happily supplied the design, knowing that the position he sought was his.

From 1605 to 1606, the emperor's artisans finished three different vaulted rooms on the first floor. Here the emperor displayed his vast collection of scientific instruments, some of the best in Europe, his books and manuscripts on scientific and arcane matters as well as his books on history and the great works of literature. In rooms beyond this library, was another library, stretching from the ceiling to the floor, known to writers across Europe as one of the greatest concerning all things scientific and philosophical. Meanwhile, exotic animals meandered through the labyrinth of corridors. In the floors below, the artisans worked in their workshop, where they made all of the wonderful machines and great devices the emperor prized so much. Beside the workshop was the alchemical laboratory, where alchemists from across Europe boiled up vats of unknown potions and experimented with secret elixirs.

The entrance to the museum was through a small antechamber decorated with images of nature, with the four elements and the twelve months of the year, all watched over and supervised by Jupiter, Rudolf's mythological self. In spite of his mythological pretensions, however, throughout the 1590s, the emperor's depression deepened. He suffered from bouts of anxiety and despair, which became worse and more frequent. Visitors noticed how sad he had become and how remote. The golden collection could only keep his madness at bay for a time, however. Suddenly, his melancholy turned to paranoia, and he began to fear his own family, believing that someone in his family would eventually murder him. Once, in a sudden rage, he threatened one of his ministers with a dagger. Fearing theft, he kept most of his golden treasure locked away in wooden chests, which was one of the reasons why Kepler had difficulty obtaining his salary. At times, Rudolf's fears swelled and he held on to his gold so tightly that there was no food in the palace. His greatest enemy, he believed, and not without reason, was his brother Matthias, the man who would eventually take over his throne.

Matthias was not as intelligent or as open-minded as his brother. He had few talents except for the applications of power, and he used Rudolf's interest in the occult to help bring him down, claiming that Rudolf was too easy with Protestants and that he dabbled in un-Christian things. Matthias did have ambition enough, and his desire for power and revenge on his older brother often choked him. Moreover, he was good at one thing Rudolf was not—he had a gift for intrigue. Rudolf hated him and feared him most of all. He took every opportunity he could to humiliate him, refusing him money and any position of power within the empire. He even forbade him to marry. After Ernest died in 1595, however, Matthias, who was next in line, became the imperial heir. Fearing the worst, Rudolf retreated more and more into his private quarters and into the galleries of his mechanical wonders, slowly abandoning the affairs of state, refusing to meet with foreign ambassadors, and flying into sudden rages.

All of his relationships gradually fell apart. His sexual life was complicated, since he was attracted to both young boys and young girls. For years, he had been scheduled to marry his Spanish cousin Isabel, but postponed the wedding year after year. His mother begged him to marry in long letters sent from Spain, but somehow he found a reason to postpone it one more year. Finally, after waiting until she was thirty-two years old, Isabel married Rudolf's younger brother Albert, the former cardinal. Hearing about their wedding, Rudolf raged about the palace in grief and anger.

⧉

THIS WAS THE PALACE that Kepler came to after Tycho's death and in which he joined the small clusters of attendants and imperial employees waiting for word on their salaries; occasionally he was summoned by the emperor to answer a question or perform some errand. A decade before, John Dee and the earless Edward Kelley, once alchemists and astrologers to Queen Elizabeth I, had haunted the castle, speaking with angels in secret and at least once, according to them, turning a pound of lead into a

165

pound of gold. Kelley had lost his ears after being unmasked as a charlatan. The team gave alchemical advice to Rudolf until Kelley announced that the archangel Uriel had spoken to him in the night, telling him that he and his partner, John Dee, should share their wives. Dee's wife, who was younger and prettier than Kelley's, put her foot down, and from that point on the partnership faded.

Ten years later, Kepler would have been standing in roughly the same place Dee and Kelley stood, in the vaulted Wenceslas Hall in the old royal palace or in one of the great rooms of Rudolf's Italian palace, waiting in line or for an appointment. Perhaps, in one corner of the room was an old man standing quietly, the only Jew ever to have had an audience with the emperor. Kepler would have known him and perhaps spoken with him. This was the mysterious Judah Löw, the great rabbi of Prague, a master of the Kabbalah. The details of his interviews with the emperor remain unknown, even today, though there are stories. One has the rabbi ushered into a wide room with a single table and two chairs. In the corner is the opening to an antechamber covered with a red curtain. The man who meets the rabbi is not the emperor, but one of his councilors, who engages the rabbi in a long, involved conversation about occult matters. He asks Rabbi Löw about his teachings, about the mysticism of the Kabbalah, and about the secrets of the universe. The conversation becomes more involved, ranging widely, until suddenly Emperor Rudolf bursts into the room from behind the curtain, where he had been listening all the while. The rabbi and the councilor stand, the emperor takes the councilor's place at the table, and the conversation continues.

This same castle, filled in Kepler's time with scientists and charlatans, mystics and philosophers, was where, seven years after Kepler had left the city, two Catholic representatives to the Diet, along with their secretary, were thrown out of a window in the old Bohemian chancellery. They survived, but the Thirty Years' War began, the war that would hound Kepler to his tomb.

ᏮᎷᎶ

*I am just completing my studies on the movements of the star
Mars, and this demands a good deal of difficult concentration.
I offer a heavenly philosophy in place of the heavenly theology or heavenly metaphysics of Aristotle. Would that you read
my work and counsel me before I publish it! Vögelin in Heidelberg will print it, though the circulation of individual
copies beforehand has been forbidden by the emperor. Besides
my physics, I am currently teaching a new arithmetic. . . . Yet
what a notion am I aiming at! It is not Mars who has incited
me to write this book, but something else. "God is in us; if
God sets us moving, then we are warming up."*

NOVEMBER 30, 1607

*You think that the stars are simple things, and pure. I think
otherwise, that they are like our earth. But experience cannot
speak here, since no one has ever traveled to the stars before.
Experience tells us nothing, therefore, neither yes nor no. In
this, I am speaking of an inference I have made about the
probable similarity between the moon and the earth. Conditions on the moon are closer to earthly conditions than we
might think. In my opinion, there is also water on the stars
. . . and living creatures as well, who exist only because of
these earthlike conditions. Both that unfortunate man
Giordano Bruno, the same fellow who was burned at the
stake in Rome over hot coals, and also Brahe, of good memory, believed that there are living creatures on the stars.*

The Copernican system

IX

Living Creatures on the Stars

Where Kepler writes the Astronomia Nova *in
Prague, a city full of magic and political intrigue.*

DURING THE YEARS KEPLER LIVED IN PRAGUE, from 1600 to 1612, he
stayed in three different residences. The first was with Tycho in a house
in Nov'y Svet, a neighborhood just west of Prazk'y Hrad, the Prague
Castle, atop the great hill overlooking the city, a short walk for Kepler.
Later, when he moved down to the Old Town, it was quite a walk up the
Steep Stair, and then a leisurely walk down a narrow lane that worms
over the hills until it ends. The houses, the buildings, the garden walls are
all taupe colored, with red tile roofs and ornate gables. Old brass
lanterns gone green are fixed to the garden walls. Tycho and Kepler's
house is No. 1 Nov'y Svet, hidden, nearly forgotten by the people dining
at the fashionable restaurant next door. Kepler's second home was across
the street from the Emmaus monastery, Na Slovanech, near the church of
St. John Nepomuk on the Rock. Both of these houses are in the New
Town, across the river from the Old, both part of the maze of communi-
ties hustled around the walls of the old castle like supplicants calling for
imperial favor.

Living Creatures on the Stars

The last five years of his time in Prague, Kepler lived at 5 Karlova Street, across the road from the Jesuit residence inside the Counter-Reformation university, the Clementinum. This house was in the Old Town, on the other side of the river, a hundred feet from the Charles Bridge, eight or nine blocks south of the Jewish ghetto, and four or five blocks west of the Old Town Square. In the same square, Kepler's friend Johannes Jessenius, the kindly anatomy professor from Wittenberg who acted as intermediary between Tycho and Kepler, eight years after Kepler's departure from the city, as an example to the rebellious would be beheaded by Ferdinand II, once the Archduke of Styria, by then the Holy Roman Emperor.

As in any medieval town, the streets of Prague are fit for pedestrians, horses, and small carts and wind a serpentine path up a gentle incline from the river. Nowadays, there are restaurants and shops selling textiles, marionettes, and Russian stacking dolls. On Karlova Street, just up from Kepler's house, one shop sells Prague crystal, while a second sells linen. It is not difficult to imagine the same street in Kepler's day, narrow, snaky, and stuffed with commerce—food sellers and wine merchants, stand-up bars, and traveling puppet shows. In the Old Town Square or near the foot of the bridge, the puppeteers played out folktales and morality plays. Traveling companies of players acted out national histories and fairy tales. Because there were no streetlights, the area around Kepler's house at 5 Karlova Street fell asleep soon after dark, after the players and the puppet theaters had all packed up and moved on, after the street musicians had wiped the river air from their violins, gathered a few copper coins, and walked home. They glanced into each pocket of shadow as they walked, like everyone in Prague, wary of petty thieves. There were cutpurses aplenty in the capital, who would steal from monks as easily as from merchants. The river air carried the sour smell of decaying water plants mixed with the nose-twitching odor of animal death. In the fall and winter and into early spring, mists wafted up from the Vltava, covering the Charles Bridge and leaving only the towers sticking out of the white lake of fog.

Prague is baroque with images, sometimes pretentious, sometimes saccharine, and sometimes mystical. A darker face hides truths that the city refuses to reveal. On the Charles Bridge is a statue celebrating the forced conversion of twenty-five hundred Jews to Christianity. The solemn, severe face of the preacher is matched by the subservience of the converted Jews, so that conversion becomes indistinguishable from conquest. The religious art haunting the city carries both spiritual and political messages, but underlying it is a grab for the numinous, an attempt to capture both the ecstasy and the torment of faith, where men and women alike are caught up in the folds of God. No art or science can in fact capture what the numinous is, for it is an encounter beyond words. Kepler tried to find it in the heavens.

If Prague is astonishing for the jaded modern tourist, so used to heroic skylines and afternoon wonders, how much more astonishing it must have been for a man from provincial Leonberg. The architecture of royalty was constructed in layers, courtyard within courtyard, the old palace dominated by the massive Gothic towers of St. Vitus Cathedral, with walls as high and imposing as St. Peter's, like fingers stretching upward, barely scraping the bottom of heaven. Kepler had made his mark in the empire, to be sure, but his situation as a Lutheran was always tenuous, and the titans who lived in Prague Castle, if only by turning over in their sleep, could create such waves as could easily swamp Kepler's little boat. One can imagine Kepler, a man born in little Weil der Stadt, beginning at the foot of the castle stair and winding up the hill past the guard post to the stone walls, to the cathedral towers, to the ornate palace of the emperors. One can imagine his staring upward, as every ordinary mortal was expected to do, filled with insecurity and fear by the enormity of the architecture. In mimicry of the heavens; in mimicry of God.

But anyone who was awake in Kepler's day would have known, after a quick tour through the Old Town, that the emperors, for all their divinity, sat upon shifting sands. The city was too diverse for a simple, uncomplicated rule. No king could command the people. He either led them or terrorized them—there was little room for anything in between. Traditionally, three nations inhabited the city, twisting the folds of its history—the Czechs, the children of Libuše and Prøemysl, who first built it; the Germans, the countrymen and retainers of the Habsburgs, who ruled it;

and the Jews, who lived as shadows, packed away in their ghetto, and suffered inside it. Of the three, the first two go back Roman times, while the Jews go back to the Middle Ages. The first two entered as settlers and conquerors. The last entered as refugees.

The history of the Czech people has been like a man carrying a stack of dishes on his head; every day the neighbors clap, shout, pound the floor, rush at him, and make faces, because they want to see it fall. Magyars, Poles, Germans, Turks—all have conquered and all have been conquered. While the Habsburgs ruled, as they did for nearly a thousand years, the German minority dominated the city, hangers-on in the imperial Habsburg court. These were no working-class Germans. All were would-be aristocracy, mixed with a burgeoning bourgeoisie. In an aristocratic society, numbers didn't matter; what mattered was power.

Prague has always been a city of mystic spectacle. Whether from the mists that cover the Charles Bridge at night, from the alchemists and Rosicrucians, or from incursions of gypsies wandering through the city, Prague has always walked halfway in shadow. Although outwardly Catholic, the nation looks back to John Commenius and the Bohemian Brethren, to Jan Hus and the Utraquists, and feels that somehow something righteously Protestant has been stolen from it. *Cuius regio, eius religio.*

Corresponding to Prague's history of near madness—Rudolf II, Edward Kelley—is its history of genius—Tycho Brahe, Johannes Kepler. Kepler wrote his first two mature works in Prague—the *Astronomiae Pars Optica* and the *Astronomia Nova.* In the former, he laid the groundwork for Newton's later science of optics by setting out the rules of refraction. He discovered that the reason the moon or the sun appears larger at the horizon is optical, rather than astronomical. In the latter, his *Astronomia Nova,* he set up the first two of his three laws of planetary motion. These laws are:

1. *The law of ellipses:* Each planet follows an elliptical orbit in which the sun sits at one focus of the ellipse.
2. *The law of equal areas:* The radius vector joining a planet and the sun sweeps out an equal area over an equal period of time.

To arrive at these conclusions, Kepler had to break with a great deal of tradition. First, the idea that the planets followed ellipses rather than perfect circles with uniform motion was near heresy. Even Kepler accepted the tradition as a reasonable starting place until he began his work on the orbit of Mars, when he eventually broke with it to make his theory fit Tycho's observations. This was a new attitude. Astronomers in Kepler's day were interested in accounting for the appearances, in creating a model that would apply geometry to motion, and the model often dominated. Although some astronomers were realistic enough to look for a relationship between the model and what actually happened, the reality they found was all too often simply an application of the model. Kepler took a different tack. He wanted to know what was actually happening, what the path of the planet actually was. In pursuing the secret of God's mind, he had to be willing to be surprised. The Aristotelians, on the other hand, had a vision of the way a well-ordered universe ought to operate, for not only was the model true, it was righteously true.

In Kepler's day, reaching back through the Middle Ages to the time before Christ, to Aristotle, people assumed in a commonsense way that the heavens were perfect. They were silent; they were orderly; they were beautiful. Standing in an open field in the middle of the night, anyone could see the great vaulted heavens arching overhead and the blue-white stars, lucid and overpowering, scintillating in the night against the black background of the sky. Such beauty harrows the soul and purifies the mind, and for Aristotle and his followers such beauty had to be perfect, for imperfection was a thing of the earth, not a thing of heaven, the place where the gods lived.

But all things for Aristotle were stacked into hierarchies. There were perfect flowers and imperfect flowers—the rose was at the top, the queen of all flowers. The diamond was the most perfect of gems. The circle was the most perfect of shapes. How could it be otherwise? The circle was simple, uncomplicated, and elegant. Therefore, since the perfect shape was the circle, the motions of the heavens had to be circular. For two thousand years, people believed this—they believed it the way we believe in democracy. There was a moral as well as a geometrical elegance to it. The vision passed from Aristotle to Ptolemy, who, being an observant fellow, realized that, although the philosopher's perfect circles were elegant, they never fit exact observation, and he set about making Aristotle's vision connect to the appearances.

One observation that had to be explained was that there were two types of planetary motions: the lower and the upper. The lower planets—Mercury and Venus—moved in an uncomplicated way. The upper planets, however—Mars, Jupiter, and Saturn—were more mysterious and more complex. Every day, Mars advanced along the plane of the ecliptic from west to east until in a period of about 780 days it completed one revolution. So far so good, but as each of these upper planets neared its position most opposite to the sun, where it could be seen at its highest point in the sky at midnight, it seemed to stop. It would hold that position for a short while and then, curiouser and curiouser, begin to move backward. After a while, it would halt again and roll forward once more. No one quite knew what to make of this. They called these meandering stars "planets," that is, wanderers, misfits who were out of step with the divine march of the heavens. The planets were oddly halfway between the perfect regularity of the stellar sphere and the frightening oddity of the comets, which appeared out of nowhere and flashed across the sky for a few months, only to vanish once again, predicting famines, floods, and the deaths of kings. Ptolemy's answer to this strangeness was the invention of epicycles, circles that turned on the larger circles of the planetary orbits, circles upon circles, perfect circles upon perfect circles.

Copernicus simplified this elegant mess by showing that the epicycles were an artifact of the motion of the earth. We need epicycles only to explain these odd motions of the planets if the earth is stationary, but if the earth moves, and if the earth moves at a faster clip relative to the upper planets, then there will come a time in each year when the earth will lap the higher planets, making them look as if they were standing still for a time, then running backward. But this only happens along a certain range of the arc of earth's orbit, when the planet is in opposition, meaning that the planet is in nearly a direct line with the earth and the sun, with the earth in between. As the earth passes each of the upper planets, the relative motion becomes less apparent, and the backward motion of the planets will cease.

But Copernicus believed, as did everyone else, in the perfect circularity of the motions of the planets. Although his brilliant simplification solved the problem of the planets' erratic motions, it didn't solve the other problem, the little one that kept hanging out in the shadows. Over the centuries, people noticed that the periods were not equal, that is, that the times between two oppositions were different, slightly different, but measurably so. But then, if the motions of the planets were in fact perfect circles, as Aristotle had presupposed, and if the centers for all the circles were the same point, that is, the earth, then the periods of the planets should have been uniform. But, as it was, the planetary orbits were not, well, perfect. How is it that the period would shift from year to year? Here, the vision warred with experience; what was seemingly right warred with what was seen. Observation of the planets over centuries had shown this.

The answer that Ptolemy gave was to offset the earth, as the center of the universe, from the orbit of the planet. This would make the orbit appear irregular. He called this offset the planet's eccentricity. Therefore, the planet still orbited in perfect circles, but the imperfection came about when the perfect circle was moved over slightly, like a hoola hoop. We, having been raised in a universe of gravitational forces, look for the mechanism that would make this possible, but astronomy in Ptolemy's day, even in Copernicus's day, did not take physics into account, because that science described the state of affairs on the earth, not in the heavens. All that mattered for the astronomer was to account for the appearances. Copernicus accepted this idea, except that he placed the sun at the center and tried to solve some of the problems by inventing a system of two overlapping circles. Tycho, who still held on to an earth-centered universe, accepted it as well. But mathematically the assumption became more and more untenable. The eccentricity of the orbit, the geometrical measure of its irregularity, was meant to explain how the orbit was irregular, but not why.

One more problem, though. It became clear in Kepler's calculations that the motion of Mars was not uniform, that it moved quicker at perihelion, where Mars is closer to the sun, and slower at aphelion, where the

planet is farther away. How to account for all of this? When Kepler received the problem from Tycho, the development of a proper Mars theory
required the calculation of the line of apsides, that is, the line joining the
real center of the circle and the offset center, then defining various points
on the orbit by extending the line out to intersect with the orbit itself. This
was best done at the opposition, where theoretically Mars would be on a
straight line corresponding to the line of apsides on one side of the earth
with the sun on the other. The problem was that this opposition did not
happen in a regular way—sometimes Mars was north of the point where
it was supposed to be, and sometimes it was south. This was the problem
of the latitudes, which meant that the planet's orbit did not perfectly coincide with the plane of the ecliptic, the single plane going through the sun
upon which the orbits of all the planets are supposed to lie like concentric
circles. But they don't coincide, at least not perfectly. Some are tilted in
one direction, some in the another. The other problem was the problem of
longitudes, the problem that led Kepler to abandon the circle for the ellipse. Longomontanus had already solved the problem of latitudes, that is,
how much the orbit of Mars tilted from the plane of the ecliptic, but he
couldn't solve the problem of longitudes. Sometimes Mars was farther
east than it was supposed to be, and sometimes it was farther west, which
shouldn't have happened if the orbit was a circle with uniform motion.

When Kepler took over the problem of the orbit of Mars from
Longomontanus in 1599, he was arrogant enough to believe that he could
solve in only eight days the problem of the longitudes that had plagued his
predecessor for months. Like Babe Ruth pointing at the home-run bleach-
ers, he told everyone so. He even made a bet on it, a bet he lost. Eight
days came and went, and Kepler still hadn't solved the problem, so he set
his paces and ran a little harder. He still couldn't solve it. Shrugging it off,
he pushed on with his calculations, sure that he could solve the problem
eventually, but frustrated that he had pushed the reason he took on the
problem in the first place, his desire to confirm his harmonic speculations,
into the background. Writing to Herwart von Hohenberg, he said: "I
would have finished my investigations into world harmony, had not
Tycho's astronomy so fettered me that I almost went insane."[1]

Kepler's problem throughout was his lack of an integral calculus, which would not be invented for more than a century. His calculations were made with high-school Euclidean geometry and trigonometry. Kepler had to invent a mathematics for himself based on old methods for calculating the perimeter of a circle by using smaller and smaller isosceles triangles, in "imitation of the ancients." Instead of triangles, however, Kepler used ovals, because he had been forced to drop the Aristotelian assumption that the orbits were circular, and therefore he hypothesized that the orbits could be calculated by taking ever smaller ovals as an approximation, thus arriving at the shape of Mars's orbit through sweaty calculation. He even complained to the readers in his text, saying that if they thought plowing through his method was boring for them, they should think of poor Kepler, who had to work through the math the old-fashioned way, with over seventy calculations to be made for each step of his long process. The abandonment of Aristotelian circles, then, came in stages, a piece at a time, and through much labor. At the end, however, in 1605, came the reward, the great flash of insight, the peek into the mind of God, where Kepler realized that the sun sat at one of the two foci of an ellipse, and that Mars's orbit, although awfully close to a circle, deviated from it just enough to be noticeable to the observant mind. When Kepler was done, a portion of perfection faded from the universe, but in doing so the universe got a little bigger.

When Kepler lived on Karlova Street, in the Old Town, on his way to see the emperor he would have first crossed the Charles Bridge. On the other side of the river, he would have seen the great, hulking castle with the spires of St. Vitus Cathedral trying to touch heaven. Inns and taverns lined the streets and above them all the palaces of wealthy men.

Once at the Imperial palace, he would all too often stand and wait for his wages and exchange news with other gentlemen who also stood and waited for their wages. There were sorcerers in Prague, alongside scholars, both real and fake. There were alchemists and astrologers, some of whom claimed to know the original language of Adam and could therefore talk to the angels. There were also a few scientists, like Kepler himself and his friend Johannes Jessenius, who performed the first public autopsy in the city. Starting with the abdomen and ending with the brains, Jessenius worked away, surrounded by well-dressed ladies and gentlemen seated in a high gallery, while he made witty comments about how fortunate the cool weather was or the stench would have been far worse.

The distinction between the two—between the alchemists, astrologers, and the angel whisperers on one side, and the true scientists—was not so clear in Kepler's day. The clear distinction is a modern one though it was incipient in the culture. Intelligent observers of the time—only a handful—could point to sets of assumptions that divided the two camps. Some believed, as Tycho Brahe and Kepler did, that regular, precise observation of the natural world, scanning for patterns of understanding that might arise out of the observations, was the only way to true knowledge. This is precisely what drives the *Astronomia Nova,* which makes it a distinctly modern text. Other natural philosophers believed first in the unity of the cosmos, that all things were mystically connected into the whole, so that while the pattern sifters such as Kepler talked of forces, the universalists such as John Dee talked of cosmic sympathies. The latter believed that because the cosmos was a single unified thing, objects within that universe could be transmuted from one form to another. Lead could be transmuted into gold, implying that alchemists, who worked for the wealthy and powerful, could also produce silver and precious stones out of lesser elements. Some believed that they could even distill small flasks of *aurum potabile,* liquid gold, which, being the most perfect of all metals, would give anyone who drank it eternal youth.

For all his title and apparent importance as imperial mathematician, Kepler's position at court was not that different from that of a court astrologer. The common folk of his time called him "the stargazer" and assumed that he was part of the troop of sorcerers and alchemists who marched up the Steep Stair to Rudolf's palace looking for imperial favor. Those who knew something of science and philosophy often feared Kepler, as if he secretly practiced the dark arts and was, if not a heretic (that would come later), then a suspicious fellow. This was true of Catholics and Protestants alike, for all feared the coming of Copernicus and the changing of the heavens.

Most astronomers of the time—Tycho and Kepler, Galileo and Mästlin—were astrologers as well. Kepler and Galileo were different from the general pack, however, because they thought that astrology was unreliable, even a bit silly, but they practiced it nevertheless. Astronomers at the time also practiced medicine, since a reading of the stars was essential for compounding cures, and therefore the practice of medicine was implicated in the art of alchemy, blending scientist and sorcerer, physician and alchemist into one strange amalgam.

At any moment, Kepler, Tycho, or any other scientists, if the politics of the city turned against them, could find themselves branded as a charlatan and trickster, just as Edward Kelley and John Dee had been a few decades before. And because Kepler was a Lutheran, the Catholic faction was always on the lookout for ways to discredit him. The only thing that made his life more stable than those of the practitioners of the occult arts may well have been the presence of the few Jesuits who understood his work and appreciated it.

The sorting process, wheat from chaff, of the seventeenth century continued on, however. Many natural philosophers had been already begun to question astrology. Kepler himself held such notions, and in 1610 he wrote a small book in response to Philip Feselius, who had roundly condemned astrologers as "absurd dunces, grunters, cyclopses."[2] Kepler too questioned astrology, but he took a middle position, saying that it still had merit. He distinguished between the old Chaldean astrology, which he considered a kind of augury, and astronomy based on physics, which he considered the royal road to knowledge of the heavens. The various rules of astrology were meaningful to the degree that they described the harmonic or disharmonic states. There are no good constellations and no evil constellations. The different aspects of the heavens affect human beings by seduction, rather than control. Their effect is psychical rather than physical, just as the sound of a bagpipe encourages the peasant to dance. He titled the book *Tertius Interveniens, das ist Warnung an etliche Theologos, Medicos, und Philosophos, sonderlich D. Philippum Feselium, dass sie bey billicher Verwerffung der Sternguckerischen Aberglauben nicht das Kindt mit dem Badt ausschütten und hiermit ihrer Profession unwissendt zuwider handlen* ("A Warning to Certain Theologians, Physicians, and Philosophers, Especially D. Philip Feselius, That by Cheaply Condemning the Superstition of the Stargazer, They Not Throw the Baby Out with the Bath, Thus Acting Against Their Own Profession").

The stars did not control life, therefore—they sang to it.

Some practitioners, such as Kepler, however—physicians, astronomers, and naturalists—stood in the breach and practiced both schools of natural philosophy without making much of the distinction. Many were fascinated by obscure Egyptian and Alexandrian texts, which they believed were written in far antiquity, but which actually came out of second-century Greek and Egyptian Gnostic cults. Possibly because of this confusion in natural philosophy, there were just as many con artists and swindlers in Prague as there were true practitioners, flimflam artists who often claimed a special knowledge acquired from some arcane text passed down to them by a magus on his deathbed or rescued from his grave. These charlatans sold their wild claims to the highest bidder, and all too often those who bought into them purchased nothing but dreams. But someone, often a member of the nobility, was always willing to buy. Such charlatans would blow into town, make extravagant claims, put on a quick séance, cheat some gullible baron out of a stack of money, and then disappear.

At first, Rudolf was tolerant of such fakes. Like everyone else, he couldn't tell the difference between the real and the counterfeit, and although he treasured those who made real contributions to knowledge, men such as Tycho and Kepler, he also wanted to know what alchemists and astrologers, real and fake alike, had to show him. In his own way, he was a universal man who wanted to gather as much knowledge as he could, so he could order it and place it into categories. Knowledge too was an essential part of his *Kunstkammer,* his museum of marvelous machines, magic stones, and exotic animals. His admirers praised him as the second Hermes Trismegistos, that fabled king of Egypt who once ordered the deep secrets of alchemy to be written on a single sapphire.

Under the influence of the Spanish Catholic faction at court, who deemed all occult matters as heretical, Rudolf imprisoned a few alchemists who had cheated some noble, but he executed none. Those who had to flee Prague, however, were often not so lucky outside Bohemia. Philip Jakob Güstenhofer, after a famous career in Prague, was finally hanged in Saxony. Another, Count Marko Bragadino, who had successfully played the part of Greek nobility, and with great theatrical flourish, set Prague atwitter as he strode through the city leading his black hounds. He was finally executed in Munich, stylish to the end, dressed in his best suit, and covered with jewelry made of fake gold. An Italian named Alessandro Scotta, for a time the talk of the city, eventually fell so far out of favor that he ended up in the Old Town Square, where he was reduced to displaying his magical wonders from inside a little wooden booth. Later on, he had a quick tryst with the duchess of Coburg and gave her a child, something she wanted desperately. Since this was not achieved by magic, but by the old-fashioned way, Scotta fled the scene before her relatives could find him and cut his throat.

What kings and emperors wanted from the likes of Kepler was knowledge that they could turn to power. From his earliest days in Tübingen, Kepler had accumulated a reputation as an astrologer, a reputation he was not really happy about, but one he had nevertheless. Rudolf II was no different. He was, to be sure, a Renaissance man who was enthralled with the natural world, but behind this there was always burning that royal need to know the future, to turn lead to gold, or to read the mystic encryptions of angels. In the early 1580s, John Dee and Edward Kelley arrived in Prague. Dee had already achieved some notoriety invoking spirits through a magic mirror, supposedly from Aztec Mexico, as well as through a crystal ball fashioned from polished smoky quartz, once given to him by the angel Uriel.[3] According to legend, he could speak the original language once spoken by Adam, the language that the angels themselves use to speak to one another. He could also understand the language of birds. In England, he achieved notoriety by gathering an astounding occult library full of ancient manuscripts written by long-dead magi, both forgotten and fabled. Because many of these books concerned old Gnostic theologies and philosophies, some in England questioned his Christianity. Eventually, he came to the attention of Queen Elizabeth and advised her on matters both naval and dental, assisting her not only with matters of military deployment, but with her bad teeth. She summoned him to Richmond on several occasions to visit with her and even visited him at his home in Mortlake, in Surrey, on the banks of the Thames. Like Kepler, Dee wanted to plumb the secrets of the universe, and at first he believed that this could be achieved through mathematics. Unhappy with the rhetorical program of a university education, he abandoned his early interest in mathematics, patterns of numbers that give structure to the world, and became interested in their arcane meaning. Soon after, in 1581, he held a séance in his house to call down the angels and to learn their secret wisdom. He believed that with the help of the right translator, or "skryer," he could seduce the angels to speak to him and could thereby read their language, much of which involved interpreting numerical codes.

For all his oddities, however, Dee was a serious scholar. It was simply his bad luck to run into Edward Talbot, alias Edward Kelley. Kelley had a hooked nose and beady, rat's eyes. Born in Worcester, he had been an apothecary's apprentice, had studied at Oxford for a time, and, while working as a scribe in Lancashire, had falsified official documents. The court sentenced him to have his ears cut off, and for the rest of his life he wore long hair and a black cap with long side flaps to hide his disfigurement. It was also designed to make him look wise and scholarly. Still afraid of the hangman, he changed his name to Kelley and then roamed about England until, while staying at an inn in Wales, he happened across a manuscript supposedly unearthed from the grave of a bishop who was also a magus. The manuscript came equipped with two ivory vials, one with red powder and the other with white. Immediately, he set out for Mortlake, appearing there on March 10, 1582. Within a short time, he had become Dee's assistant, and in one séance after another, he convinced Queen Elizabeth's astrologer that he was indeed speaking to angels and bringing to earth a wealth of arcane knowledge.[4]

Likely it was Kelley's idea that they both travel to eastern Europe, ostensibly to study occult knowledge at the court of the emperor, but also to make a pound or two. They traveled to eastern Europe at the invitation of a Polish nobleman, the palatine of Sieradz, Olbrecht Laski, who traveled through England in June 1583, where he visited them at Mortlake. During one séance, a spirit spoke through Dee's mirror predicting that Laski would succeed Stephan Báthory on the throne of Poland. Of course, this had to be true, because the angels had said it, and besides, it was what Laski wanted to hear. Dee and Kelley traveled to Cracow on Laski's invitation. Dee brought his young wife, Jane Fromond, who people said was more than pretty, and his son, Arthur. While Dee and Kelly were in Poland, the spirits continued their predictions of Laski's ascendancy, right up to the time that Dee and Kelley decided to move on to Prague. The Spanish ambassador Guillén de San Clemente arranged for them to meet with the emperor. The audience did not go well, however, because Dee, in a fit of enthusiasm, prophesied that a glorious new age would fall upon the empire, starting with the conquest of the Turks, if only the emperor would repent and change his sinful life. Rudolf was dubious. Later they wrote a letter to him, intimating that they had achieved a prior success in transmuting metals, which they would be willing to do for him as well. It didn't help.[5]

Rudolf ordered one of his secretaries to investigate the pair. The papal nuncio, supported by the Spanish faction at court, eventually convinced Rudolf to banish Dee and Kelley from his lands. They never made it out of Bohemia, however, because Vilém of Rozmberk offered them asylum at his estate in Třeboň, where Rozmberk spent an astonishing amount of money on them as they continued their experiments in secret knowledge.

Somewhere in there, Kelley decided that he no longer wanted to interpret angelic scripts, with all their numerical codes, and told Dee that he wanted to leave. Dee was convinced by that time that he would be lost without Kelley, and so, after some fighting and much blackmail, he signed an agreement with Kelley that they should hold all their assets in common, which ostensibly included their wives. Jane Fromond put a stop to that, and then Queen Elizabeth recalled Dee to England, where he died penniless, having been rejected by James I, Elizabeth's successor. In the years before he died, he sold off what remained of his library, book by book, just to pay the bills.

Kelley's end was more tragic. For a time his star seemed to rise, and he took Bohemian citizenship, gathered powerful protectors about him, and was even knighted by the emperor, which included a title, "de Imany," referring to his supposed Irish heritage. He married a rich, well-educated Czech woman who gave him a son and a daughter as well as a sizable dowry. In 1590, Rožmberk aided him in acquiring the town of Libeřice, an estate in Nová Libeň, and several villages, including the peasants who lived there. Then from his dowry he purchased a brewery, and then a mill, and then a dozen more houses in a gold-mining region. He bought two more houses in Prague, in the New Town, one near the Emmaus monastery. This house is known today as "Faust's house."

Then he did it. He got himself into a duel on the hospital field outside the Poříč Gate, and killed a Bohemian officer. Rudolf had forbidden all dueling and ordered Kelley thrown into prison at Křivoklát Castle. Agents stood by to question him, by torture if necessary, and to drag out of him the truth about his tinctures and his séances. They wanted to know especially about the *aurum potabile,* the liquid gold that bestowed eternal youth, and about the hidden meaning of strings of numbers they found written down among his papers, supposedly taken down during angelic séances. During his incarceration, Kelley attempted an escape by jumping out of a high window, but he crushed his leg on the rocky ground where he fell. Eventually, they released him so that he could get medical treatment, which didn't help much, because they amputated his leg.

But then Rudolf, goaded on by the Spanish faction, ordered Kelley's imprisonment once more and had him sent to Castle Most, in the northern part of Bohemia. Kelley tried to escape again and leapt once more from a high window, this time into a carriage driven by his son, breaking his other leg. Knowing that he would certainly be caught and, if caught, would spend the rest of his life in prison, Kelley mixed up a poisonous potion for himself and committed suicide.

This was the world that Kepler came to—civilized, urbane, Byzantine, dangerous. Four years after Kelley's death, Kepler arrived in Prague and began his work on the orbit of Mars, struggled on with the problem, nearly despairing of it in 1604, and falling into another depression. A year later, during the Easter season, he suddenly came upon the insight he needed and formulated his area law.

That same year, he wrote a short book on the new star that had appeared in the constellation of Ophiuchus, near the conjunction of the three upper planets, Mars, Jupiter, and Saturn. What more auspicious event could have occurred? On the night of October 17 the weather cleared, and Kepler saw it, the new star, nearly as bright as Jupiter, shimmering with color. Everyone, from the emperor on down, waited for Kepler to speak. His little book was funny, intelligent, and thoughtful, full of astronomical and theological reflection on the ways of God. God teaches mere humans by such signs. The new star was no accident, but a way in which God made his will known. While many others predicted Armageddon, the defeat of the Turks, revolution, and the coming of a new king, Kepler wrote about the conversion of America and of vast migrations of peoples out of Europe, as people had once migrated westward into Europe. Still, this kind of speculation was painful for him, and his arguments disputed with themselves. Kepler finally did not know the significance of the new star. He was no prophet; that wasn't his job, and he told people that they should just go on with their lives and examine their consciences.[6]

Reflections on the new star did not immediately raise Kepler's spirits, however. As they had done so many times in the past, his thoughts turned toward death and fixed there. What if he could not finish his manuscript? What if he died beforehand? He even planned to have his unfinished work sent to Tübingen for deposition in the archives. Eventually his depression lifted, and he pushed on toward publication, dropping his plans for Tübingen. Finally, the manuscript was complete. He titled it, rather forwardly, *Astronomia Nova αἰτιολογητοσ seu Physica Coelestis, Tradita Commentariis de Motibus Stellae Martis* ("New Astronomy Based on Causes or Celestial Physics Treated by Means of Commentaries on the Motion of the Star Mars"). Indeed, it was a new astronomy, for it was the first truly modern work in that field.

In his dedicatory letter to the emperor, Kepler joked that his book was the result of his long war with Mars and that he had brought that most noble captive to the court of His Majesty. "He has been constrained by bonds of computation," Kepler wrote. Mars was a captive because up until that time, its orbit was so unpredictable that it could not easily be caught. Kepler quotes Pliny, saying that "Mars is the elusive star." In the middle of his self-deprecating humor, Kepler could not resist the temptation to complain a bit about his suffering. "Meanwhile, in my encampment, has there been any sort of rout, any kind of catastrophe that has not taken place? The overthrow of my most distinguished master [Tycho Brahe], revolution, epidemics, plague, pestilence, household matters both good and bad, destined in all cases to use up time . . ."

Then Kepler set forth two schools of thought that existed among astronomers. The first was led by Ptolemy and the second by Copernicus and Tycho, and though this second school was more recent, it had its roots in the work of the ancients. Ptolemy's theory treated the planets individually and hunted the causes of each planet's heavenly movement separately, while the second school treated the planets as a system. The second school was then divided again. Copernicus, who led the first division, tried to treat the earth, sun, moon, and all the planets as a single system, with the sun at the center and all the planets, including the earth, moving around it, thus showing that the motion of each planet was in relation to all the others. The retrograde motion of Mars and the other upper planets—Jupiter and Saturn—was therefore the result of the relative motion of earth in relation to those planets. The other division was led by Tycho Brahe, who created a two-tiered system that kept the earth at the center, following Ptolemy, and sent the sun spinning around the earth, with the planets spinning around the sun.

Kepler spent a good deal of his time in the *New Astronomy* both praising and criticizing his old master's theory, for he recognized Tycho as a great leader and innovator and as the man who had made the most perfect observations up until that time. On the other hand, he criticized Tycho's planetary system as too complicated. In his search for a celestial mechanics, he assumed, with Ockham's razor, that the order of the cosmos had to be simple, for the mind of God was not compatible with wasteful and irrational motions. He found that by formulating a preliminary notion of gravitation based on magnetism, where all objects naturally attract all other objects, and emanating primarily from the sun, he could explain the planetary system in a much more simple way. What's more, this *vis motoria*, this vital force, diminished with the inverse of the distance. He had therefore come terribly close to Newton's later formulation of the law of gravity, which described the force diminishing inversely with the square of the distance, which made the force fall off at a much faster rate. His instincts were dead on; it was his mathematics that was undeveloped.

Tycho's system required that the sun and all the other planets orbit the earth, which would mean that the earth would have had to exert some powerful gravitational forces to keep it all in line. The earth simply wasn't big enough for that, but the sun was. Here Kepler took a huge step, probably larger than even he knew. He decided that he no longer needed to explain the movement of the planets through some kind of animate faculties, souls or living beings. Rather, he could explain it all through the action of physical forces, forces that he had identified with magnetism. If the earth was merely one of the planets, as Copernicus had said, and as opposed to Tycho, the power of planetary motions had to reside in the sun, and not in the earth. Certainly, the earth had the power to attract, as all matter did, but only the sun was large enough and great enough to hold the entire system of planets in check.

Kepler set forth several axioms to support a new theory of gravitation: every body, to the extent that it is bodily, is naturally suited to rest where it is when it is outside the influence of a like body. This axiom, therefore, showed that he had no modern idea of inertia, first formulated by Galileo, which said that a body in motion tends to stay in motion. Gravity, therefore, for Kepler, was a mutual corporeal quality existing among the various bodies, uniting them, with the more massive bodies exerting the greater force. The earth, therefore, attracts a stone more than the stone attracts the earth. This force, Kepler argued, is also the cause of the tides. Galileo believed that the tides occurred because, as the earth spun on its axis, the water in the oceans sloshed around like water in a bathtub. Kepler argued that it was because of the attractive pull of the moon, which causes water to leave some places and pile up in others, creating the ebb and flow of the sea.

LETTER FROM KEPLER TO TOBIAS SCULTETUS
APRIL 13, 1612

ᏻᎷᏌᎧ

*I had a partner, I don't want to call her the most loved, for
that is always true, or at least it should be. She was a woman
whom public opinion presented with the crown of honor,
righteousness, and purity. She combined these attributes, in
undisputedly rare fashion, with beauty and with a happy disposition. Not to mention the qualities that are less obvious—
her belief in God and her charity for the poor. She had given
me children who flourished, particularly a six-year-old boy,
very much like his mother. Whether you looked at the blossom of his body or the sweetness of his behavior or listened to
the promising prophecies of friends, in every sense one could
call him a morning hyacinth in the first days of spring, who
with tender fragrance filled the house with the smell of ambrosia. The boy was so close to his mother, people would not
simply see their relationship as merely love, but as a deeper,
more lavish bond. Now I had to watch how my wife, in the
prime of her life, having been subjected to three years of repeated attacks, slowly shattering her nerves, became often
confused and was rarely herself. Just when she started to recover, however, the repeated illnesses of her children brought
her down again. Her soul was deeply wounded by the death
of the little boy who had been half her heart. Numbed by the
terrorism of the soldiers and by the bloody war in the city of
Prague, despairing of a better future and consumed with grief
for her dearest children, she finally contracted the Hungarian*

spotted fever, and died. She was a victim of her compassion, for she could not be convinced to stop visiting the sick. In melancholy and hopelessness, in the saddest state of spirit, she took her last breath.

Emperor Matthias

X

Who with Tender Fragrance

Where Kepler's marriage is troubled,
Rudolf II dies, and the Counter-Reformation
comes to Prague in force.

IN THE TIME OF Libuše, old customs were dying, as old customs do, even
the good ones. According to the legends, the balance between men and
women had tipped, and the old prerogatives of women were fading away.
At one time women had the right to choose their husbands, and the husband was expected to move in with the woman's family, not the other way
around. For one reason or another, perhaps foreign influence or perhaps
just because of a weakness of memory, these old customs gave way to
new, where the rights of men were seen everywhere and the rights of
women, nowhere. Libuše defended the rights of women, and her husband,
Přmysl, defended the rights of men, and while they both lived, the balance
lived as well. Then Libuše died, and Přemysl, in his grief, nearly went
mad. When he returned and sat on his throne, however, he had lost half
his wisdom with the death of his wife, and he sided with the men from
that day on without care for the women. The women had lost their de-
fender, and they seethed with resentment. There was no one to prophesy

for the people anymore, and in their fury the wisdom of women gave way to witchery.

One woman, a handmaid to Libuše named Vlasta, stood up and mocked the men, saying that they had mead dribbling from their beards and had all fallen into a drunken stupor. She walked out of the great hall into the night and gathered young girls and women to her. She would not submit to foolish, drunken men, she told them, and she would fight anyone who tried to make her submit. The women liked what she said, each one vowing to do the same. They would set up their own land, with their own castle, which they would call Devin, the Women's Castle, and it would be a nation of women who would not submit.

They built this castle on the opposite bank of the Vltava, in plain sight of Vyšehrad, where Prince Přemysl could see it and worry. He knew what his lack of wisdom had brought about, and he knew that the power of women, once unleashed, was fearsome, so he called his warriors about him, and said: "There is a new castle across the river. You can all see it. The girls are building it even now, and they call it Devin. I have had a terrible dream, in which I saw a young woman drenched in blood charging through the countryside, her face mad with rage, her hair flying. As she passed, blood flowed into the streams, and she climbed down from her horse to drink the blood. Then she came to me with a bowl of blood taken from the river and offered it to me to drink. I awoke in a terrible sweat."

But the men would not listen. "They're just girls," they said. "What can they do?" And the prince knew that they were fools.

Meanwhile, Vlasta and her women fortified Devin and trained an army of Amazons. They lured men into the forest and then slaughtered them. There were no men living in Devin, though women lived in Vyšehrad and all the villages around. Many of these women became spies for Vlasta, stealing weapons and horses. The women of Devin knew everything the men were doing, but the men knew nothing of the women. Finally, the men gathered themselves together, thinking that if they marched out to confront the women, the sight of all those warriors gathered together would terrify them, and they would run away. Přemysl begged them not to do this, but they would not listen. They approached the women's castle, but it was silent, apparently empty. Congratulating themselves, they approached the silent gates, little aware of the number of archers that lined the battlements. Suddenly, the gates opened and the women's cavalry poured out. Arrows fell upon the men, and the cavalry, led by Vlasta, chased them back across the river.

After that, a new pride grew among the women. Some loved their husbands and would not leave them. Some disappeared in the night, only to appear among the women in Devin. Other women changed suddenly, and their men left their homes, fearful of being murdered in their sleep. For some time, the women's army swept through the land, breaking up happy marriages with cunning where they could and murdering where they could not. With time, the men began to gather once again, this time in earnest, arming themselves for the fight.

Using cunning, the women captured and killed one of Přemysl's men, Ctirad, who had been sent out to settle a dispute among the clans. A woman named Sárka was tied to a tree to make her look helpless, and when the men stopped to help her, she plied them with drugged mead until they passed out. Then the women came out of the forest and slaughtered them in their stupor, all except Ctirad, whom they took back to Devin and tortured on a wheel in front of the castle. At that point, seeing what the women did to their fellow, the men lost all doubt about attacking them, gnashed their teeth, and swore revenge. The men rode into the forest, looking for Vlasta's gangs of warriors and cut their throats when they found them.

This infuriated Vlasta, and she gathered her army for an assault on Vyšehrad itself. The men saw them coming and, full of hatred, rode out to meet them. The two sides crashed into one another before the gates of the castle. Blood ran into the Vltava in torrents, and death was the only victor. Vlasta was the most daring of all the women, and in her battle fury she rode out ahead of her army, so far out that she was soon cut off and surrounded by seven young men, warriors who hated her most of all. She was soon cast to the ground and pierced with swords. Then the women retreated in confusion inside the gates of Devin. The men set fire to the castle with the women still inside and shouted curses at them from the fields beyond the walls. The castle burned through the night and into the day, and the light of it could be seen for miles. The wound between the men and the women, brought about by the loss of women's wisdom, half of the whole, would take centuries to heal. And perhaps it never would.

⟨~~~⟩

KEPLER LIVED IN PRAGUE for eleven years, and during that time he heard stories. He most likely heard this one, for it was an old fairy tale, a Czech story going back a thousand years, to pre-Christian times. His marriage to Barbara Müller in those years was troubled, with little peace and little companionship, and somewhere in there, when he heard this story, he must have wondered if men and women were always destined to hurt one another.

Johannes and Barbara Kepler were not well matched. They lived in two different universes with two different sets of physical laws. It is not easy to be married, even for the most compatible couples, but when a simple girl marries a genius it is nearly impossible. Johannes's mind was always elsewhere. He was the kind of man who would rather burrow in his study and work out his calculations than do anything else in the world. Even as he walked across the Charles Bridge and up the Steep Stairs to the imperial court, his thoughts swam through a sea of numbers, reeling in the movements of the heavens. Even while he was standing and waiting in the Bohemian chancellery, in the high-vaulted Vladislav Hall of Prague Castle, where they held banquets as well as jousts, making light talk with barons and imperial secretaries, or while attending the emperor himself in some alcove of his *Kunstkammer,* his mind was never far from his study, his papers, his computations. And yet the work was hard and taxed him greatly. He struggled through the rough emotional seas that creative minds must navigate—first up, then down, first elation, then depression. And yet he loved it so. He was, after all, a man who rummaged for God in the balances of his own mind. For him, the geometry of the heavens, the dances

196

of the planets, the secrets of the universe were more real than the twists of human politics, for the first was complete with mystical joy, while the second was full of fear.

Barbara knew nothing of his work and cared nothing about it. His fame and his position as imperial mathematician gave her honor, but she would have preferred a more normal man. Her father, Jobst, had spent his life tending to his wealth and his position, and his daughter was much the same. She was in a sense an uncomplicated woman—uneducated, a small-town girl who saw little real value in book learning. She was, like her father, a practical woman. To outward appearances, Johannes and Barbara had, if not a perfect marriage, at least a good one. They were, to all those around them, comfortable German citizens, though some noticed how Barbara's melancholy seemed to grow worse with each year. They had sufficient money to live on, though never quite enough for Barbara. Unlike the lives of many of their friends in the nobility, their life was simple, yet was adequate to the position Johannes held as imperial mathematician. Kepler joked about it, comparing himself to Diogenes, who sent his works to the king from a tub. He often referred to his home as his comfortable "tub," though they had not rejected wealth, nor were they poor by any means. They were good middle-class people with good middle-class values and a good middle-class lifestyle. And with good middle-class uncertainty. When they first moved to Prague, Kepler spent between 400 and 500 gulden annually, but with newborn children adding to the family every few years, and with Barbara's health in constant decline, his expenses jumped to between 600 and 1,000 gulden a year. Barbara's own inheritance was sacrosanct, however; fearful of poverty, she would not let him sell or pawn so much as a pewter drinking cup to pay for firewood.

There were money problems, of course. The emperor kept making promises about Kepler's salary, but because of his collection fetish and imperial shopping sprees, he almost never had enough money in the exchequer to cover his imperial debts. Depending upon whether Rudolf was flush or not, the Keplers' lifestyle, like those of so many others at the imperial court, fluctuated from nearly comfortable to nearly impoverished. Barbara was constantly terrified of running out of money and saw her inheritance, which was mostly in land, as her hedge against an uncertain future. And uncertain it was. Who knew what the emperor would do next, because he was growing more unstable by the day? And who knew what disease would come burning through the city next, costing money for physicians and medicine? The children, some of whom actually survived into adulthood, came down with strange coughing sicknesses or burned with summer fevers, and everyone prayed that it was not plague or smallpox.

The Keplers had numerous friends as well, many from wealthy families with important positions, people who had plenty of money. These people had to be entertained. This was a sore spot for Barbara, who yearned for a taste of that noble life. Kepler's house seems to have been a gathering place for men, along with their wives, who were interested in the stars—imperial secretaries and representatives of the Estates, the local Czech nobility; some were barons, some even dukes. There was Johannes Jessenius, the great anatomist, who was later executed by the Counter-Reformation; Johann Georg Gödelmann, the ambassador from Saxony, who was also a part-time expert on witchcraft law; Jost Bürgi, the imperial watchmaker and the man who first used logarithms in his computations; and Johannes Mathäus Wackher von Wackenfels, an imperial adviser and a relative of Kepler's. Most of Kepler's patrons and intellectual friends had solid positions at court or were nobility from the Estates and had independent sources of cash. Kepler even counted a few Jesuits among them, a fact that his fellow Lutherans did not overlook. Such men surrounded Kepler and sought his advice on astronomical and astrological matters. Many of them visited him in his home, showing off their wealth with the unconscious flourish that only the very privileged can manage.

One of Kepler's friends was Johannes Pistorius, a former Protestant who later became a Catholic priest and the emperor's confessor, and later bishop of Freiburg. Imagine the relationship between these two! They both loved to argue religion and did so every time they met. Imagine the effect this had on Barbara, who had no theological background and spent her life in simple piety. Later, when Pistorius became ill, he wrote a sweet letter to his friend Kepler, saying that soon he would shrug off the vanities of this world and find peace in the arms of his Savior. Kepler, however, responded in a most un-Keplerlike way, taking Pistorius's letter as an opportunity to attack the Catholic church as an enemy to religious freedom because it staged an assault on free conscience and tried to rule salvation itself, making itself the single doorway through which to find Christ. Pistorius responded kindly, saying he didn't really want a theological conversation, and that he had nothing but fondness for Kepler and wished him God's blessing. This is the kind of argumentation that Kepler opened his doors to during his years in Prague. The bishop was undoubtedly sickly during that last exchange of letters; when he was healthy, he must have given as well as he received.

This was the house that Barbara lived in then, filled with Kepler's intellectual circle of like-minded men and their wives, all of whom admired him and his work and wanted to be close to the flame. It must have seemed to Barbara that these people appeared out of nowhere, showing off their finery. On the other hand, Barbara herself spent a good deal of her own time and money keeping up appearances. She was, to all those who knew her, a woman of charity and generosity. People from many parts of the city admired her as a model of Christian virtue.

In private, however, Barbara was not so pleasant or so generous. She had survived much in her life. Her father yielded control of his children to no one, and all of them, especially his daughter, he wrapped inside his schemes for higher social standing. He was a demanding, unforgiving, towering presence in her young life. At his behest, she had married two older men, because of their standing and reputation, but also because Jobst wanted to add their fortunes to his family's. Her one chance of marrying a man closer to her own age, that is, Kepler, her father disputed and resisted for years.

Barbara was prone to depression and therefore prone to a wealth of diseases that the tides of melancholy carry with it. She was an unhappy woman, made unhappy partly by her husband's short temper and his obsessive work habits. Likely, however, she would not have been happy had she been married to a king. She was embarrassed by her husband's work and often wished that he had a regular job. In response, Kepler, who had little patience, called her simpleminded, naïve, and silly. He hated the fact that she would disturb him at his work, bringing trivial household matters to him while his concentration was focused on the mathematics at hand. Kepler was often short with her or ignored her altogether. The two of them would then fly into rages or sulk, sometimes for days. She envied the wives of Kepler's many associates, not realizing how much regard they had for her husband, that though they enjoyed a higher position at the imperial court, they valued Kepler's brilliance even more. She could not help noticing, though, the difference between her station and theirs: her financial struggles, her one elderly, bandy-legged maid, her ordinary middle-class house compared with their grand coaches with four horses, their footmen, and their battalions of servants. When one of them had left the house, she must have watched him leave in grand style and compared him to her husband, who retreated to his study to calculate angles and scratch on pieces of paper.

Kepler admits in his letters that he was no great treasure. He was irascible and often unkind, and he vacillated between bouts of anger and windstorms of guilt and repentance. He never tried to understand Barbara, because gentlemen did not feel the need to understand their wives, only to provide for them. When he brought her to tears, he was immediately sorry, which never solved anything exactly, because what he considered important and what Barbara considered important were so very far apart.

It galled her that some of the people referred to them as "Mr. and Mrs. Stargazer." Although Kepler thought this was highly amusing and often referred to himself as "Mr. Stargazer," Barbara took these things to heart. Few women in her day had much education. For all of her complaining, she wanted nothing extravagant—freedom from poverty, social respect, a warm family, and her husband's attention. Mr. Stargazer, the man whose mind was forever turning toward the heavens and to his endless calculations of the planetary orbits, was all too often absent, off in the world of his books.

In contrast, Barbara Kepler read almost nothing—no novels, no stories, and certainly no mathematics or astronomy. The one consolation she had was her prayer book, for she was intensely pious. But this too created a difficulty for her, for though Johannes was also deeply religious and his spirituality and hers ran along parallel lines, religion and piety are not the same thing. The religious person wishes to experience the full range of the faith, to understand its traditions, and to face its weaknesses without flinching. Kepler wanted to study the heavens as his contribution to the faith. In the course of his studies, he had developed a theological position that was his own, and even as a strict Lutheran, which he remained all his life, no matter what the Tübingen consistory (the duke's council of advisers, both religious and secular) whispered about him while he lived in Prague and said openly later after he had moved to Linz, he had charted his own religious course through the troubled waters of the seventeenth century.

Barbara, on the other hand, was pious and would not have dreamed of doing what Johannes had done. Pious people often take their religion in narrow slices and are little interested in the grand sweep of its history or the range of its theological opinions. Indeed, there are religious people who are also pious; the two categories are not mutually exclusive. There is a sliding scale, however. In their intense search for a relationship with God, the strictly pious person often avoids intellectual challenges and cannot abide even the most favorable critique of the faith. For Barbara, her faith was in her Lutheran prayer book, and she could no more chart an independent theological opinion than she could understand the movements of the heavens.

What troubled her most was her own husband's reputation within the Lutheran community. Pious Lutherans who knew them well could see that Barbara's melancholy was growing, and they blamed Johannes for leading her away from the true Lutheran faith. Without trying to understand Kepler's exact position on Calvinism and predestination, they had heard rumors that he was a crypto-Calvinist, an idea he thoroughly rejected, and so they believed that Barbara's melancholy was the direct result of dark thoughts brought about by a belief in predestination and a fear that her soul was in jeopardy. This was not really the case, but given the times and Kepler's reputation, one could see how some people might have believed this. In truth, the idea of discussing his theological opinions with Barbara never even occurred to Kepler. This did not mean that Barbara was stupid, not by any means; it meant she was ordinary and lived in an age when women were not expected to be educated or to have rational opinions. Barbara just had the bad luck of being married to a man who was not ordinary.

Almost all of Barbara's side of the story has been lost, sad to say. Four hundred years later, all we have are Johannes's letters and Johannes's description of Barbara, but nothing of her description of him. The he said–she said is all too one-sided. With the single exception of one complaining letter that Barbara wrote to Johannes while he was visiting Graz, a letter that seems to back up his less than flattering hints about her, Barbara herself remains mute. One can at least say in all fairness, however, that Barbara and Johannes were constantly zipping past each other like shooting stars, constantly missing one another in understanding, constantly fighting over trivial things, a fact that drilled itself into Johannes's soul. He knew that Barbara's melancholy, which was perpetual, had sickened her body as well, and he felt a great empathy for her struggle, for he too had suffered bouts of depression. Often during an argument his comments would cut her to the bone, and immediately he would realize what he had done, that he had gone too far, and would pull back from the brink. He apologized profusely then, no matter who was right or who was wrong, which showed the depths of the fellow feeling he had for her. At those times, he would have plucked out his own eye rather than say something to embitter her further. For this reason, though there was no love between them and little passion, their marriage never sank to the point of open warfare; neither took the other to court or carried difficulties beyond the bounds of the family.

Surprisingly, however, in spite of the lack of love, the Kepler family grew at a fair clip. Barbara had difficulty in childbirth, but she still gave birth to three children in the years that they were in Prague. On July 9, 1602, she bore a daughter that they named Susanna, possibly in memory of their first little Susanna, who had died so soon after birth in Graz. On December 3, 1604, she gave birth to a son, Friedrich, and then on December 21, 1607, she gave birth to another son, Ludwig. Given the religious climate of the times, Kepler chose to have all three children baptized in an Utraquist service, the Czech Reformed church that followed Jan Hus and proclaimed that the people should receive Communion *sub utraque specie,* under both species, rather than by a Lutheran minister. By Rudolf's decree, only Catholic and Utraquist clergy were allowed inside the city, and therefore the Utraquists were the closest thing to Protestants that Kepler could find, and besides, one of his friends, a vastly rich young man who had squandered fortunes on his interest in alchemy, Peter Vok von Rozmberk, was also a leader in the Utraquist church.

Kepler's had his suspicions about alchemy, something Rozmberk's misfortune may have had a hand in. In his last years, he got into a bit of a controversy with an English alchemist and practitioner of occult science, Robert Fludd. Kepler's critique of Fludd's work was devastating. He took Fludd's ideas and analyzed them factually and rationally as a modern scientist would, while Fludd, who saw himself called to be a priest of secret knowledge, responded that Kepler saw only the outside of things and not the inside. This later controversy acts as a window into Kepler's view of occult sciences, a view he may well have developed while surrounded by alchemists and theosophists in Prague. It also may account for Isaac Newton's later coolness and his refusal to acknowledge Kepler, for Newton was a great devotee of alchemy and the secret sciences.

Barbara and Johannes chose godparents for all three children from the highest ranks of Prague society. Susanna's godmothers were three wives of imperial guards, while her godfathers were members of the nobility that the Keplers had met in Graz—Baron Ludwig von Dietrichstein, Baron Herwart von Hohenberg, Baron Weickhard, and Baron Dietrich von Auersperg. Friedrich's godfathers included the Baden ambassador, Joseph Hettler, the imperial treasurer, Stephan Schmid, and the venerable scholar Johannes Mathäus Wackher von Wackenfels, Kepler's distant relative and the imperial adviser. Little Ludwig's godfathers included Philip Ludwig and his son Wolfgang Wilhelm von Pfalz-Neuburg, both Protestant counts of the Palatinate, whose prince elector was Frederick, the Winter King, the man who accepted the throne of Bohemia after the Protestant rebellion in 1618, starting the Thirty Years' War.

Kepler loved his children, doted on them, and showed them a level of patience he rarely showed anyone else, including his wife. Little Friedrich was his favorite, though the boy took up much of his father's time. Just after Friedrich was born, Kepler wrote a letter to Herwart von Hohenberg, admitting that the noise around the house was keeping him from his work and from maintaining a proper correspondence with his friends and colleagues. Troops of women kept marching through the house to visit Barbara in her bed after the difficult birth. Children ran about, demanding attention. And Kepler had to play the host to all those who came to visit, so he couldn't easily slip away into his study.

Johannes didn't complain much, however, about the other troops, the battalions of his relatives, who kept visiting him from Swabia. He was, after all, their celebrity relative, and a trip to Prague was a pilgrimage for many of them. His mother, Katharina, arrived in 1602. Two years later, his sister Margaretha showed up, expecting to see the sights and hanging on Johannes's every word. Margaretha later married Pastor Georg Binder of Heumaden in 1608, a man who became a central figure in Katharina's trial for witchcraft. Eventually, even Heinrich, the unlucky brother who wandered through the world with a dark cloud over his head, came to visit, stayed, and joined the imperial guard. He even married and had two daughters.

And then there was travel. In 1601, after Jobst Müller, Barbara's father, had died, Kepler traveled to Graz to settle her estate. That was when Barbara sent her letter, the one that got Kepler in trouble with Tycho Brahe. In October 1606, another round of plague mysteriously boiled up in the city, and Kepler fled Prague with his entire family to Kunstadt in Moravia. Kepler must have had some money set aside at the time, for not everyone could leave the city. Only those with enough money to travel could do so, which meant that the middle and upper classes could escape, while the poor had to stay behind in the city to suffer and die by the score. Because of the plague, the emperor at that time remained in Brandeis. In November of that year, he summoned Kepler to attend him at his court there, which meant that Kepler had to return first to Prague by himself, and then from there go on to Brandeis. By the beginning of the next year, the plague had burned itself out, and those who had left returned to the city. Kepler's family returned with them and took up residence once again in their home in the New Town, near the Emmaus monastery.

By the spring of 1609, Kepler traveled to the Frankfurt book fair, and from there to Heidelberg to supervise the final printing of the *Astronomia Nova*. The book was finally finished and the dedication to the emperor complete, but all along the way Kepler had struggled with Tycho Brahe's heirs for the right to publish his own work, because so much of it was based on Tycho's observations. Science had not yet evolved to the point where data could be exchanged freely. The emperor had promised 20,000 gulden to Tycho's family to purchase his instruments and his observations so that Kepler might be able to finish the *Rudolphine Tables*. However, as with so many other imperial promises, the emperor did not have the money to cover his intentions, and while Kepler and Tycho's family waited for the promised pay, the emperor's bureaucrats dithered. Over the next year, the Brahe family received a few thousand talers, but this did not come close to paying the emperor's debt.

Meanwhile, Kepler's own research led him further away from the Tychonic system. Because they had not yet received the promised price for the observations and instruments, Tycho's family, led by Franz Gansneb Tengnagel von Camp, Tycho's son-in-law, resisted any publication on Kepler's part that would use his old master's observations. They did this because they didn't want to see any diminishment of Tycho's glory, and there was money in it. Tengnagel was not a very accomplished astronomer, but he had enough of a reputation as one of Tycho's former assistants to promise the emperor a publication of his own based on his father-in-law's observations. But no publication was forthcoming, because Tengnagel was simply not competent to do the work. All that Tycho's family could see was the potential profit that they could gain by taking control of the *Rudolphine Tables*, but there was no one in that family with the dedication, training, and mathematical acumen to do the necessary work to compile the tables, not to mention the ability to take those observations and create a proper theory of planetary orbits, which was something far beyond their intellectual means.

Kepler's luck did hold out, however. The emperor appointed Johannes Pistorius, his father confessor and a friend of Kepler's, to supervise Kepler's work. Because the two men had been friends for some time, this was a fairly easy relationship. Meanwhile, Tengnagel busied himself bolstering the family's demands. In 1604, the Brahe family forced Kepler to sign an agreement in which he promised not to publish anything using Tycho's observations without Tengnagel's personal approval. A few years later, Tengnagel, who had been a Lutheran like Tycho, sniffed the political winds and converted to the Catholic church, so he could join the imperial council. This made Kepler even more dependent upon him. However, Tengnagel had promised the emperor he would finish his own version of Tycho's work, but in four years time he had done almost nothing. Now, as a member of the imperial council, he could stop Kepler from publishing without ever having to publish anything himself. However, this did not stop Kepler from researching his own theories using Tycho's observations. The agreement allowed him to continue to work on the *Astronomia Nova*, even if he couldn't publish it without permission. When Tengnagel did not produce his own version after four years, Kepler believed that he was free of any obligation to Tycho's family, but Tengnagel still tried to stop him from publishing.

Finally, the two sides reached an agreement that allowed Kepler to publish the *Astronomia Nova*, as long as Tengnagel could insert a short preface to the work, in which he would explain that Kepler's book was based on Tycho's observations, that it did not follow the Tychonic planetary theory, and that Kepler had used Tycho's observations for his own purposes. Nevertheless, Kepler agreed to this insertion so that the book could be published. Now, after many centuries, Kepler's *Astronomia Nova* is accepted around the world as one of the foundational texts in modern astronomy, while Tengnagel's inserted preface has become a mere tidbit of history.

The other problem in publishing the work was money. The emperor promised him, and actually delivered this time, 400 gulden for printing, but because Kepler had not been paid in quite a while, this money quickly disappeared into the family coffers. This meant that Kepler had to scrounge together enough money to print the book himself. Eventually, another 500 gulden came his way, but suddenly the emperor forbade him from selling the book or distributing it to anyone, not even one copy, without his explicit permission. It seems that Kepler had written the book in his role as imperial mathematician, which meant that the emperor wanted to hold on to any works produced by him as imperial property. Rudolf could see the value of the work and wanted to distribute the *Astronomia Nova* himself. But once again his finances had become precarious, and the emperor had bigger problems to solve, so he dropped the idea of distributing the *Astronomia Nova* himself, allowing Kepler the chance to sell the entire edition back to the man who printed it, who could then offer it for sale on the open market.

ᏺᎳᏃ

MEANWHILE, THE TENSION between Protestants and Catholics in Prague, and indeed all across Germany, had reached the boiling point. The German princes had been taking sides in the conflict since the days of Luther—some were Catholic and some Protestant—with the Catholics gathering around the Habsburg family and the Protestants gathering around Frederick IV, the elector Palatine, a Calvinist. The power of the princes had become too confused with their choices of religious confession, so that a strong defense of each ruler's own religious beliefs became synonymous with legitimate rule. We must avoid the temptation to ascribe this kind of behavior only to the Catholic Habsburgs, though they were certainly the most vigorous defenders of the Counter-Reformation. Arch-Duke Ferdinand of Styria, who would later become Emperor Ferdinand II, had already outlawed Protestantism in his territories in Austria, including Graz. The same was true in Bavaria. Nevertheless, the Lutheran Duke of Württemberg and the Calvinist elector Palatine practiced the same kind of intolerance, for it was an intolerant age, and many read any concession granted to other religious confessions as a sign of weakness. For this reason, across Germany, the tensions between Protestants and Catholics flashed into civil wars. In Donäuworth, the violence had become so great that Duke Maximilian of Bavaria used it as an excuse to march his army into the region and enforce his own solution. Everyone, Protestant and Catholic alike, had forgotten the unifying teachings of Jesus, that anyone who is not against you is for you, and had abandoned forgiveness altogether, taking up with their own clutch of theologians, who forgave nothing.

In May 1608, the Protestant princes gathered around the elector Palatine to form the Protestant Union. The next year, in July, Maximilian of Bavaria gathered Catholic princes around himself to form the Catholic League. Both sides began to conscript armies. Meanwhile, Rudolf was slipping further and further away from reality, sinking into paranoia about his brother Matthias, and leaving the rule of his own kingdom to his often corrupt ministers. Although most of the Thirty Years' War occurred after Rudolf's death, his melancholy and shyness had contributed greatly to the political tension that produced it by creating a power vacuum, permitting the squabbling princes to carry on without interference. Instead of trying to solve the problem, he let his paranoia toward his brother rule his actions, which made matters worse by inciting dissension in the Habsburg family. Moreover, Rudolf had no official heir, since he had never married, and none of his illegitimate children could ascend the throne. The leading contenders were his brother Matthias, the man Rudolf hated more than anyone else in the world, the man who Rudolf in his madness believed was trying to poison him, and Rudolf's cousin Ferdinand, the ultra-Catholic Archduke of Styria.

Rudolf's own relationship with the Protestants had been checkered. In 1602, he had once again banned the Bohemian Brethren, closing their churches and schools. Meanwhile, in 1604, the Hungarian Protestant nobles, led by István Bocskay, rebelled against the Habsburg empire. This was not surprising. The emperor's army, led by a clique of Italian generals, had been rampaging through Hungary and Slovakia, slaughtering peasants and generally terrorizing the people. Bocskay was quite good with light cavalry, returning the favor to the emperor by laying waste to the Moravian countryside and eventually threatening Vienna itself.

By this time, Matthias had become the official head of the Habsburgs in spite of the fact that Rudolf was still emperor. Some of the members of his own family thought that Rudolf might be possessed; others thought that he was mad, which came down to the same thing. All in all, they were pretty much in agreement that dear cousin Rudolf was no longer good for the family business and had to go. In order to strengthen his position against his brother, Matthias had to make concessions to the Protestants, confirming the rights and privileges of the Estates. He then signed a peace treaty with Bocskay, and another one with the Turks, who were once again threatening the empire's eastern frontier. Matthias then formed a confederation of Estates, including both Hungary and Austria, to ensure against future rebellion.

Suddenly, Matthias was at open war with his brother, Rudolf. He marched on Prague, leading an army from his newly formed confederation to try to head off any attempts by Rudolf to sabotage his new peace agreements. But Rudolf wasn't dead yet. Thin and sickly looking, he called a meeting of the Bohemian Estates and actually showed up in person. Somehow he managed to arouse Bohemian national sentiments. The Estates formed their own army and set out to fight Matthias's invasion force. Now that Rudolf had an army of his very own, Matthias discovered a new flexibility. In June 1608, in the town of Libeò, now a suburb of Prague, he signed an agreement with his brother to slice up the empire. Matthias got Austria, Hungary, and Moravia, and Rudolf got Bohemia, Silesia, and Lusatia. Rudolf also got to keep the title of emperor, which was nice. Matthias, on the other hand, became the imperial heir, which galled Rudolf no end.

Everything seemed to be over. Peace fell on the empire, except for one little problem. The Protestant Bohemian Estates had saved Rudolf's kingdom for him, perhaps even his imperial skin, or at least his freedom. The Habsburgs had been known to do some pretty nasty things to one another. Rudolf owed them hugely, and he didn't like it. A true Habsburg to his core, he wanted revenge, revenge first of all on his brother, that miserable traitor, and second of all on the Estates, who had forced concessions out of him when he didn't want to give them. For months, he perched in his hidey hole in the *Kunstkammer* and schemed. His plan, however, was not thought through. He contacted his nephew, the twenty-three-year old Archduke Leopold, bishop of Passau, Bavaria, who was a notorious adventurer, a man who had no political and even less military experience, a typical young man who longed for glory. Leopold gathered an army at his estate, which he entrusted to a Colonel Ramée, a mercenary soldier, ostensibly to fight the Protestants in Württemberg and the Palatinate.

The real, sneaky reason for this army, however, was to march on Prague, to crush the Protestant Estates, and to return to Rudolf all his previous lands and control of the Habsburg universe. But this was a nightmare army, smaller, but in character not too different from the horde that sacked Rome a century before. Both the Spanish ambassador and the papal nuncio had their doubts. Leopold's army traveled through Upper Austria and then turned north toward Prague, occupying the Minor Town, south of the New Town, on February 15, 1611. And of course, they had to burn and pillage along the way, first going after the rich houses of the nobility and then, after everyone else. They tried to cross the Stone Bridge to the eastern side of the Vltava to take the Old Town, and then went back across the Charles Bridge to take the New Town. The Passauer cavalry finally beat its way up to the Old Town Square, a short walk from Kepler's house, where the mechanized statue of death was banging out the waning time on the astronomical clock every hour on the hour.

The Protestants, however, were waiting for them. They pulled the cavalry soldiers from their horses one by one and slaughtered them. Then the Protestants went mad in their turn, burning down monasteries and assassinating as many monks as they could find. From Kepler's house, his family could have heard the riot easily, heard the gunfire and the cannonade. They could have heard the screams of death and the cries of rage. It must have been difficult at that moment to know who was friend and who was foe. When the most obvious monks were dealt with, the Protestant mobs stormed the Jewish ghetto, sacking the homes of the wealthy and killing any Jew they came across just for good measure. The Jesuits managed to escape, however, but only just.

Matthias's army was nearby, and he rallied the Estates, who brought up artillery to blast the Passauers out of the Minor Town. Finally, on March 10, Archduke Leopold led his ruffian army out of the city, and Matthias moved into Prague Castle with his army. He rounded up any local supporters of the Passauer invaders and tried them as criminals, then crowned himself king of Bohemia on May 23. Both Protestant and Catholic Estates supported him, because Rudolf in his madness had lost the support of everyone and, well, because Matthias had an army in town. Surprisingly, Matthias was generous with his brother, after a fashion. He didn't imprison him and allowed him to remain in the Prague Castle surrounded by his glittering collection of wonders for the rest of his life. He could also keep the title of emperor, though he had no power whatsoever. Rudolf finally had the solitude he yearned for, but he didn't live long enough to enjoy it. In January 1612, he took sick. He had a liver inflammation, his lungs were failing, and his body broke out in gangrenous sores. On January 20, Rudolf II, emperor of the Holy Roman Empire, onetime king of Bohemia and Moravia, onetime ruler of Silesia and Lusatia, once leader of the Habsburg world, died quietly. They buried him respectfully, but without much fanfare, and thereby ended the golden age of Prague.[1]

⟨∽∾∽⟩

While this was still going on, back in 1609, Kepler was in the midst of the last touches on his *Astronomia Nova*. Rudolf was still alive, and the Passau invasion had not yet happened. After all his struggles at the Frankfurt book fair, he had finally negotiated its publication with Tengnagel and a publisher in Frankfurt, and after a swing through his native Swabia, including Tübingen, he returned to Prague and presented his work to the imperial council. For himself, Kepler was overjoyed by the emperor's decrees of toleration. Protestants could now practice freely across much of Germany and Austria, that is, if Rudolf kept his word, which, unknown to Kepler, he had no intention of doing. But Kepler was no fool and, despite the general elation among the Protestants, he could see that the emperor's power had been severely weakened and that this would not necessarily lead to peace and freedom for the Protestants. The more enmity there was between the denominations, the worse it was for everyone. The more each creed circled the wagons by defining all other creeds as the enemy, the more inevitable war became. When that war would happen in Bohemia, Kepler was not certain, but it would happen, and it would happen soon.

Meanwhile, in order to keep his family safe and to please his ailing wife, Kepler decided to look for a new position, somewhere outside Prague, somewhere in a town similar to Graz, where Barbara had been happy. But because Kepler had been a stipend student in the duchy of Württemberg and had received his education as part of the duke's educational program, he was sworn to the duke's service, even after all those years away. If he were going to seek a new position somewhere, he would need the duke's permission to work for another ruler outside the duchy. Kepler was quite faithful to this. Perhaps he still yearned after a professorship in Tübingen, but the one that was most suitable for him was still held by Michael Mästlin. Still, Kepler longed for his homeland and would have flown there like a cannonball had they offered him even a meager position somewhat worthy of his reputation as imperial mathematician.

While Prague was between battles, Kepler traveled to Stuttgart in May 1609 and petitioned the duke in person to consider the troubles that he was undergoing in Prague, without hope of finding sanctuary in his own homeland. Once again, he asked the duke for permission to work for another ruler, which the duke's councilors granted with the stipulation that, should the duke need his services, he would drop everything and enter the duke's service.

But once again, he stuck his foot in it. In a second letter to the duke, he promised that he would be willing to sign the Formula of Concord, but only conditionally. He would not struggle against it or speak out in any way, and he would seek to find some kind of unity between himself and the strictest members of his own church. Then, he naïvely tried to explain himself once again, not realizing that the decision against him had already been made. He wrote that from his youth he leaned toward the Calvinist doctrine of Communion, and that this was the single article on which he agreed, even partially, with the Calvinists, and that for years he could find no reason why a person who held the Calvinist notion of Communion could not be called a true brother in Christ by a Lutheran. Now was the time, he said, to try to make peace with the Calvinists. They had made changes in their own doctrine, themselves rejecting predestination. Frederick IV, the elector Palatine and the leading Calvinist in Germany, was not at all happy with Calvin's predestination, which he thought cruel. So why not make peace now?

Whether this last part showed Kepler's naïveté is uncertain. It is possible that because he was getting his news about Tübingen from Michael Mästlin, who often softened the blow for his student, he did not have a complete reckoning of the true opinion of the theological faculty at the university. This would change, however, after his excommunication in Linz. Old professors like Hafenreffer, who had been kind to him as a student, suddenly became cold disciplinarians. It is likely, though, that Kepler was simply the kind of man who needed to explain himself at every opportunity, believing that if they only understood him and understood that he meant no harm, then they would accept him. He knew that his position was at odds with the official doctrine of the Württemberg Lutheran church, a church that was influential all across Europe, and so he took whatever opportunity he could to prove to them that he was not a bad Lutheran and that his deviation from the Formula of Concord did not constitute a break with the church. He would soon learn, however, that church leaders believed otherwise.

On his return to Prague, Kepler fell into another depression. The city that had been his intellectual sanctuary for ten years was falling into chaos. His reception at his home university was less than warm, and he felt as if a long winter had settled on his soul. For a time, he stopped sending out feelers for a new position, though he knew that some of his important friends in the Estates of Upper Austria were trying to make room for him there as district mathematician and a teacher in the Protestant school. Instead, his depression dragged him down and froze his brain. He stopped working.

Then a wonderful thing happened, setting fire to Kepler once again. Galileo Galilei down in Padua had taken the newly invented *perspicullum*, or telescope, and during the winter of 1609–10 had applied it to the heavens. By itself, this was no great thing, since several other astronomers around Europe had been thinking along the same lines. And his telescope was not spectacular; it had little more power than a modern pair of binoculars. However, somehow Galileo knew where to look and what to look for. One discovery led to another. He found strange bodies near the planet Jupiter, bodies that no one had ever seen before! Not knowing what these were precisely, he named them the Medician Stars, after his patron the Grand Duke of Tuscany. Then, he turned his telescope to the seven-starred constellation of Pleiades and counted at least forty stars there, thirty-three more than anyone else had yet seen. He found that the planet Venus passed through phases, like the moon. He also found that the Milky Way, that fuzzy band of light stretching across the sky, was actually made of tiny individual stars. Galileo, as careful an observer as Tycho Brahe had ever been, marked all of these discoveries down in his notebook. He invited famous and important men to peek through his telescope to confirm his great discoveries by their own authority.

Kepler first heard about this when, on a cool day in March 1610, his friend and relative Johannes Mathäus Wackher von Wackenfels, the imperial adviser, stopped by Kepler's house and called him out to his carriage. There Kepler stood on the street while the two men discussed the news. Wackher told Kepler that he had received a report that Galileo had used the new telescope and had found four new bodies near the planet Jupiter, bodies no one had ever seen before. Wackher agreed with Giordano Bruno that, if the report were true, the stars could all be suns like our own sun and that the number of the stars would be infinite, existing in an infinite space. He thought that these bodies of Galileo's must be stars not yet discovered. Kepler, however, disagreed with Bruno, for he could not believe in an infinite space and told Wackher that these new bodies may be moons orbiting Jupiter, just as our own moon orbits earth.

Needless to say, Galileo's discoveries created quite a stir all over Europe. Galileo visited the great astronomer and mathematician Giovanni Antonio Magini in Bologna and set up his telescope in the garden so that Magini and his followers could peek through, but no one saw much of anything. Astronomers from both Catholic and Protestant camps shouted objections, some based on preconceived Aristotelian beliefs and some based on Scripture. Martin Horky, an acquaintance of Kepler's and the son of a Bohemian Protestant pastor, was a partisan of Magini, who kept a low profile himself, but encouraged men such as Horky, who had also been there to look through the glass, to write against Galileo. Horky published a short tract called *Peregrinatio Contra Nuncium Sidereum*, or *The Sojourn Against the Starry Messenger*. In his tract, Horky claimed that Galileo's reported discoveries were mere fables, mere fairy tales, that the bodies Galileo had seen around the planet Jupiter were only reflections in the glass, and that talk of the Milky Way being resolved into individual stars was a story as old as civilization. Father Scheiner, a Jesuit who would debate Galileo over the question of sunspots and would come off the worse for it, also raised objections.

Suddenly Galileo was at the center of a firestorm. Unlike Kepler, Galileo had many enemies. While Kepler was a peacemaker, Galileo was a controversialist. Like Tycho Brahe, he protected his reputation by maintaining constant vigilance against those who might wish to diminish his glory, even if it meant making enemies. Galileo was an arrogant man who not only knew he was right, even when he wasn't, as in the controversy over his explanation of the tides, but who made sure you knew it as well. This was a trait that would eventually get him into a sea of trouble with Pope Urban VIII Barberini, who had been told by Galileo's enemies that Simplicio, the fool in his *Dialogue on Two World Systems*, was spouting the philosophy of the pope himself. This did not make the pope happy, and when the pope isn't happy, no one at the Vatican is happy either. Including the Inquisition. Especially the Inquisition.

Kepler waited anxiously for Galileo's full report, which came to him on May 8, barely a month after he had first heard the news. Galileo's short work the *Siderius Nuncius*, or *The Starry Messenger*, detailed his discoveries about the Medician Stars, which he had sent to Kepler by way of the Tuscan ambassador, Julian de' Medici, along with a request for Kepler's reaction. This was a one-way bargain, however, because Kepler had been waiting to hear from Galileo about his own *Astronomia Nova* for some time, and Galileo, forever haughty, had kept silent. Over the next few months and even years, Galileo did not show much heart in his relationship with Kepler, probably because the imperial mathematician was his only real competition. Perhaps the difference may also have been between a theoretician and a practitioner. Galileo believed that one had to go out and look for oneself, a true empiricist, and yet astronomy has always had need of its theoreticians who grab raw data and make universes out of it. Moreover, Kepler never quite managed to keep his metaphysical and aesthetic speculations out of his astronomical work. And the fact that he was also a Lutheran might have been a factor as well, a source of prejudice for Galileo, the pious Catholic. Still, the temper of the times affects even men of science.

No doubt with all the voices rising up against him Galileo needed the imperial mathematician's support, and he was not disappointed. Kepler was alive again; his imagination, his sharp reason, his love of learning had quickly thawed. He would have written to Galileo no matter what, for the love of God and for his own wonder about the universe. "Who could be silent before the knowledge of such great things?" asked Kepler in his response, titled *Dissertatio cum Nuncio Sidereo,* or *Conversation with the Starry Messenger.*[2] "Who would not be overflowing with the riches of God's everlasting love?" It took Kepler only eleven days to write his response, just in time to send it back to Italy with the courier from the Tuscan ambassador.

Kepler's little book was a voice of quiet reason in a room full of shouting. Too many of Galileo's contemporaries had been stung by his wit or been treated to his hauteur, as if he were the only one with any brains on the entire continent of Europe. The Aristotelians branded him a charlatan, a flimflam artist with a couple of glass lenses. The Jesuits fretted about the appearances. Wasn't astronomy about accounting for the appearances? And yet Galileo claims to have found something new, to have found that the moon was a rough body just like the earth! He claims that it has mountains on it, and valleys as well! Ridiculous, when everyone knows that the moon is a perfectly smooth body composed of the fifth element, *quinta essentia,* or quintessence. People have known this from the time of the Stagirite, Aristotle himself. And what about that time when—April 24 or perhaps 25—Galileo visited Magini and tried to show him and his important guests the Medician Stars he had discovered. Did they see anything? Of course not! Is this surprising, when this Galileo, this Italian, doesn't even know how the telescope works? These images—aren't they somehow unnatural? Somehow ghostly? Everyone knows that you can shape pieces of glass into lenses to see more clearly, but using lenses to make things appear that weren't there before? Spooky. Perhaps even—heretical? Galileo promises to write something about the optics of his *perspicullum* sometime in the future, but until then, who can trust him?

Kepler too had been a victim of Galileo's silent scorn, but that did not stop him from praising the truth where he saw it. He called Galileo's discoveries "highly significant" and said that anyone who was a true philosopher would want to study and reflect on them so as to know God's universe better. Then he went after Galileo's enemies, those "sour opponents of new things, who reject anything they don't know, and call wicked anything that goes beyond the traditional boundaries of Aristotle's philosophy." After that, he put Galileo's discoveries into historical perspective, laying out what Galileo owed to others, placing his discoveries into a wider history of astronomy, and softening some of Galileo's more extreme claims. Finally, Kepler looked to the future, letting his imagination out for a run. Certainly, he said, the telescope was in its infancy, and new refinements in the instrument would lead to further discoveries. If Jupiter had moons, could there be moons around Saturn as well? Someday, he speculated, "Ships of the air, with sails designed for the atmosphere of heaven, could be made, and then people would arise who would not fear the vastness of space."

But if there are other planets like our earth, then how are human beings special? How can we claim that divine providence is watching over us if we are only a part of the universe, if the stars and the planets were not created for our sake, if we are not masters of creation? Wasn't that promised us in the Garden of Eden? Kepler feared this thought as much as anyone. Here is the committed Lutheran staring the modern age in the eye. Actually, his answer was not that different from Einstein's or Chandrasekar's. It was a Platonic answer—geometry rules everything in the universe, regulating the motions of the stars and the planets, fixing the order of the universe on immutable laws, for geometry flows out of God's mind, is an outpouring of divine reason. It is eternal. "That we human beings participate in it makes us an image of God." It is human reason and human consciousness that makes us special. It is the human ability to find the superstructure of the cosmos that makes us part of the divine.

Of course, everyone read Kepler's tract as they wanted to. Galileo's enemies thought that Kepler had put Galileo in his place. George Fugger, the emperor's ambassador to Venice, said that Galileo's mask had been torn. Mästlin misread the pamphlet entirely and congratulated his old student on pulling out Galileo's feathers.

Galileo, for the most part, read Kepler rightly. On August 19, four months after he received a copy of the *Dissertatio,* he wrote a letter thanking Kepler for being the one man to have the intellectual ability to see the truth when he had not himself seen the Medician Stars. Later he wrote to the Tuscan minister and told him that Kepler had agreed with everything he had written in *The Starry Messenger,* without a single doubt.

It was not enough for Kepler, however, to simply read Galileo's report. As much as he supported the Italian's discoveries, he wanted to confirm those discoveries and to see what he could find out for himself. To do this, he needed a decent telescope, something he could not get in Prague. He asked the Tuscan ambassador, Julian de' Medici, to request a telescope for him from Galileo, which the ambassador did, but Galileo ignored the request. Not that Galileo had run out of telescopes. By that time, in order to curry favor with the great and powerful lords around Europe, he presented telescopes to anyone who was important enough to help his cause.

For some reason Galileo did not want Kepler to have a telescope of his own, however. It is likely that he was afraid that Kepler, who was unused to the instrument, might have the same problems that Magini had. Perhaps he was afraid that Kepler might join the voices of his critics, having, in an inexpert way, peered through the eyepiece to see nothing but a blur. On the other hand, Galileo's personality was such that he may have refused to send Kepler a telescope out of bald fear of competition. If Galileo could find wondrous things, could not the imperial mathematician, the author of the *Astronomia Nova,* also find other, even more significant things? So Kepler was caught. Without a telescope of his own, he could not confirm Galileo's reports, and yet scholars from all over Europe were complaining to him about his support of Galileo, hinting that because no one of any consequence had been able to confirm Galileo's story, then who could believe him? That Italian fellow might have been making up the entire thing.

Kepler wrote to Galileo requesting the names of those who could confirm his discoveries. Armed with such a list, Kepler felt that he could safely confront Galileo's enemies without fear of being accused of naïveté. Galileo wrote back, but he did not supply the requested list. Instead, he wrote that Cosimo II, the Grand Duke of Tuscany, and Julius de' Medici, the brother of the Tuscan ambassador, were his witnesses. But neither of these men was expert in astronomy, so that although they could say what they saw through the telescope, they could not tell what it meant. Then Galileo told Kepler about how much money he had made on his discoveries. The grand duke had given him over 1,000 ducats, with 1,000 more ducats per year added on as his regular salary. Kepler, who had trouble getting any money at all out of Rudolf II, must have felt as if Galileo were bragging about his wealth rather than supporting his own achievements.

That August, however, Kepler was finally able to observe the newly discovered moons of Jupiter. The most recent war between the Habsburg brothers had just ended, and the elector Ernst of Cologne, the Duke of Bavaria, had come to Prague from Vienna, where he had been negotiating with Matthias. He had been one of the important lords whom Galileo had graced with a telescope and, knowing that Kepler was in Prague, he brought it along to lend to the imperial mathematician so he could make his observations. While in the capital, the elector was kept busy meeting with the other princes to help them settle the war, so Kepler got a chance to use the telescope while the meetings continued. Gathering his friends around him, Kepler spent from August 30 to September 9 observing Jupiter and recording the results. To ensure accuracy, the group set up a protocol, a kind of controlled experiment, with each man taking his turn observing through the eyepiece and then drawing what he saw on a tablet without showing his drawing to any of the others until all had taken their turn. Then they met and compared their drawings. There was enough agreement among them that Kepler published the results in a pamphlet he called *Narratio de Jovis Satellitibus, The Story of the Satellites of Jupiter.* Thus Galileo, for all his silence, had his discoveries confirmed by a man whose support he treated with such unkindness.

But what about the telescope itself? How did it work? How could it make things that were small to the naked eye appear large? How could it make things that were invisible to the naked eye become visible? Up until that time, the instrument was a mystery. Galileo had promised a report on how it worked, but he had not produced it. He was, understandably, too busy using it to make his observations. However, no one could really trust this new instrument, nor could they trust any of the discoveries that were made with it without some understanding of how it worked.

In September 1610, soon after he had finished his own observations of the planet Jupiter, Kepler wrote a short work he called the *Dioptrice,* which was the first theoretical analysis of the workings of the telescope. In this book, he set down the basic laws of optics, explaining how the passage of light through a series of lenses produces the effects it does. After his long depression, Kepler had come alive again, and as he so often did after a melancholic period, he suddenly possessed a furious kind of energy. His desire to know overwhelmed him; his desire to figure things out drove him on. In the *Dioptrice,* he described the double convex converging lens, and explained how using two or three lenses extends the effectiveness of the instrument. He also demonstrated how a set of two convex lenses can make an object appear larger while inverting the image. This led him to an explanation of Galileo's telescope, which used a converging lens in the body of the instrument and a diverging lens in the eyepiece and provided a higher magnification of objects, a serious improvement over a telescope that employed only a single converging lens. By way of appreciation, he offered his new short work to the elector Ernst of Cologne, the man who had made available to him his own copy of Galileo's telescope.

Meanwhile, Galileo continued with his observations. His telescope was not strong enough to resolve the rings of Saturn, but he was able to see them as two fuzzy blobs sticking out on either side of the planet. He thought that Saturn was actually built that way, one large blob with two smaller blobs stuck to it like Mickey Mouse ears. Then he turned his telescope to Venus and noticed that it passed through phases like the moon. This was a proof of Copernicus, since if Venus passed through phases, then those phases likely were the result of its position relative to the sun, just as the moon's phases are the result of its position relative to the sun and the earth. It was highly likely, Galileo concluded, that the light of Venus came from the sun and was reflected off the planet's surface.

Kepler received word of these new discoveries not from Galileo, but once again from the Tuscan ambassador, Julian de' Medici, who kindly passed the reports on. He heard about Galileo's Saturn observations in the early part of August and about his Venus observations on December 11. Perhaps it was because Galileo was afraid of his enemies or perhaps because it was fun to play word games, but he cast the meat of his reports into complex puzzles. Of course, unless Galileo had already sent a key along to his correspondents in an earlier letter, none of them would have been able to read his reports. His discovery about Saturn he wrote in the single Joycean word: "Smaismrmilmepoetaleumibunenugttaurias."[3] And later, when he reported on his discoveries about Venus, he took pity on his readers and cast his secret into a Latin anagram: "*Haec immature a me jam frustra leguntur oy,*" which is translated, "I already tried this in vain too early," which, like all codes, means almost nothing, or almost everything.

Kepler loved ciphers and puzzles, and Galileo's way of reporting his discoveries revved him up. A secret message demands revealing, and any mathematician worth his salt would feel called upon by reason itself to tackle the problem. Kepler played with the first one, fussed with it, and tried different combinations with the same fearful determination and intense love of his work that he had brought to bear on the problem of Mars's orbit. He sorted the letters into words and the words into phrases, thus building a message for himself out of Galileo's first meaningless code. His solution was: *"Salve umbistineum geminatum Martia proles,"* which, oddly enough, means "Greetings, twin protuberances, Mars's children."[4] Kepler was wrong about the puzzle, but not about Mars. Galileo wasn't talking about Mars at all, and yet, two hundred years later, Asaph Hall, an American astronomer working at the Naval Observatory, discovered that Mars did indeed have two moons and named them Phobos and Deimos.[5]

Galileo, however, would not reveal the secret of his puzzle, that is, until the emperor took notice and sent word to Italy that he wanted to know the secret himself. Galileo would not respond to the emperor's mathematician, but he would respond to the emperor. His own solution to the first puzzle was a Latin phrase: *"Altissimum planetam tergeminum observavi,"* which means, "I have seen the highest planet three-formed." This meant that instead of Mars having protuberances, Saturn had them. Kepler couldn't stand it anymore, so after he received the second puzzle, he wrote a letter to Galileo, begging him to reveal the second puzzle's meaning as well. "Think of the anguish you put on me by keeping silent," he said. On January 1, 1611, Galileo relented and sent back the answer *"Cynthiae figures aemulatur mater amoram,"* which means, "The mother of love emulates the shapes of the moon."[6]

Galileo never responded to Kepler about his little book, the *Dioptrice*, though it solved his problem of the telescope for him. Perhaps it is too much to ask of Galileo to expect more than that, but if it had been the other way around, Kepler would have written at once. Over the next year, Kepler wrote at least six letters to Galileo. Galileo wrote none, except a terse note recommending a young friend of his to the imperial mathematician. After that, their communication ended forever.

Then 1611 came, red with blood. Just at the end of 1610, Barbara, whose health had never been good, suddenly took sick. Illnesses came in rapid succession. First, she came down with Hungarian spotted fever, a new disease that had invaded the city along with one of the raiding armies. Then she began having epileptic seizures, which, given the medical prejudices of the time, must have been terrifying to everyone in the household. Imagine Kepler pacing the floor, calling for physicians, instructing her maid to watch her constantly to make sure she didn't have a seizure in her sleep. With time, however, the fever, perhaps a form of typhus, retreated, and Barbara crept toward recovery. But then came sorrow after sorrow.

Within a few weeks, all three children were in bed with high fevers, exhaustion, and body aches. Small red spots erupted on their tongues, inside their mouths, and down their throats. The spots turned to sores, which broke open and oozed liquid. A rash appeared on their faces and invaded their arms and legs; by the next day their little bodies were covered. After three days, two of the children, Susanna and Ludwig, began to feel better. The rash changed to bumps, which filled with a thick, viscous fluid, and then formed little calderas on their skin, raised rings with depressions in the middle, like thousands of belly buttons. The pustules scabbed over, and as the scabs broke open and fell away, the children rallied, chattered, ate again, whined to leave their beds to play. Kepler knew what this was, for he had suffered it himself as a child, the disease that had mangled his hands and ruined his eyes. It was smallpox, and as he watched over little Susanna and little Ludwig, praying for their recovery, he also prayed for little Friedrich, whose fever had not abated and who grew weaker every day, and he knew that Friedrich, only six years old, would not survive.

Friedrich died on February 19, in a winter full of war. "Whether you looked at the blossom of his body or the sweetness of his behavior or listened to the promising prophecies of friends, in every sense one could call him a morning hyacinth in the first days of spring, who with tender fragrance filled the house with the smell of ambrosia,"[7] Kepler wrote the next year. The boy was his heart, as he was his mother's. As little Friedrich lay dying, outside in the streets, soldiers marched into the city, mercenaries from Passau, sent by Archduke Leopold, that rash boy. Houses were burned and looted, families turned out onto the streets. The soldiers had not been paid, and unpaid soldiers all too quickly become beasts. Young girls screamed, raped by soldiers in back streets and alleyways. From time to time, a child was run down by horses.

In the midst of this, little Friedrich died in his bed. Cast down with grief, Barbara's recovery faltered. "The boy was so close to his mother, people would not simply see their relationship as merely love, but as a deeper, more lavish bond." Kepler stood by, watching his wife sink into despair. After three years of illness, her mind broke down and she sat in her bed, confused and unaware. Meanwhile, outside the door, on the streets of Prague, the soldiers of Archduke Leopold, bishop of Passau, had been unleashed on the citizens. In her broken state, Barbara quivered with each gunshot and wept with each rolling boom of the cannonade. Her world had finally collapsed along with her wits. Prague had fallen to barbarians; her little boy had died in her arms—what was there left to live for? This world was altogether vile.

Running for physicians, calling upon neighbors, asking for help for his wife, Kepler must have seen much of the battle. Did he watch the Passauer cavalry slash their way across the Stone Bridge and up the street to the Old Town Square? Did he hear the horses screaming in pain, shot through the neck or through the lungs with musket fire? Did he hear the mercenaries shouting to one another, see their rage, their cruelty, and wonder if the grief inside himself over Friedrich's death would ever match the grief of the city over its dying people? Day after day of slaughter. Bohemian troops, Protestants this time, entered the city from the other side, attacking monasteries and burning churches. Protestant citizens in the Old Town Square pulled the slashing Passauer cavalry off their horses and killed them to a man. The representatives of the Estates called to Matthias for help, and he responded with another army, an Austrian army to throw the Passauers out of the city. As they arrived, however, Rudolf paid off the Passauer cavalry and the men quietly slipped out of Prague and returned home. When he arrived, Matthias was the last man standing, the only one with an intact army. He could dictate terms. Matthias forced Rudolf to abdicate in 1611, and the next year in Frankfurt, the princes elected him emperor.

Throughout these events, both sides asked Kepler for astrological advice, but knowing the damage that astrology could do to a king if used deceitfully, he gave Rudolf what advice he could without guile. Rudolf was desperate and would have believed anything. However, whenever Rudolf's enemies asked for his advice, he was glad to tell them that the stars had lined up firmly behind the emperor. He would not say that to Rudolf, though, for the emperor's own good.

With Rudolf deposed, Kepler's need to find a new position became acute. He had been in communication with his friends and patrons in Linz for some time, mainly because he wanted to find a city about the size of Graz where Barbara could finally be happy. But in the back of his mind, there was always the hope of returning home. He wrote the Duke of Württemberg to help him obtain 2,000 talers the emperor had promised him from the House of Silesia, and in the letter he begged the duke to help him find a professorship in philosophy or to give him some other task in the duchy. What he wanted most of all was peace, peace to finish his work. His life had been anything but restful, and up until that point Württemberg had been an island of Lutheran safety in an increasingly uncertain Germany. The duke sent Kepler's request to his chancellor, who could not find anything wrong with it. Here was a good Swabian fellow who had graduated from Tübingen *summa cum laude* out of the duke's own stipend system, who had become the emperor's own mathematician, and who had become by almost everyone's accounting a great man. What better addition could he find for the duke's university? Mästlin was old and would need to be replaced soon enough. Kepler was young; he had years left in him.

Then the chancellor sent the application off to the consistory, where tongues began to wag. They searched through the files and found another letter, written only two years ago, in which Kepler said that he couldn't see why anyone who ascribed to the Calvinist doctrine of Communion could not be called a brother in Christ by his fellow Lutherans. He could not see it, but they could. This was false doctrine. This was heresy. This was criminal. This Kepler fellow was a furtive Calvinist, a heretic in Lutheran clothing. He was opinionated, with original and eccentric ideas about philosophy and, worse, about doctrine. He would poison the minds of the students, filling them with dangerous ideas. There would be controversy. Besides, all the teachers at Tübingen University had to follow the

Augsburg Confession, which meant that he had to sign the Formula of Concord. No waffling. No sly reservations. If a man could not bend to the church, then there was no place for him in Württemberg.

Personally, the duke wanted Kepler, as did the chancellor. Neither of these men was a theologian, but they were statesmen who kept their eyes open to whatever advantage they could find to help the duchy through difficult times. But to override his consistory meant that he was giving a no-confidence vote to the leaders of his own church, and that would cost him far more than having Kepler at Tübingen would gain him for the school. The duke, following the lead of the consistory, denied Kepler's application.

A slight possibility arose about a job opening up in Padua, filling Galileo's spot when he traveled to Rome. Galileo had actually recommended Kepler before the Venetian council, but nothing happened, and Kepler didn't have his heart set on the job anyway, preferring to stay in Germany. Linz looked better and better, even though Barbara was sick and failing. He traveled to Austria to arrange things, hoping that this might bring Barbara around, but when he returned to Prague, she had fallen sick once more. A new wave of illnesses had fallen on the city with Matthias's army, and Barbara, whose immune system was likely compromised by grief, had succumbed to it as well. "Just when she started to recover, however, the repeated illnesses of her children brought her down again. Her soul was deeply wounded by the death of the little boy who had been half her heart." On top of that, the wars drove her closer toward death. "Numbed by the terrorism of the soldiers and by the bloody war in the city of Prague, despairing of a better future and consumed with grief for her dearest children, she finally contracted the Hungarian spotted fever."

Inexorably, Barbara slipped toward death. As she was dying, her maid put a clean white shirt on her, and Barbara asked, "Is this the robe of salvation?"[8] "In melancholy and hopelessness," Kepler wrote, "in the saddest state of spirit, she took her last breath."

*What would be more reasonable than that I, as a philosopher
who has passed and is nearing the end of his prime, who has
muted passions, who is soft of body and dried up by nature,
should marry a widow long known to myself and to my wife,
a woman who was recommended to me not too subtly by the
latter. At first she appeared to agree; she certainly contemplated the matter, but finally excused herself most humbly.
With their mother I was offered her two daughters, along
with an unfavorable prognosis, if violation of honor may be
portrayed this way. Moving from widows to virgins, the looks
of the current one and her pleasant face caught my attention.
Her education was more brilliant than necessary for me. She
had been given more than her share of intellectual pleasures,
her age not ripe for domestic concerns. Finally, after evaluat-
ing all arguments, the mother decided that the daughter was
too young. This matter took up a month, and then I left
Prague, for I had decided and explained to the mother that I
would either get a bride or give up the city. This was the second. Now about the third.*

*On the way to Linz I made a detour to Mähren by resum-
ing my plan. Here my soul grew warm. I liked the girl, for she
was well brought up, the way I prefer it. She cared about my
children with extraordinary willingness. I left them with their
future mother to collect them at a later time, at my expense:
But the good girl had promised to be faithful to another before the year was through. Then came number four, the first of*

the Linzer women. She could sufficiently present herself by
beauty and her mother's distinction. I fell for her because of
her tall build and athletic body, and it would have been
settled, had not both love and reason forced a fifth woman on
me. This one won me over with love, humble loyalty, econ-
omy of household, diligence, and the love she gave the
stepchildren. I also liked her loneliness and the fact that she
was an orphan. After listening to Helmhard Jörger's wife, I
began to decide on the fourth one, angry that the fifth one was
put aside. However, I continued: a sixth one came recom-
mended by my stepdaughter and her husband, while friends
played matchmaker. She came from nobility and was not
without money, which was attractive. On the other hand she
lacked the years; her nobility made her suspect to pride, and I
was hesitant about the extensive costs of such a wedding.
When the fifth woman was already happily alive in my heart
and my words, a sudden rival made her company, whom I
counted as woman number seven. Friends praised her sophis-
tication and sense of economy. She had a face that made her
worthy of love. When courting her family and the girl, I in-
cluded warnings and negative remarks. What consequence
other than a refusal could have come out of this?

To quiet the gossip, I now turned to folks who were com-
mon, but aspired to sophistication. Among those I chose, with
the advice of a friend, an eighth woman. Beauty was not one
of her assets, but the mother was honorable; respectable edu-
cation, modest habits, and some money made them stand out.
Destiny, however, sought revenge with my restlessness and
doubtfulness by facing me with a being of the same unsteadi-
ness. At first she and the relatives were willing, but neither I
nor she herself knew whether she was willing or not. Finally, I
became more careful; the rest, for there will be three more, I
kept quiet about it.

Pretending to have a woman (number eight) whom I couldn't let go, I talked to one, which I will call the ninth, and looked for a sign of her affection. In light of my unsure behavior, the girl declined. With number ten the difference in our bodies was too obvious: myself lanky, insipid, thin; she short and fat. A friend brought number eleven onto the scene. I went to meet her, I liked her personality, but everything was arranged rather secretly. Money, sophistication, and economy were present again, and she was quite young. A friend managed the matter; I waited four months, and finally we received word that the girl was too young. While preparing to travel to Regensburg, I returned to the fifth women, declared myself, and was accepted.

Susanna is her name, the parents Johann Reuttinger and Barbara Bürger of the city of Efferding, the father a carpenter, both already deceased. Her education, in lieu of a large dowry, took place in the girls' school in Starhemberg, which receives high praises throughout the area for its discipline. Frame, habits, and body conform with mine. Not a trace of pride, no extravagance, patience at work, an average knowledge of how to keep house, middle-aged and enough sense to learn what is missing. By decree of the honorable Herr von Starhemberg I will marry her this coming October 30 at twelve noon in the presence of the Efferding congregation. The wedding celebration dinner will be held in the House Moritz, whose emblem is the golden lion.

FROM KEPLER'S JOURNAL
1614

The year 1614. I was bothered by bodily pains brought on by insipid fluids that constricted the folds of the body.

EFFIGIES DNI DANIELIS HIZLERI
THEOLOGI IN AUSTRIA ET SUEVIA
FAMIGERATISSIMI

Symb:

AH DEUS IN QUÆ NOS SERVASTI TEMPORA EN EM FAC IN MUNDO

MUNDO TEMPORIBUSQUE MALIS

IMUNDO MUNDO

Hoc ore, his oculis, octo cum lustra feneret,
Hac fronte HIZLERVS conspiciendus erat:
Dispeream, si usquam quicquam conspexeris isto⁕
Suavius, aut gravius, Candidiusve Viro!

Iob: Wagner

Pastor Daniel Hitzler, who excommunicated Kepler

XI

To Quiet the Gossip

Where Kepler, after moving to Linz, Austria,
must come forth to defend his mother on
charges of witchcraft.

THE STORY OF KATHARINA KEPLER'S AGONY begins with a business deal gone badly. There was a time when she and Ursula Reinbold, her accuser, had been friends, a couple of conspirators meeting over the backyard fence or along the road to pass the time of day and to chew over all the little happenings in the village. No doubt they talked about their children, the price of beef and pork, the price of onions and cabbages. No doubt they talked about their neighbors and all the little scandals, real and imagined. No doubt as Kepler's star rose Katharina regaled Ursula with his success and his growing fame. Her son, the *imperial mathematician*. Her son, who just a month ago was consulting with the emperor. Her son Johannes, you remember little Johannes, who is now rubbing elbows with kings and dukes. No doubt Ursula grew weary of hearing it, and no doubt so did just about everyone else in town.

For her part, Ursula was not the kind of woman to be trifled with. She lived inside her resentments, and when someone insulted her, she generally found a way to make them suffer. Ursula had a troublesome reputation, for she had once been punished as a public prostitute. She had had a series of abortions, boiling up her own potions to induce early labor—all in all, Ursula Reinbold had a bad smell. But strangely enough, the people of Leonberg were willing to forgive her much, because her husband, Jakob, the glazier and glass maker, was a successful man, and because her brother, Urban Kräutlin, was the barber of the duke himself. Moreover, people often forgive true malefactors things that they would never forgive other people who are merely irritating. Katharina Kepler had committed the ultimate sin—she made everyone nervous. She was a little bit mad, but only a little, which was far more dangerous than being an abortionist and prostitute.

The trouble began with Katharina's son Christoph, a respected tin-smith and pewterer in town. An artisan and a tradesman, he too was a successful businessman, a thing one would not expect from the child of a mother who was a little bit mad. But Katharina was only half mad—part naïve child and part serpent. The details of the business deal between Ursula and Christoph are murky, but one can imagine that it had some-thing to do with a dispute between the glazier and the tinsmith, because both of whom worked in household goods. Someone made promises; someone reneged. Words were said. The fight started. Christoph brought up Ursula's unfortunate past. Then, unwisely, Katharina got into the fight and upbraided Ursula for her bad reputation. Stung by Katharina's in-sults, Ursula from that time on looked for some way to make Katharina pay for what she had been made to suffer.

Unfortunately for Katharina, Germany was at the pinnacle of its witch mania. More than six women had already been condemned and executed in Leonberg. Witch hunts were popping up all over central Europe, with hundreds of trials and thousands of victims—75 percent of whom were women. Within twenty years, however, the tide would turn. Subtly, almost without fanfare, more and more people would start to pick at the flaws in the witch-hunters' arguments. Too many upstanding citizens had died without substantial proof, without justice. But in 1613, the year that Ursula first leveled her accusation against Katharina, the windstorm of fear had already become a hurricane. Katharina's trial was typical, for it was built mainly on gossip and innuendo.

To understand the witchcraft mania in the seventeenth century, one must first understand village life. This is hard for twenty-first-century people. We cannot understand a world in which people think that twenty-five miles is a great distance, half their world, really, and where they walk everywhere they go, and where they breathe the same air with the same people each day, each year, each decade. Small-town life in our time is almost dead, and where it exists it is unrecognizable.

People in villages have believed in the demonic for millennia. Soon after people settled into towns and villages, after they had created the first controlled environments, they noticed the shadows in their houses, the spooky places under the beds, and the wicked glint in a neighbor's eye.

Stories about witches made sense of the unfathomable misfortunes of life—why children die, why we sicken, why we grow old. They incorporated the hope that whatever sufferings the evildoer, the witch, had dropped upon the village, she might have the power, and the will, to undo. Farm life was insecure enough in Kepler's day, as it is today. Too much rain or too little, and people starve. A sudden outbreak of plague, a series of bad crops, the threat of famine, an explosion of disease among the herd animals, children suddenly breaking out in unknown rashes or burning with a fever, and then, like candles blowing out, dying. Something evil must be at work, and someone evil must be to blame.

There was plenty of disaster to be had in Germany in the seventeenth century, and not all of it was of human origin. Starting in the twelfth century and lasting on into the Enlightenment, Europe entered what climatologists have called the Little Ice Age, a period in which there were various-length spans of unusually bad weather—cold winters, strange thunderstorms, and year after year of bad crops. One of the reasons astrologers were so well paid at the time is that they were the meteorologists of the day. By reading the stars, they could predict the weather and perhaps give kings and princes foresight on whatever new catastrophe might be coming. Mostly, however, people in the villages suffered, and suffered without understanding.

Witches were the terrorists of the seventeenth century—unseen, moving about society disguised as ordinary citizens, with malevolent wills hidden behind smiling faces. This was not a superstition, as some Enlightenment writers would claim—it was a worldview. All that was good came from God. All that was evil, all that was malevolent, all that caused death and disease and despair bubbled up from the devil. But the devil did not work alone. Like any potentate, he had his minions, which he drew to him by promises of wealth and power. Not that he ever made good on his promises, because after all he was the devil, but those who had turned themselves over to him, for whatever goods they believed that they were going to get, had become his slaves. These were the witches, the practitioners of poison, potions, and other pharmacological horrors. They were, according to the common village fantasy, members of a secret society who traveled through the air to unclean mountaintops or ancient pagan glades, where they copulated with the devil, ate feces, and practiced one unholy blasphemy after another.

Some have claimed that the witchcraft trials of the seventeenth century persecuted an underground minority of secret believers in leftover paganism. The witchcraft trials were therefore a byproduct of Christianity's intolerance, rooting out the last vestiges of a once noble Celtic nature religion. The historical record, however, does not bear this out. Some individuals did practice magical healing rites, either as a part of herbal cures or as charms to protect the crops, and the church universally condemned such practices. Few such people would have considered themselves pagans, however, for most of their beliefs were an amalgam of Christianity and animism. They were accused of witchcraft simply because they were odd, and because they were known to be odd by the members of their own village. Most of the time, they were persecuted not because they were pagans, but because they were at war with their neighbors.[1]

Historically, there is no evidence that witches as we have imagined them ever existed at all. The typical image of the witch was that of "a creature with long, straight hair, a very sharp nose, and long slender fingers. She has a big mouth with pointed teeth. She dresses in black and she wears a pointed black felt hat on her head. A witch usually sails through the air on a long broom and is always accompanied by a fierce-looking cat."[2] By this account, most of the women of that time might have been witches, for the dress described was common enough. It was essentially the description of an old woman, and little else. Except for the flying part.

Writers of manuals for witch-hunters, such as the *Malleus Malificarum*, published in 1486, were sometimes openly misogynistic, describing women as more vulnerable to the wiles of the devil because of their sensuous sexuality. But most of the fear welled up from the villages. To look at the witchcraft mania from the top down, as the acts of governments or churches, is to see only the reaction, for the fear of unseen evil permeated the entire society—men and women, kings and peasants, popes and village idiots. Village life was such that people developed reputations over long periods of time, reputations that stuck to them, and that often had little to do with their actual behavior. The friction between people mounted and, through the power of gossip, became heavy, like water behind a dam; if released, such water could sweep everyone away. Often, the people who were accused of witchcraft had been suspected of being witches for many years, but it was always a dangerous business to accuse a neighbor of witchcraft. Ursula Reinbold could easily have become the next victim of the witch mania that she rode so successfully against Katharina Kepler. Her only protection was her family's friendship with the local magistrate, Luther Einhorn.

Luther Einhorn came to Leonberg in 1613 to take the position of *Vogt*, or magistrate. He was, even said diplomatically, a suck-up. He was the kind of man who scouts out those who are connected to power and befriends them. Although Katharina Kepler had connections with the imperial mathematician, Johannes's influence was far away, with people who lived in Prague or Linz or Bavaria. Ursula Reinbold, however, was well connected closer to home. Several members of her family worked as servants of the duke. Her brother Urban Kräutlin was the duke's barber; another relative was the duke's *Forstermeister*, his forester. Einhorn had befriended the Reinbolds fairly quickly and was closely associated with their camp from the beginning.

For some time before she had met Einhorn, however, Ursula had been declaring publicly that she had become sick, both mentally and physically, after drinking a potion given to her by the Kepler woman. She even went to the hospital, where she lay sick for a time. There, she met Donatus Gültlinger, who testified at Katharina's witch trial that the Reinbold woman had told him that as soon as she drank the Kepler woman's brew, she said "Good Devil! What is this? What did you give me to drink? It is as bitter as gall!" The fact that this was hearsay evidence didn't seem to matter much. After talking to Herr Gültlinger, Ursula went up and down the hospital floor, spreading rumors and complaining to everyone she met about the Kepler woman and what she had done to her. Ursula's health was likely not very good; she may have had gynecological problems stemming from her sexual history, and there are hints that she had recently aborted another child through one of her potions, so her story might have carried some credibility.

At first, Christoph thought all of this was just women's folly and visited Einhorn to ask that he stop Ursula from gossiping all over town, but he didn't get very far. Ursula's accusations would not have been heard at all, had the people of Leonberg not been predisposed to believe that Katharina was a witch in the first place, predisposed to accept whatever fanciful tale anyone might tell about the Kepler woman.

In 1615, Prince Friedrich Achilles, of Württemberg, visited the Leonberg *Forsthaus*, a hunting lodge. Something like a restaurant out in the woods surrounded by thick trees, the *Forsthaus* was often used for public gatherings, for it was a cool refuge from heat of late summer, just after the harvest. The prince was there to hunt, but that almost always meant a big party. Einhorn was there on that occasion, as were the Reinbolds, probably because the *Forstermeister* was her relative, which meant that he was also in charge of the *Forsthaus*. Einhorn and the Reinbolds had been drinking all evening, drinking to brotherly love by entwining their arms and then downing the beer together. This meant that they were either very drunk or they had passed the point of formality and were now bosom companions. Somewhere in there Ursula brought up her complaint about the Kepler woman. They drank on into the evening, while the Reinbolds worked on the magistrate.

Eventually, Ursula, her husband, Jakob the glazier, and her brother, Urban Kräutlin, the duke's barber, convinced the magistrate of Katharina's wickedness and incited him to take action. Still intoxicated, Einhorn led Kräutlin and the Reinbold couple to the courthouse and summoned Katharina to him. Depending on how early they had started drinking and when they had stopped, this summons might have come in the evening or possibly even at night. The entire meeting was against the law, for Einhorn had summoned Katharina without also summoning her son Christoph, her legal protector, or without summoning the two *Kriegsvögte,* the war magistrates, who, because she was a widow, were assigned to look after her welfare. Now thoroughly enmeshed in the Reinbold camp, Einhorn abandoned legalities and joined in the persecution of the old woman.

When Katharina arrived, Einhorn accused her "most seriously" of being a witch and ruining the health of Ursula Reinbold and demanded that she should use her powers at once to remove the curse. No excuses. Standing and shouting drunkenly, Kräutlin swore that if Katharina refused to remove her curse on the spot, he would draw his sword and stab her through the heart. Then, without waiting for a response from Katharina, he drew his sword from its scabbard and, staggering toward her, shouted curses at her while he held the blade against her breast.

This was a trap. It may not have been an intentional one on the Reinbolds' part, because they were drunk and wanted to terrorize the old woman just for the fun of it. She was there alone, after all, defenseless. If Katharina refused, she might be murdered on the spot, but if she agreed, then they could accuse her of using magic powers and of being a witch. Murder was common enough in witchcraft accusations, when the aggrieved family turned to violence. In some cases, the witch accommodated them, according to the stories, and if the sick person were healed, that might have been the end of it. But if the sick person died, then the cure would be one more example of the witch's malevolence.

The folk tradition held tons of stories about people who had taken sick and afterward accused a local woman of witchcraft. In the stories, the sick person's family often beat the witch until she agreed to perform the magic rites. In some stories, the witch would tell them to lay a chicken on the sick person's chest to take away the illness. The chicken would soon die, but the sick relative would then live. Other times the witch refused, and they beat her to death. Things that night had gotten out of hand in just that fashion, for Kräutlin was drunk and angry enough to have run Katharina through.

Katharina, for her part, held to her story and, shaking all over with fear, denied any wrongdoing. She said that she had never made the Reinbold woman sick and could therefore not help her regain her health. Enraged by Katharina's stubbornness, Kräutlin grabbed his sister by the arm and threw her against Katharina, for everyone knew that witchcraft could be cured by "counter-witchcraft." At this, Luther Einhorn sobered up enough to realize the legal danger he was facing. He warned Kräutlin against using force against the Kepler woman and then sent Katharina home.

As Johannes Kepler later pointed out in his letter to the Leonberg Senate, the entire affair was dodgy, for by forcing Katharina to perform this rite, Einhorn and the Reinbolds had solicited witchcraft. If the devil had actually appeared during the witchery, then everyone involved would have been in a lot of trouble, to the point of torture and death. Within a few days, the Kepler family filed a suit against the Reinbolds for slander. By the end of the year, Johannes learned of his mother's accusation, torment, and shameful treatment and immediately wrote the first of his letters to the duke.

Apparently the Kepler family's lawsuit was the first official legal action taken in this case. The Reinbolds had been gathering allies in the village, people such as the magistrate and the Haller woman, and orchestrating a whispering campaign, using gossip to poison Katharina's already shaky reputation. They had not yet made an official accusation of witchcraft, but complained about it only on the side to Einhorn, who then pressed the case forward.

FOR ALL HIS DIFFICULTIES WITH MONEY, Kepler had loved his time in Prague, for Prague was one of the great centers of the world, the home of emperors, and a meeting place of the greatest minds in Europe. After twelve years at the emperor's court, Kepler had been changed by the city, and he was no longer the provincial boy who had arrived in the city at death's door in 1600. After the deaths of his son and his wife, Barbara, he planned to leave the city to return to another provincial capital, where things were smaller than they had been in Prague, but he had remained in the city for a while after Barbara died. Barbara had not written a will, but she had made it clear that she would leave Kepler nothing. All her money went to the two surviving children and to the stepdaughter, Regina. Regina's husband caused trouble, for Kepler had charged the estate for Regina's upkeep and for her dowry. The husband, Philip Ehem, from a prominent Regensburg family, thought that Kepler had charged too much and wanted a larger share of the estate.

Rudolf, now deposed and sickly, had asked Kepler to stay on in the city, and since Kepler was loyal to his emperor even when the emperor was an imperial shell, stripped of power, he stayed on. Finally, Rudolf died on January 20, 1612, and Kepler had nothing left to stay for, so he took his children and all his household goods and rode out of Prague. He left the children in Kunstadt, Moravia, under the care of a certain Frau Pauritsch, a widow. Traveling on alone, he arrived in Linz, the capital of Upper Austria, in May.

Linz was a provincial capital, and the people in Linz lived, worked, and fought like provincial people. Although that may have seemed comfortable in a way, Kepler's own mind had been expanded far too much by life in Prague, and the spiritual freedom he had first established for himself by questioning the Formula of Concord had grown in him as well. Not that he had doubts about the faith. He simply had a mind of his own.

His next few years in Linz were a lonely time for Kepler, for unlike Prague there were few people in the city who could match his erudition and his intelligence. Instead of welcoming him as the great man, the city resisted him as outlandish, even foreign. The problems he had experienced with the Württemberg Lutheran church had festered and become dangerous. In his forties, Kepler was no longer the inexperienced, brash young man, and by this time in his life, after being the imperial mathematician for twelve years, he knew who he was. He knew his place in the world and the shape of his own life. Because of this alone, he became "a sliver in the eye" of some of his fellow Lutherans.

Instead of sending him into a depression, for once, this new isolation spurred him on. It called forth qualities in his personality: "the blind self-assurance, the demonstration of both piety and compassion, the grasping after fame, with surprising new plans and remarkable works, while the rest was a constant searching, interpreting, analyzing of the most of diverse causes, along with his tortured uncertainties about his salvation."[3] Intelligence carries its own torments. Perhaps Kepler would have been happier had he taken a position at a university such as Wittenberg or Bologna, rather than returning to the job of district mathematician and teaching in a district school, but no place in the world would have quieted his great yearning desire for God.

He first started looking into the position in Linz to please Barbara, because she had never been happy in Prague and longed to return to Austria, to the small-city life she knew as a young woman. But now that Barbara was dead, the position chafed. He was responsible to a school administration, both civil and religious, once more, and no one among them could quite forget that he was the great man and that he had been brought to their city by other great men. Unlike in Graz, where his teaching position had become available by the death of another teacher, the position in Linz had been created for him, just to bring him to the city.

But there were compensations. Within a short time, Kepler made new friends—Baron Erasmus von Starhemberg and Baron Erasmus von Tschernembl, two of the most important Protestants in the region. He also made friends with several of the local lords: Herr von Polheim, Helmhard Jörger, and Maximilian von Liechtenstein. These men were kindred spirits who offered him their protection and their fellowship. Moreover, Kepler retained his position as imperial mathematician, since Emperor Matthias had confirmed him in that job soon after Rudolf's death with a salary of 300 gulden a year along with 60 gulden added on to pay for his housing and firewood. Kepler was therefore able to maintain his status in the world, though the emperor retained the right to choose where Kepler would live. But Matthias was not Rudolf. He did not suffer from depression, as Rudolf had, but then neither did he have his brother's intelligence or his Renaissance love of learning. His need was for an imperial astrologer, not a man of science. He therefore permitted Kepler to live in Linz, far away from the court, and required his attendance only occasionally. This gave Kepler the time he needed to complete the *Rudolphine Tables* and to work on the composition of his masterwork, the *Harmonice Mundi, Harmony of the World,* which would include his third law of planetary orbits.

Six years later, in 1618, the religious pressure cooker that was Europe finally exploded. The Second Defenestration of Prague (the first having taken place during the Hussite Rebellion), then the Battle of White Mountain, and the flight of Frederick IV, the Winter King, led to the worst religious war in European history. The religious air in Linz, as it had been in Graz, as it had been in Prague, was electric with war fever.

The school in Linz where Kepler was teaching was about as old as the one in Graz, but it was smaller and less prestigious. It had been another Württemberg foundation, built in the latter half of the sixteenth century, but then it was closed by the Counter-Reformation in 1600 as part of Rudolf II's new series of religious rules. The school opened once again in 1609, however, as part of Matthias's concessions to the Protestant Estates. It had a rector, a co-rector, and only four full-time teachers on the faculty. Kepler's presence there gave the school an extra shot of prestige. The degree of installation, promulgated on June 14, 1611, declared that he was hired because of his "celebrated ability and commendable virtues." He was ordered to complete the *Rudolphine Tables* in order to give honor to the emperor and to Austria, and he was expected to continue his own astronomical work and to do anything he believed to be useful and appropriate in astronomy, physics, or history.

In spite of his newfound freedom, Kepler was lonely in Linz. There is one quality that Johannes Kepler and his mother shared—neither of them fit in. Both of them were stubborn. Katharina did not want to leave Leonberg, even when the townspeople brought her to court, even when they threatened her life. During one phase of her trial, she lived in Linz with Johannes for nine months, where she was safe from accusations. But she was never happy there and longed to return to her little house in Leonberg, where everyone hated her and where the tiger of a corrupt magistrate was waiting for her in the grass. Even after her trial, when the town rejected her utterly and threatened to stone her to death, she left Leonberg with great sadness.

Johannes, likewise, would not leave the Lutheran church in spite of its many rejections. He had been cast off by Tübingen; the theological beliefs he had developed as a young man had carried him further and further toward the frontiers of his own church, which, as his own fame grew, got the Tübingen faculty buzzing like a hive of bees. His own faith, his own reason, his own conscience taught him to think in ways about the ubiquity doctrine that differed from the Formula of Concord. The ultraorthodox church leaders at Tübingen accused him of being a hidden Calvinist, though he would never become one. Nor would he become a Catholic, in spite of all the pressure the Counter-Reformation brought to bear on him. He was a man caught between two iron orthodoxies, neither of which would bend, both of which required absolute obedience and conformity, and he could give neither.

The Counter-Reformation papacy had long since lost its Renaissance openness and had once again become enamored of the Inquisition. To be Catholic meant an unquestioning obedience to the church, and this obedience had to stretch across one's entire life into one's actions and beliefs. This was part of the Tridentine reform preached by the Jesuits, who had themselves taken a special vow of obedience to the pope. On the other hand, the Tübingen consistory demanded an equal level of obedience, of complete acquiescence to the Formula of Concord, the document that was meant to bring peace to the diverse factions in Lutheranism. Instead, it became the means of excluding men such as Kepler who could not bring themselves to conform to it. This conflict came to a head in the first few weeks of Kepler's life in Linz.

The Württemberg Lutherans were really a very tight group. Everyone knew everyone else's business, and they wrote letters back and forth across Germany about those suspected of heresy. In an informal way, they had invented their own inquisition, based on the village more than the city. Where the Catholic Holy Office had professional inquisitors, tribunals, and royal armies to back them up, the Lutheran inquisition had winks and nods and secret letters. In 1610, two years before Kepler himself arrived in Linz, Daniel Hitzler had become its chief pastor. He was a Württemberg-trained minister, another graduate of the duke's stipend program who, five years after Kepler, had taken the same course of studies. He was about as different a man from Kepler as one could imagine. He was certain of his orthodoxy and of the need for absolute conformity within the body of the church. Kepler believed that a man's conscience was inviolable, a matter between himself and God, and that one had a religious duty to reason through the complexities of the faith and to understand them. Hitzler believed that the individual conscience needed to submit to the church and that wild horses such as Johannes Kepler needed to be reined in.

Soon after Kepler arrived in Linz, he met with pastor Hitzler and requested Communion. During the conversation, he set forth the entire contents of his faith, including where he agreed with the Formula of Concord, which was nearly all of it, and where he disagreed. After all, self-destructive honesty is always the best policy. However, unknown to Kepler, Pastor Hitzler had already received a communication from someone in Württemberg, possibly Hafenreffer, informing him of the consistory's suspicions about the imperial mathematician, who had been a source of controversy from the time he entered Tübingen. Because Linz, like Graz, was a missionary colony of Württemberg, most of the people already knew about the suspicious nature of the imperial mathematician. Unlike with Galileo, whose struggles with the Inquisition had more to do with angry astronomers and philosophers and finally the proud intransigence of Pope Urban VIII, the persecution of Kepler, like his mother's later witch trial, was a matter of gossip. The Lutheran consistory in Stuttgart and the faculty at Tübingen considered Kepler to be an "unhealthy sheep" who might infect the flock with his unorthodox ideas. This was not even close to the truth, for Kepler kept his concerns about the ubiquity doctrine within a narrow circle of friends. Ultimately, he believed that all Christians—Calvinists, Lutherans, and even Catholics—should respect one another as Christians, and that was the most heretical thing of all.

Hitzler took the letter about Kepler he had received from Tübingen at face value and demanded that the astronomer sign a copy of the Formula of Concord and stipulate to all of its provisions without reservation. This was the price of receiving Communion. Kepler told him that he accepted the Formula of Concord in every way, except that he had one small reservation. The pastor told him in return that it was all or nothing, and he demanded absolute obedience—for Hitzler, Kepler's personal faith was not as important as his compliance. In Kepler's mind, this meant abandoning his own conscience, his own faith, the thing that Luther had taught was the only way to salvation. It was his most immediate connection to God. Did not the Lutherans rebel against the Catholic church on the same issues? To abandon his conscience would be to abandon his faith.

Worse yet, Hitzler did not treat this affair in the way a wise spiritual leader should have. He broke all confidentiality and gossiped shamefully, spreading word of his dispute with Kepler out to the entire community. For most of the people in Linz, their minister had accused their mathematics teacher—that strange fellow from Prague—of being a heretic, and that was the end of the matter. Ignorance has its own kind of logic. "The chief pastor of the church and the inspector of the school have branded me publicly as a heretic."[4]

With amazing naïveté, Kepler wrote a letter to the Stuttgart consistory appealing his pastor's decision. Either he did not know about the Tübingen communication, or he believed that his personal relationship with some of the members of the consistory, and with the duke himself, would counterbalance this. These were men he had spoken to, men who had congratulated him after he received the title of imperial mathematician. Some of them were men he had spent time with while staying at the duke's palace. They were people he thought admired him, or at least respected him. But on September 25, the consistory wrote back, answering Kepler's appeal and the mask was torn. They had sided completely with the pastor and joined him in the condemnation.

In his appeal, Kepler had argued that Pastor Hitzler did not actually have the power to exclude him from Communion, because the dispute between them was over a question of conscience, and that any attempt to exclude him from Communion on the basis of his conscience would override Kepler's faith. He presented the consistory with two choices—either he be admitted to Communion in spite of their concerns or he would receive Communion elsewhere.

The consistory's answer to Kepler's first point was without exception—no true pastor of the church could accept anyone into Communion who outwardly pretended to be a follower of the true evangelical faith, but who could not follow that faith precisely, who followed his own path, wandering away from true doctrine, obscuring the faith with specious questions and even more outrageous speculations. No pastor could accept a man who was confused in his soul, for his confusion could spread to others like a contagion, nor could he accept a man who followed his own judgment on the faith and on matters of God. Nor could any pastor accept a man who could not commit himself strictly to clear statements of doctrine, by not ascribing to the entire Formula of Concord, without exception, for it was the symbol of the orthodox Lutheran church, founded in Scripture. Thus, they said, Pastor Hitzler was correct in refusing Communion to Kepler, who had placed himself outside the true religion by the things he had written and said. Kepler could not be admitted into Communion unless he abandoned his erroneous speculations and followed the true religion. They said that Kepler had shown that he was at odds with the orthodox doctrine of the faith on many occasions, dating back from the time when he was a student in Tübingen, and that he had denied the omnipresence of the body of Christ, that is the doctrine of ubiquity, and that he showed some sympathy, if not agreement, with the Calvinists. It did not matter that Kepler disagreed with the Calvinists in most things; it only mattered that he agreed with them in some things. This was a response worthy of the Holy Office in Rome.

The heart of the dispute was that Kepler had actually taken a critical stance toward Lutheran orthodoxy, and the church could admit no such criticism. But it was Kepler's personality to weigh and measure all things, to argue all things, and to come up with his own opinions. He could no more abandon that than he could abandon the Lutheran church.

Finally the consistory said: "You want nothing to do with our confession: how then could you ask for admission to Communion?" As to the scandal, they laid that completely at Kepler's door. If only Kepler would acknowledge the authority of the church, return to a complete observance of the faith, and thereby follow the will of Pastor Hitzler, the scandal could be avoided. And if Kepler tried to receive Communion outside the Linz community, then he would be furthering that scandal and would be carrying it into the heart of whatever Lutheran community would accept him. If this community opened its doors to Kepler and admitted him to Communion, it would then be as suspect of Calvinism as he was, as would the pastor who received him. Kepler should keep to his mathematical studies, they said, and not involve himself in theology, for this was not his profession. What's more, he should not bother the consistory with any more useless appeals. "Don't trust your own mind too much, and make sure that your faith rests not on human wisdom, but on God's strength."

In its own way, the Lutheran church of Württemberg was undergoing a recapitulation of the Catholic experience. Catholics always believed that the faith was a matter of individual conscience, because one had to stand alone before God to be judged. But for the faith to be preserved through time and not dissolve into a battle of confusion, with each individual coming up with a personal interpretation of the Scripture, the church needed to establish some uniformity of doctrine. This meant a unified interpretation made by an exclusive group of official interpreters. This also meant that the church had to have authority, which included the power to compel its members to comply with its teachings, rather than their own consciences. In keeping with this line of reasoning, only three hundred years after the founding of Christianity, Augustine of Hippo encouraged the persecution of the Donatists, because they would not conform to the church's teaching about baptism. They insisted that all those who had fallen away from the faith during the waning years of the Roman persecutions needed to be rebaptized. Augustine disagreed. Subtly, in any organization, religious or otherwise, solidarity becomes ossification, the faith becomes orthodoxy, and compliance becomes more important than conversion of spirit. By the time of the Reformation, Christianity had gotten to the point where authority itself had become the problem.

To reform Christianity, Luther and his followers returned to the idea of individual conscience and individual faith, ideas that had been asleep in Christianity for a thousand years. Luther and his followers, and the pastors who followed them, took the title "Doctor" rather than "Father." They wanted to be seen merely as educated men rather than men who possessed semidivine authority. Kepler was very much in line with this Reformation thinking. The Lutheran church of Württemberg, however, in order to survive intact in the noisy marketplace of religious ideas, had to begin to develop an orthodoxy. Without it, what Luther and his followers had taught would have disappeared, dissolving into thousands of smaller denominations, each one based upon some private interpretation of Scripture. Lutheranism would have dissolved into the religious landscape altogether.

The very nature of human organizations creates orthodoxy, and orthodoxies, in turn, give birth to reformers and mavericks, men such as Luther and Kepler. Kepler, as a good Lutheran, found himself at odds with the Lutheran church, but as a thinking Lutheran he almost had to. He believed that to be a good Lutheran, he had to follow his faith, which meant attending to his own conscience, which also meant that if he did not agree with the Formula of Concord in every detail, then he must not sign it. This did not mean that he stood against his church; it meant that he participated in it more fully. For the Württemberg consistory, however, if anyone, especially a famous man such as Kepler, were to be allowed that kind of freedom of conscience, it could eventually spell the end of the church itself.

Kepler and his own church were at odds, and there was no solution in sight. But his position was not completely eccentric. He had searched the works of Christian antiquity, the writings of John of Damascus, Gregory Nazianzen, Fulgentius, Origen, Virgilius, and Cyril, and he could find no trace of the ubiquity doctrine. It was, in his opinion, not part of the Christian heritage. Oddly enough, while the consistory condemned Kepler as a dangerous innovator, Kepler himself believed that the Formula of Concord, by including the ubiquity doctrine, was itself dangerously innovative.

Therefore, Kepler's fight with the consistory continued. He informed them in a return letter that he would avoid scandal and cause no further troubles for Pastor Hitzler. However, he would not abandon his request for admission to Communion, stating that he would return to it at a later date, perhaps in a different community. Neither the consistory nor Pastor Hitzler would let the matter rest, however. Much of Europe was heading blindly toward the Thirty Years' War, and the tensions between the Christian confessions drove the authorities in each church to demand rigorous compliance from all of their members. Kepler the famous mathematician could not be allowed to step out of line, which of course was the very attitude that led to the Thirty Years' War itself. Accusations against Kepler mounted daily:

> I have been denounced as a man without principle, approving everyone, incited not by an honest heart, but by a desire to have the friendship of all parties, whatever may happen, today or tomorrow. I have been called a godless scoffer of God's word and of God's holy Communion, who cares nothing about whether the church accepted him or not, and who, instead of being eager to receive Communion, decided that it should be kept from him. I have been attacked as a skeptic who in his old age has yet to find a foundation for his faith. I have been condemned as unsteady, now siding with this group, now with that group, as each new and unusual thing is brought into the arena.[5]

Accusations grew up all around like fungus. Some people accused him of taking sides with the Catholics on specific points just to help his career. Others said he was a Calvinist because he believed in their ideas on Communion. He was like a weathervane turning in the wind—Calvinist in some things, Catholic in others. So he rejected the Calvinist doctrine of predestination, calling it "barbarous." This didn't make him a true Lutheran. He wouldn't accept Dr. Luther's great book on the captive will either. He even agreed with the Jesuits on the doctrine of ubiquity. So was he Catholic or Lutheran or a Calvinist? He wasn't any of these things, but a man alone, unchurched, a man who wanted to start his own confession, the Church of Kepler! He was a newcomer, and perhaps an atheist and a heretic. So much for the great mathematician.

Kepler was hurt to the quick by all these accusations. How little those who made them understood him. He struggled daily with his own conscience and could find no way out, for his conscience was the bellwether by which he made all his decisions:

I bear witness before God that I am not happy with this position I find myself in, nor do I want them to judge me as a new man, separate from others. It hurts my heart that these three great blocs have ripped at the truth so terribly that I am left collecting it piece by piece, wherever I can find them. I have no misgivings, however. Instead, I work at reconciling the divisions, if I can do so with the truth, which allows me in all honesty to agree with several of them. So some call me a mockingbird, when I say I often side with two of the divisions against the third. See! I am always hoping for an agreement, agreeing with either all three parties or with two of them against the third one. My antagonists, on the other hand, are happy with only one party, and they imagine for themselves a single irreconcilable quarrel. God willing, my hope is Christian; what those others imagine, I cannot say. God already has rewarded our warring Germany with lamentation.[6]

Kepler could see what was coming down the road for Germany. This was not prescience, for just about anyone with an honest mind and open eyes could see that what he said was true. The true enemy of Christianity was sectarianism, and what Kepler wanted was peace and unity. He wanted the clouds of war then gathering to dissipate. "I bind myself to all simple Christians, whatever they call themselves, through the bond of Christian love; I am the enemy of misunderstanding, and I speak kindness wherever I can."[7] In this, as in so many things, Kepler was a man out of time. "My conscience commands me to love an enemy and not harm him, to avoid adding new causes for separation; it tells me that I ought to be an example of moderation and mildness for my enemy; perhaps through my actions, I might encourage him to do the same, and then at last may God send us the dear desired peace."[8]

For Kepler, it was the authorities of his own church who were beating the war drums, who had broken the peace. "My argument about religion is that the preachers are becoming too haughty in their pulpits. They do not live by the old simplicity. They arouse dispute; they bring up issues that hinder devotion, accuse one another wrongly, stir up the nobles and lords against each other, are too malicious in their interpretations of the actions of the pope, and cause many to fall away when a persecution begins." In writing to Mästlin, he said "I could quiet the entire dispute by signing the Formula of Concord without reservations. Yet I cannot be hypocritical in questions of conscience. I would sign, if they accepted the reservations I have already presented. I want no part in the fury of the theologians. I shall not judge brothers; for even if they stand or they fall, they are still my brothers and brothers of the Lord. Since I am not a teacher of the church, I should pardon others, speak well of others and interpret favorably, rather than indict, vilify and distort."[9]

But this did not end it. The war with his own church had begun and would not pass away that easily. The whispering campaign had had its effect. People glared at him in the streets. People warned one another against him. Those in authority tried to remove him from his teaching position, so that by the autumn of 1616 Kepler's position as a teacher was in question. The commissioners for the school suddenly discovered that they were paying too much for their mathematics teacher and that the money could be better spent elsewhere. The matter came before the representatives, and some of the members raised all their old objections to hiring Kepler in the first place. There were four groups in the body of representatives: the prelates, the lords, the knights, and the city officials. Old class resentments rose to the surface. The barons were for Kepler, and therefore the knights were against him. Baron von Starhemberg and Baron von

Tschernembl defended him once again, and their side won the day, but the matter stayed precarious.

Suddenly a letter arrived from the University of Bologna. Giovanni Antonio Roffeni, a philosophy professor at the university, had written Kepler about the death of the astronomer and mathematician Magini, offering the chair to Kepler. He quickly refused, because Bologna was a Catholic city and too closely allied with the papacy. At heart, he was a German and would never leave his country for another. Also, how could he give up the fight?

A year after Kepler's arrival in Linz, in 1613, in the middle of his entire struggle with the church, Kepler married again. She was Susanna Reuttinger, a twenty-four-year-old woman from the town of Efferding. Her father and mother were both dead, but they had been upstanding citizens—her father had been a cabinetmaker—so Susanna had a good reputation. Baroness Elizabeth von Starhemberg, the wife of Kepler's patron, adopted her after her parents had died and for the next twelve years acted as her guardian.

Just before his wedding he wrote to an unknown nobleman, possibly Peter Heinrich von Strahlendorf, and told him the entire sordid story of his courtships. The attacks from the church authorities were in full swing, and Kepler typically wrung his hands over his unhappy conscience. "Can I also discover God inside myself," he asks, "the God whom I so easily grasp when I consider the universe?"[10] Were all his failed courtships not a sign of moral weakness? Maybe they were right about him. Maybe he had so many false starts because of lust, or because he was a fool and lacked judgment, or because he knew nothing. Then Kepler finds his feet and the rest of the letter goes on tongue-in-cheek. One can almost hear him: If I am a fool, then I might as well show it. Perhaps it was all God's will anyway.

Kepler had decided that he wanted to marry again only a few months after Barbara's death. There was a widow in Prague whom Barbara had introduced to Johannes before her illness. The widow had been a close acquaintance of Barbara's, perhaps even a friend, and as her illness deepened Barbara apparently recommended the woman to her husband as a suitable replacement. Shortly after Barbara's death, out of respect for his dead wife, Kepler presented his case to the widow, who considered it quite seriously: "At first she appeared to agree; she certainly contemplated the matter, but finally excused herself most humbly."[11] This was probably a wise move on her part, because Kepler's interest in her was founded mostly on grief and on a sense of obligation to Barbara's memory. This is rarely a good basis for a marriage.

Apparently, there were enough people around who wanted to play matchmaker, so there was another attempt to fix him up. A woman in Prague offered him one of her daughters for marriage, a girl who was quite attractive, "the looks of the curent one and her pleasant face caught my attention."[12] Still, there would have been hell to pay if honor had not been satisfied to the letter. Oddly enough, however, this pretty young girl was too educated for him. "Her education was more brilliant than necessary for me. She had been given more than her share of intellectual pleasures, and she is too young to take on household matters."[13] Apparently, Kepler was not looking for an intellectual equal, but a woman to take care of his children and to run his domestic affairs. In the end, however, the choice was taken out of his hands, when the girl's mother decided on her own that her daughter was too young for marriage. After that, Kepler left Prague. On his way to Linz he stopped at Kunstadt with his children, and there he met a young girl who turned his head. "Here my soul grew warm. I liked the girl, for she was well brought up, the way I prefer it. She cared about my children with extraordinary willingness."[14] They struck an agreement, and Kepler left his children with her, promising to return from Linz at a later time to collect them and to finalize their arrangements. Before the year was out, however, the girl had, not too surprisingly, become engaged to another man.

In Linz, the process started all over again. The first of the Linzer women was something of a dish. "I fell for her because of her tall build and athletic body, and it would have been settled, had not both love and reason forced a fifth woman on me."[15] We should all have such problems. The fifth woman was Susanna Reuttinger, and he would marry her, but he did not know that yet. "This one won me over with love, humble loyalty, economy of household, diligence, and the love she gave the stepchildren. I also liked her loneliness and the fact that she was an orphan."[16]

But then, he actually listened to people's advice, which he shouldn't have done. Helmhard Jörger's wife preferred the dish athlete to Susanna and convinced Kepler to drop Susanna in favor of the other girl. Kepler was angry at this, but he did it anyway. After all, women know about these things. Then his stepdaughter, Regina, and her husband got into the act and recommended another woman, who came from the nobility and had money to boot. As Kepler said: ". . . and not without money, which was attractive." But once again, she was too young for him, and he worried about her noble status, which he feared might make her proud and difficult. Besides, the wedding would have cost him a fortune. All the while, Kepler was still thinking about Susanna and remembering her presence, her voice, the way she carried herself, her kindness. She was in fact perfect for him, because she was intelligent enough to appreciate something of his work and supportive enough to help him continue on with it. But that didn't stop him from courting a seventh contestant, a woman presented to him by friends, who praised her sophistication and her skill in running a household. Another pretty girl, but Kepler's heart wasn't in it. Even as he courted her, he warned her about himself and revealed the doubts he had about her. He wasn't surprised when she refused him.

Then there was an eighth woman, and he liked her well enough. She wasn't pretty, but she would have made a good mother; she was honorable, with a good education, and had some money of her own. At first, things went well, but then the woman had doubts, and no one could figure out whether she wanted to marry Kepler or not, so he quietly backed away and found woman number nine. He was unsure; she was unsure— nothing happened. Number ten he didn't find attractive at all. He was thin as a stick, and she was short and round. Then a friend introduced him to number eleven. Once again, she was a young girl who was not ready for marriage, but the friend wanted to act as go-between and carried on the negotiations in secret. This went on for four months, but once again nothing happened. The girl was too young, as Kepler already knew.

Sick of endless rounds of dating, he prepared for a trip to Regensburg and all the while thought of Susanna. Before leaving, he rode over to see her, declared his love, proposed marriage, and Susanna accepted. One can imagine that she knew he would come back all along.

The wedding took place in Efferding on October 30, 1613. The reception took place at the Sign of the Lion. One story has it that during the reception Kepler got distracted by the problem of inventing a new way to measure the amount of wine in a barrel and spent half the time in a mathematical haze. Because he was the imperial mathematician, he invited Emperor Matthias, who sent his regrets. He also invited the representatives of the Estates, who sent him a wedding gift of a goblet worth about 50 gulden.

When the festivities were over, everyone started complaining again. People did not approve of his choice. Regina wished that he had married the aristocratic girl, and she did not think that Susanna was capable of properly caring for Kepler's children, who were eleven and six. With time, however, Kepler and his new wife proved them wrong. His marriage to Susanna was far happier than his marriage to Barbara, for she understood him. For the first time, perhaps in his entire life, with his wife and children around him Johannes Kepler found peace.

Over the next fourteen years, Susanna Kepler gave birth to six children. The first three died as babies. Margareta Regina, born on January 7, 1615, struggled with epilepsy and died two years later on September 8 of a simple cough that turned into consumption. Katharina, born July 31, 1617, died eight months later, on February 9. She too caught a cough that turned into consumption or possibly pneumonia. Sebald was born January 28, 1619, and died from smallpox on June 15, 1623, on the Feast of Corpus Christi, the Body of Christ.

The last three children survived. Cordula, born on January 22, 1621, came into the world while her mother was visiting Regensburg. Then, two years later, Susanna gave birth to Fridmar on January 24, 1623, and finally, after another two years, Hildebert on April 6, 1625. Kepler liked the name Hildebert and chose it for scholarly reasons. There had once been an eleventh-century theologian by that name who wrote beautifully on the nature and importance of the Eucharist and was the first to use the term "transubstantiation."

ᕮᎭᏗᏂᎧ

Highest Honorable Dear Duke and Master:

Recently, the wife of Jörg Haller, a poor townsman and day laborer, forwarded a complaint to me. Accordingly, on October 18, her little daughter, a girl of about twelve years, helped the brick maker's daughter carry bricks and limestones to the kiln. That is when the Kepler woman passed them and hit Haller's girl on the arm. The girl felt pain immediately and her agony grew by the hour, so that she couldn't move hand or finger. Although the plaintiff (Haller's wife) does not openly accuse the Kepler woman of magic making or witchcraft, she respectfully requests an investigation as to why the Kepler woman hit her daughter on the arm. Since I saw the girl's injured arm and hand, I summoned the Kepler woman to the courthouse and had her confronted with the charges.

The Kepler woman did not want to confess and instead called her accuser a liar. She said she did not touch the Haller girl, who was coming toward her at the time. She said she merely passed the girls, but I say that was when she turned around and hit the Haller girl on the arm, which was witnessed by the brick maker's daughter. So I summoned the brick maker's girl and, since she is only eleven years old, also her father and mother. The witnesses stated that their daughter, right after it happened, came running home and told her parents that the Kepler woman had hit Haller's Catarina on the arm and that the Haller girl is in pain. They told their daughter to keep quiet about it. They also stated that the Haller daughter is a pious girl who would not give anyone reason to hit her.

The Kepler woman is now over seventy years of age and her husband, who according to her is still alive, left her twenty-eight years ago. She has been under the intense suspicion of witchcraft for some years now. The wife of the local glazier, Jakob Reinbold, in her defense to the civil charges brought against her by the Kepler woman, swears to her death that the Kepler woman gave her a magic potion four and a half years ago. That potion caused her to suffer inhuman pain that could not be relieved by any remedy or cure. The trial date (for the slander case) had been fixed for last Monday in order to examine the witnesses; however, it was then canceled in light of the above developments.

Therefore, I seriously questioned the Kepler woman under threat of imprisonment and also told her that I had no choice but report the matter to you. She, however, did not admit to any wrongdoing, and I let her go. She returned to my office with her son Christoph, the pewterer, who has an honest and good reputation. The Kepler woman said I should not believe her accusers. After I had explained to her son the details of the accusations against his mother, he sighed deeply and said that he wishes to God he could leave this town overnight with his poor belongings because of his mother. He said I should proceed with what I believe is in my jurisdiction, that God must want it that way.

After this, the Kepler woman returned to the court for a third time solely to see me and beseeched me to refrain from reporting her to you, or at least to cause a delay or a halt. She wanted to give me a nice silver cup (unknown to her creditors, etc.), which she would promptly deliver.

Now, having advised you of my modest opinion and having

submissively reported to you the situation or at least having given you a better understanding (foremost because named Jörg Haller, a hardworking and respected lad, is also accused by the Kepler woman. Meanwhile, the girl is enduring great pain in her arm, which is still lame), I also want to advise you that the Kepler woman has gone to be with her son in-law, the pastor of Heumaden, and is there at present. I dutifully and expectantly submit my report and submit myself obediently to your grace.

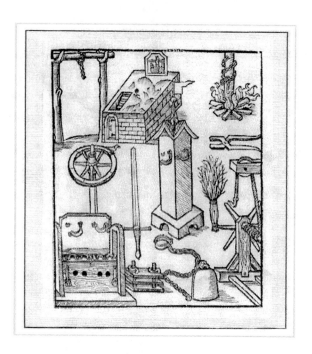

The instruments of torture

XII

If One Practices the Fiend's Trade

*Where Katharina's trial begins, and she is horribly
mistreated, and where Johannes must leave the
emperor's service for a time to come to her aid.*

AFTER THE SHAMEFUL ATTACK ON KATHARINA, Einhorn moved to
quash the Kepler family's slander case against Ursula. The Keplers expected their case, *Kepler v. Reinbold,* to come to court within a reasonable
period of time, but because he was the magistrate, Einhorn was able to
postpone it for over a year. His continuing disregard of the duke's order to
proceed with the Kepler family's case against Ursula Reinbold during that
time was surreal, a parody of justice, and can make sense, even politi-
cally, only if we consider that through Kräutlin, Einhorn enjoyed a secret
behind-the-scenes channel of communication with the duke and his
councilors. They were willing to forget his insubordination because
Katharina's main supporter, her son Johannes, was himself in a stink with
the Stuttgart consistory over charges of hidden Calvinism. There was talk
that Johannes might be as guilty of witchcraft as his mother. Some industrious soul had dug deep into the Tübingen archives and found a disputation that Kepler had written while in school there, which later became his
Somnium, or *Dream,* illustrating Copernican theory by imagining a

fanciful flight to the moon so the universe could be described from the perspective of moon people. Because in this tale, perhaps the first actual piece
of science fiction, it was his mother who, through magical powers, summoned the spirits of the air to carry him to the moon, the wagging tongues
of Leonberg assumed that this proved that the son too was a practitioner
of dark magic.

Although nasty tongues could threaten a simple, stubborn old woman,
there was not much they could do against the emperor's mathematician.
In many ways, despite his loyalty to the duke, Kepler had outgrown his
homeland and was too big a fish for them to spear. When he heard about
this new round of gossip from his sister, he fired off a hot salvo to the
Leonberg councilor and to the Leonberg Senate. In his letter (quoted in
full before Chapter 1) he attacked the "devilish" people who had threatened his mother and used illegal means to terrorize the poor old woman
into admitting that she was a witch when she was not. He rejected the talk
about himself as one more intrigue to bring down his mother and then
blatantly referred to his many years of imperial service, hinting not too
subtly that they really should be careful, because he had the ear of the
emperor himself. He demanded that the court send him copies of any documents they had gathered to date, and exhorted, or possibly even instructed, the councilor to do his job properly and to look after his family's
interests. Then he soundly reprimanded the magistrate without mentioning any names. But Einhorn had no shame, and because of his powerful
protection by the Prince Friedrich Achilles and Kräutlin, Kepler's admoni-
tions meant little to him. He had little choice but to carry on, for he was
afraid of being implicated in the events of August, when he and his dear
friend Kräutlin had intoxicated themselves and threatened the life of
Katharina Kepler with a sword.

After a year of Einhorn's stall tactics, the court in Stuttgart grew nervous. When the Reinbolds tried to postpone the case for another twenty
weeks by "forgetting" to send their defense brief to the clerk, the judge,
who had finally hit his limit, stopped the postponements and set the date
as a *terminus peremtorius,* that is, no more delays.

Then the Reinbolds got a break. There was a woman in town, the wife of Jörg Haller, who lived almost strictly on Ursula Reinbold's charity; Ursula threw her odd jobs and passed her name around town as someone who would work cheap. The husband, Jörg, was the son of the long-dead *Wasenmeister*, what the British would call a knacker, the man in charge of killing sick horses and disposing of them. Jörg Haller was a day laborer and alcoholic who had "chased his money down his throat" and from that point on had tried to support himself through theft, so that his family had the reputation of being "loose, thievish, and godless riffraff."[1] Jörg had only recently stolen oats and hay from a local dyer and had run confidence games; he would often contract for labor, get paid, and then not show up.

The sad condition of the Haller home caused two of their children to suffer from chronic illnesses. Frau Haller was terribly superstitious and spent a good deal of her time measuring other people's heads, which was illegal because it involved magic and fortune-telling. Her brother also dabbled in fortune-telling and sold snake oil on the side.

In October 1616, during the harvest, while the days were growing shorter and cooler and the townspeople prepared for winter, the local brick maker, Endriss Leibbrandt, needed some day labor. He was hiring young girls in town to carry baskets full of bricks to his new kiln and had already conscripted his daughter Barbara and about ten other girls, when he stopped by the Haller house to ask if their daughter Katharina could join them. Frau Haller agreed, because work was work and she could at least be assured that her daughter would get a hot meal that day. The girls gathered at Leibbrandt's kiln on October 15 and carried bricks, probably from around five in the morning to past sundown. Katharina Haller had been with them, carrying heavy bricks and limestone, and her muscles must have been tired.

At nine o'clock in the morning, the Kepler woman walked by, just as the group of girls had finished unloading their baskets of limestone at the kiln and was returning to the lime hut. The Haller girl was startled when she saw Frau Kepler and circled around her rather than get too close. Either her mother had said something to her or the gossip about Katharina Kepler had permeated everywhere, because the Haller girl was afraid when she saw her and tried to run away. After Frau Kepler had gone, the Haller girl returned and carried bricks for another hour and a half, when her arm hurt. Likely, she had tendonitis and her arm did hurt, or possibly she was lazy and wanted to skip off work early. That was when she claimed that Katharina Kepler had hit her on the arm.

The brick maker didn't believe a word of it, but his wife did. He argued that the girl had just been frightened when she saw the Kepler woman and there was no evidence that old Frau Kepler had done anything, so he questioned the other girls to find out what they saw. All of them said they saw the Kepler woman walk by, but only one girl saw Frau Kepler hit the Haller girl on the arm. The Haller girl then said that the other girls didn't see the Kepler woman hit her, because they had already gone. Frau Kepler had returned after she had walked by and that's when she hit her. The brick maker refused to attach any significance whatsoever to the Haller girl's story, but his wife, possibly because he was so skeptical, believed every word. She went to Ursula Reinbold and told her the whole story.

Ursula saw her chance. She suggested to the superstitious Haller woman that the pain that the poor girl had suffered was the *Hexenschuss,* the witch's shot.[2] When her daughter's pain did not subside, Frau Haller took her case to Luther Einhorn.

Two days later, Katharina Kepler was carrying an armload of hay to her home. Suddenly the Haller woman stood in front of her in the street, just before the gate to Katharina's house, and demanded to know by what right she had beaten up her daughter, the poor girl who had never done her any harm. Katharina denied hitting Haller's daughter, saying finally that perhaps she might have brushed her lightly with her basket as she passed her on the narrow path, but nothing more. Suddenly the Haller woman became violent, screaming, "Help the poor girl! I demand you help the girl!" Still shouting, she advanced on Katharina, who slowly backed away toward her home, which made the Haller woman scream even louder, accusing Katharina of witchcraft and demanding that she come at once to heal her daughter. Katharina was finally backed up against the wooden gate of her house, still denying that she had done anything to the girl, so the Haller woman pushed her so hard that Katharina fell against the gate, springing the latch, and then she stumbled backward, nearly falling.

The Haller woman stood over her, demanding that Katharina come at once to cure her daughter, but Katharina insisted that she had done nothing to hurt the girl, which only made the Haller woman more violent still. Finally, she drew a knife from her belt and screamed at Katharina that she would help the girl or die on the spot. Saying this, she advanced on Katharina once again and held the knife at her throat. To save her life and to calm the mad Haller woman, Katharina agreed to follow her home. Standing, she followed her out onto the street and then suddenly turned and called out for her son Christoph to come: "Help me please!" she said. "Just one more time!" Christoph appeared from the house and stood between the two women. He spoke sharply to the Haller woman and told her that if she had any accusations to make, she should do so in a court of law.

Luther Einhorn, spotting an opportunity, summoned Katharina to him two more times. On both of these occasions, the Keplers made some unfortunate mistakes and gave ammunition to the Reinbolds' camp. The first time Katharina and Christoph appeared, Christoph muttered some foolish remark to the magistrate about this being only a silly matter between women; he said that he knew that his mother was a pain, and if he could, he would just leave town and be quit of her. He never thought his mother was a witch, but neither did he seriously reckon the forces that were gathering against her. He did not realize when he spoke with Einhorn that the magistrate was set against his mother and had no scruples about misusing his position.

Katharina, however, was not much help. As with Johannes, it was her nature that once she believed she was in the right, she would pursue a dispute until she was vindicated. But she was also naïve. All this time, Einhorn was trying to keep Katharina's slander case against the Reinbolds from coming to court and was looking for any opportunity he could find to turn the tables on the Keplers. During her second visit, Katharina begged the magistrate not to report the situation to Stuttgart and to allow her trial against the Reinbolds to come up. At this point, she did something incredibly stupid. She offered Luther Einhorn, who was already dead set against her, a bribe, a silver cup worth about 50 gulden, an act that would come back to haunt her throughout the trial.

On October 22, Einhorn reported the affair to Stuttgart in a letter, for this was his chance to divert the duke's attention from the slander case to the witchcraft case, which up until this point had gotten little notice. He related the Haller woman's claim and placed himself into the story as a witness, saying that the girl's agony started immediately and increased to the point that her hand was paralyzed. He also said that the Haller woman requested an investigation, though she fell short of accusing Katharina Kepler of witchcraft. Later on in his letter to the duke, he quite simply lied, making claims about the brick maker's story that were not so, saying that he called the brick maker and his family in to testify and that the brick maker and his wife told him that their daughter had run home with the story, but they told her to be quiet about it.

Einhorn here gives little indication that the brick maker had found the Haller girl's story dubious or that he believed she had made up most of it. Charging on, however, Einhorn relates how the Kepler woman had been under intense suspicion of witchcraft for some years, which was not entirely true. Moreover, a week after the incident, people saw the Haller girl in the forest cutting wood.

Einhorn then gets to the point—his point. For the first time, he has the opportunity to tell Ursula's story in a legal setting, to bring her accusations out of the shadows of gossip where they had been festering for years and into the well-lighted courtroom. "The wife of the local glazier, Jakob Reinbold, in her defense to the civil charges brought against her by the Kepler woman, swears to her death that the Kepler woman gave her a magic potion four and a half years ago. That potion caused her to suffer inhuman pain that could not be relieved by any remedy or cure. The trial date (for the slander case) had been fixed for last Monday in order to examine the witnesses; however, it was then canceled in light of the above developments."

This was it. This is what Einhorn had wanted all along. At this point, he played his trump card. "After this, the Kepler woman returned to the court for a third time solely to see me and beseeched me to refrain from reporting her to you, or at least to cause a delay or a halt. She wanted to give me a nice silver cup (unknown to her creditors, etc.), which she would promptly deliver."

Einhorn needed to stop the slander trial and to insert the witchcraft trial for several reasons. First, the slander trial would have inevitably brought up questions about his own behavior; second, he was thoroughly entrenched in the Reinbold camp; and, third, if Katharina Kepler were convicted of witchcraft, then her property would be forfeit and, as magistrate, he would have control of its dispersal. On October 24, 1616, two days later, Einhorn got his wish. The *Oberrat,* the superior adviser to the courts, having been influenced by Prince Friedrich Achilles, who was in turn influenced by Urban Kräutlin, ordered from Stuttgart that the magistrate, Luther Einhorn, arrest the Kepler woman on suspicion of witchcraft, leave her in jail for a few days to let her fears work on her, and then seriously examine her. The use of torture was not specifically mentioned, but the threat of torture always remained in the background, a shadowy threat in any court examination.

After the order for Katharina's arrest came out, Christoph sent a letter to Johannes in Linz, informing him of their mother's dangerous situation. The two brothers conspired to get the old woman out of town and as far away from the Leonberg authorities as they could. They wanted to talk Katharina into immigrating to Linz, where Johannes and Susanna could watch over her and where she would be far outside the Württemberg authorities' jurisdiction. This would not have made Johannes's life in Linz any easier, to have a mother accused of witchcraft move into a house where the son had been accused of heresy. Nevertheless, to have Katharina move to Linz was the best of a number of bad choices.

But none of her children had reckoned on Katharina's stubbornness. She wouldn't have it. She believed that she was in the right, and that was that. To move out of Leonberg would be to abandon the field to the Reinbolds, something she could not do, before God. Katharina refused to go to Linz for fear that they might think she had a guilty conscience.

Instead, she moved to Heumaden, which was close by, to the house of her daughter Margaretha and her son-in-law, Pastor Georg Binder.

But living in Heumaden meant to Katharina that she would no longer be living in her own house or managing her own properties, and this burned her like acid. After she left town, the Reinbolds claimed that she had given evidence against herself by running away, which was exactly what Katharina feared. What she couldn't see was that her life was in danger, whether she was a witch or not, and getting out of town was the only way to save it.

After a great struggle within the family, Katharina's children managed to convince her to lease her home to Christoph for three years by promising that they would support her during that time. Since Christoph, who had just become a citizen of Leonberg, was still living in Luther Einhorn's jurisdiction, the three children agreed that Margaretha and Johannes would take turns inviting Katharina into their homes. Christoph Kepler was adamant about his mother's move to Heumaden, if not Linz, because he had a feeling that things were going to get very bad indeed for Katharina. The lies had been repeated so many times by so many people that they had begun to be accepted as the truth. By this time, Katharina was nearly under house arrest. Einhorn had warned Christoph to keep his mother from visiting other people in their homes, because this would cast even more suspicion onto her.

On December 2, 1616, the duke's consistory in Stuttgart ordered Einhorn to inventory all of Katharina Kepler's assets. Einhorn did what he had been ordered to do—closed up her house and counted all her assets, with the expectation that he would be able to make payments out of it to the Reinbolds and the Hallers for their pain and suffering. After doing that, he bragged about town that it was he who got the old woman arrested, he who got her estate closed up, and he who would be in charge of her fortune.

A few weeks later, in January 1617, Jakob Reinbold wrote to the duke himself and requested that the "fugitive" Kepler woman be arrested and brought back to Leonberg. He claimed that because his wife was still suffering and no one knew the kind of poison Katharina had put into her potion, the Kepler woman should be returned to Leonberg so that she could be forced to confess and to hand over the information they needed to help Ursula. Oh, and by the way, he asked the duke if Katharina's estate could be confiscated rather than just inventoried and if he and his family could have a share of that estate for their pain and suffering.

Meanwhile, the Kepler family found out about Jakob Reinbold's petition, and in January 1617 Christoph and Pastor Georg Binder petitioned the duke to dismiss Jörg Haller's request that Katharina's estate pay for the support of his children. For once, the Keplers were successful. The duke, perceiving the corruption behind the whole affair and yet not knowing what to do about it, was embarrassed and avoided taking sides. However, as more petitions fluttered his way, his embarrassment grew. He really wanted the whole thing to go away.

In February Emperor Ferdinand, who succeeded Matthias after his death, summoned Kepler to Prague, and so Kepler hurriedly left for the capital. He didn't stay long, however, because he wanted to return to Linz to take his mother to Leonberg. Meanwhile, in Leonberg, Einhorn had refused to show Katharina Kepler's advocate the duke's order and instead read them excerpts—"a little of this and a little of that." On March 9, Christoph Kepler and Pastor Binder wrote to the duke directly to complain about Einhorn's behavior, especially his refusal to show them the duke's order.

From February to August, the Keplers fought with Einhorn, trying to get him to allow the slander case to come to trial. The duke ordered Einhorn to do so, but on more than one occasion he ignored the order, and then bragged that if the duke forced him to obey his order, he would petition the Royal Chamber Court. The town bailiff of Leonberg, Jakob Kern, set the date for the slander trial on at least three occasions, but each time Einhorn managed to find a pretext to worm out of it.

At the end of August 1617, the weather was still hot, the air still humid. The farmers were preparing for the harvest, and the gossip of Leonberg was afire with the terror of witches. Katharina, certain of her innocence, insisted on her day in court and, in spite of all that had happened to her, returned to Leonberg and tried to move back into her house. She still believed she would be vindicated in the end and that the court would restore her good name.

On September 1, Johannes Kepler wrote to the duke to inform him of all the events in Leonberg and to request a proper administrator for his mother's assets. Up until this time, the administrator would have been Einhorn, which would have been a serious conflict of interest. In his letter, Johannes explained to the duke: "I have doubts and I worry that before the examination of those witnesses, by whom she will prove her innocence, that because she is a careless old woman (who has already let go of the keeping of her own house and has rented her estate), it will lead presently to additional useless costs, and also subject her to the machinations of her adversaries." He then asks the duke to authorize "that her property and estate be managed by someone who will administer, advice, and provide expert opinion to the war magistrates of Leonberg, and will save her estate from ruin and reduction, and will take from it only as much as is necessary for her legal personal use and also for the payment of her debt."

In October, Katharina arrived in Heumaden. She had been gone over nine months in Linz and complained bitterly about being away from Leonberg. Johannes left Linz soon after, but on the way he suffered another tragedy. He stopped in the little town near Regensburg, on the Danube River, where he found out that his stepdaughter, Regina, by then a married woman with children, had recently died. Her husband, Philip Ehem, had no one to care for his children, so he asked Kepler to leave his fifteen-year-old daughter, Susanna, behind to watch the children while Kepler traveled on to Leonberg. Finally, on October 30, Kepler arrived. It should have been a time of celebration, for it was the Lutheran jubilee. All around him, however, there were tragedies. His stepdaughter had died. His mother was on trial for her life. And fire had destroyed Vaihingen, a nearby town.

Kepler had traveled all the way to Leonberg hoping that his presence would help his mother's case and that he could run political interference for Katharina in ways that Christoph could not. The chief counsel set a new date for the slander trial in November, but once again Einhorn managed to slip by the order. On November 20, without consulting with the bailiff, Einhorn summoned Katharina to his office once again to fish for some new "development" that might allow him to evade the new slander trial date, set for November 24, even though the date had been announced and all the parties informed. Frustrated and angry, Kepler finally asked the duke if his mother could return with him to Linz without loss of honor on her part, and the duke agreed. But Katharina would not go. She would have her day in court, even if it killed her. Even so, the high court in Stuttgart ordered the magistrate to allow Kepler to take his mother back to Linz with him. Also, they told Kepler that the duke would not allow the Reinbolds to postpone the slander trial any longer.

Throughout 1617–18, Einhorn managed to sidestep everybody. Jakob Kern, the bailiff, tried to schedule the slander trial in spite of Einhorn, but failed. Johannes Kepler, realizing that his mother's case against the Reinbolds was stalled, returned home in December 1617 only to find his daughter Katharina sick in bed.

Nothing much happened after that until May 1618, when the examination of witnesses on the slander trial began. No one knows what day that trial took place or what the result was. Somehow the documents disappeared. Kepler claimed that his mother had won the day, because four honest old town councilors, elder burghers with some standing in Leonberg, had come forward to set the record straight, saying that Katharina Kepler's reputation was impeccable and that the Reinbold woman's reputation was nowhere near as clean. The Kepler family waited for the judgment against the Reinbolds to be made public, but it never was. That judgment was quickly lost, because Einhorn and the Reinbolds, possibly with the help of the Prince Friedrich Achilles, worked furiously to bury it.

Suddenly there was a bill of indictment against Katharina consisting of forty-nine articles, finally putting on paper what up until that time had had merely been gossip. Johannes was furious. This started off as an evil fantasy, and now, because of Einhorn, his mother was fighting for her life.

Then on May 23, 1618, the Thirty Years' War started with a comic opera in Prague, and Germany paid for its years of sectarian hatred.

LETTER FROM KEPLER TO HERZOG JOHANN FRIEDRICH
VON WÜRTTEMBERG
NOVEMBER 1620

ᏮᎭᎭᎧ

Highest Honorable, Merciful Duke and Sir,
Your Ducal Highness is aware of my modest obedient
service.
Merciful Duke and Sir,
As much as I did not want to bother your Ducal Highness,
the misfortune and the misery of my mother, under arrest at
Güglingen, is becoming so great that all of us, including our
relatives, fear what is becoming obvious—her utter ruin. Her
suffering is accompanied by the suffering of the three of us,
her children, but we do not in any way assume that your
Ducal Highness has taken joy in this judgment, but rather, as
the trusted father over his subjects, that you would like to
help all of us escape from it.

For our mother, who in her seventy-fourth year and with
present maladies of body and soul, was foremost never con-
victed before this arrest, now four months long, is heartbreak-
ing. And if she, poor thing, considers that she has withstood
four months of torment, with neither judgment nor legal pro-
ceeding, it is all the more painful for her when she remembers
that those accusations she was charged with are as incredible
as they have ever been. She has not knowingly committed
even the least visible injustice, nowhere close to any of the
charges brought so far. For this reason, they want to assume
her to be culpable on the basis of the gathered suspicion. We
would rather that these same legal proceedings should be

praised as Christian love in many parts, and that those who have suspected others unfairly—only and solely her adversary, those who slandered her among the people six years ago and was lawfully charged for it—is to carry blame.

But much harder for us, and much more dangerous for our mother, is that our mother is being guarded by two keepers, men who themselves are deeply in debt, whose common thinking and conscious efforts are solely intended to extend their duty as long as possible. To achieve this, they think that no act is too audacious or too low. They take the unfortunate utterances, despondency, fickleness, impatience, and whatever else this troublesome situation may exact from such an old woman and interpret it maliciously, comment and portray it in a manner that when, finally, were she to be released, she would be burdened with even more suspicions than were put upon her before her arrest. To this end, we can doubtlessly interpret the purpose of our adversary's journey to Güglingen.

Thirdly, these useless and wasteful guards have caused unmanageable expenses by unnecessarily burning wood so they did not have to sleep too close to the fire, to such an extent that within a few weeks, after keeping house in this manner, my mother would have nothing left of her income. Even her income is in danger, since the magistrate of Leonberg, in order to obtain alimony for her, sold all her fields and then raised an untimely dispute between us, the children, who have our own expenses. In addition, my brother fears that I will cause him to be thrown into poverty in the future, since my arrival in the country has led to such an extension of the court case and has incurred such great expenses. I have to lament to God in Heaven, since I have been forced to abandon my poor wife and children on the road in a foreign place near Regensburg, without food or money. And still I will soon have to leave, task unaccomplished, with shame, ridicule, and heartache, for I have no credit in the country.

The Second Defenestration of Prague

XIII

With Present Maladies
of Body and Soul

*Where the Thirty Years' War begins with the
Second Defenestration of Prague,
and where Katharina Kepler is tried
and convicted of witchcraft.*

ON MAY 23, 1618, in Prague, spring was slowly edging into summer, the spring rains were giving way to summer's heat, and the capital was nearly at war with itself. The Protestant Bohemian Estates had come to the point of murder, because the Habsburgs, who had once made promises of religious concessions to them, had reneged.[1] After the Habsburg brothers' war, Matthias replaced Rudolf as emperor. He stayed on in Prague until after his brother's death and then transferred his capital to Vienna, where it remained. He passed the rule of Bohemia on to his nephew Archduke Ferdinand of Styria, who with the negotiated blessing of the Estates took the crown in 1617. Before leaving, Matthias confirmed all the promises he had originally made to the Protestants, promises that he had made under duress when he needed their help to fight Rudolf. Protestants could practice their religion as they saw fit; they could hold public office; they could live without fear of losing their land, their titles, or their lives. But then the Catholic faction, like a boulder in the river, refused to move. This was the same group that Rudolf, by nature more open to new ideas than Matthias, had to deal with for years. Led by the Spanish but with cousin Ferdinand's not too secret support, they would not sign any document that promised equality to the Protestants, because for them there was only one Christian church, one heir to Peter, and one true doctrine. There could be no compromising with heretics.

This was not that different from what the more radical Protestants believed in their secret hearts, though in the Habsburg lands, being the underdogs, what they settled for was freedom. They did not get that either. Under the new archbishop of Prague, Johannes Lobelius, the Catholics turned up the heat, and one by one the concessions to the Protestants evaporated. Anger swelled in the ranks, and the Bohemian Estates turned from petitions for imperial concessions to secret meetings and whispers about violence. Civil war was brewing all over the city. Protestants could no longer hold the offices that they once held, while the Prague towns had lost their right of self-rule. The cap was the day the Catholics tore down two new Protestant churches that were built on lands once belonging to a Benedictine monastery, claiming that the land was still Catholic land. The Protestants stamped about, their whispers becoming shouts. All church lands belonged to the king, they said.

The Protestant Estates held a meeting at the largely Hussite Carolinum University to discuss the situation. Their resolution was to send one more petition to Matthias requesting that he honor his promises. One evening, the radical faction, led by Václav Budova of the Bohemian Brethren and Count Matthias Thurn, known for his hot temper, met in the Minor Town in the home of one of their members, Jan Smiřický, to plot murder. Their specific targets were the royal administrators who lived in the Prague Castle, whom they accused of subverting Matthias's honest promises to them.

But there were deeper reasons hidden behind these. Even in the age of absolute monarchy, kings had limited power. As rulers of national governments, they still had to deal with the local nobility, town councils, and estate representatives, since most of the actual work of government was done by them. Kings rarely had standing armies, and what they did have was meager. They relied on taxing the local governments to maintain their armies and their way of life, and if a king did not have a healthy relationship with his local governments, then his support, military and financial, could dry up. This was especially true after the Reformation and the Peace of Augsburg. In spite of the old formula "whose the land, his the religion," if a local ruler belonged to a religion different from that of his people, most especially from that of those who formed his local governments, then his position was precarious. Even in a world that operated on the divine right of kings, new radical theories were popping up around Europe, even in Prague, that held that the people could legitimately depose their own king, if that king practiced the wrong kind of religion.[2] But even the most radical would rarely shout that theory from the rooftops, for it was too new and frankly dangerous. They would rather make their revolt while gravely proclaiming they were saving the rightful king from bad advisers. That was the position that Ferdinand II found himself in, in both Austria and Bohemia.

Before they sent their petition to the emperor, radicals such as Thurn and Budova were in the minority. But that soon changed. Matthias responded to their request, but his response was not what the Estates had hoped it would be. His answer was haughty and condescending, the answer of a man who had achieved the power he had wanted and planned on using it. Matthias's refusal set fire to the city. The radicals had suddenly become the darlings, and the political faction of the Protestant Estates had become a movement.

On May 23, a hot crowd of angry Protestants, all members of the Estates, followed the radical faction up the hill to the Prague Castle, where the Catholic Diviš Černín opened the gate for them. He would later die on Emperor Ferdinand's scaffold for that mistake. From there, they stormed the staircase leading to the offices of the royal administrators; they brawled their way in and confronted the bureaucrats along with their secretaries. There were only four of the administrators there that day, the other six having left town rather suddenly, for their health. The remaining four, Jaroslav of Martinic, Adam of Sternberg, Dìpold of Lobkovic, and Vilém

Slavata of Chum, either had important business to conduct, were very brave men, or were slow on the uptake.

The mob confronted them and accused them of intrigue and of being responsible for the king's bad faith. The tension in the room nearly exploded, teetering on the edge of riot. The four administrators pleaded with the Estates men, assuring them that they were not the ones responsible for the tone of Matthias's answer and that they were only servants of the emperor, like the Protestants themselves. This did not mollify the crowd, however, for they had come ready for blood. The radicals had been planning for this day for some months and would not back down. The fact that the four men had little to do with the king's answer to the Estates did not matter. The fact that the answer had actually been written by Melchior Cardinal Khlesl, the gray eminence of the imperial court, meant nothing. These men were Catholic, they were intransigent, and they were there.

The radicals would have done away with all four of them, but the rest of the Estates men were more level-headed. They separated out the moderate Catholics—Dìpold of Lobkovic and Adam of Sternberg—from the ones who were more intransigent and pushed the moderates into an adjoining room, where they let them go. Then Count Thurn incited his fellow rebels, shouting encouragement. The time for talk was over, and the time for action was beginning. They should act now, decisively and quickly.

The remaining administrators—Jaroslav of Martinic and Vilém Slavata—had been two of the most anti-Protestant members of the king's court in Prague and had been deeply involved in the Catholic faction for twenty years. They had consistently dragged their feet when ordered to approve Protestant rights, had argued against religious tolerance, and had worked tirelessly against the Protestant cause. Count Thurn, in good Bohemian fashion, hearkened back to the first defenestration of Prague performed by the followers of Jan Hus in the fourteenth century and ordered the two men thrown out the high window of the Bohemian chancellery, a fall that was a good fifty feet. Out went Jaroslav of Martinic without much ceremony. Then they picked up Vilém Slavata, but he grabbed on to the window sill and called out for his father confessor. Apparently, his conscience was not as clear as his compatriot's. This stalemate went on for a few long painful seconds until one of the rebels took the hilt of his dagger and pounded at Slavata's hands until he let go. But that was not enough, not quite. One of the secretaries, Johannes Fabricius, was edging along the wall toward the door when one of the Protestant men saw him; the crowd took him and threw him out the window too. Well, why not?

Oddly enough, all three men survived the fall and managed to crawl off in spite of the pistol shots that fell all around them. Later the Protestants claimed that the men fell on a pile of horse manure that had been left at the bottom of the wall under the window. Catholics, on the other hand, claimed that the men lived because the Blessed Virgin had spread her mantle under them so that they fell as lightly and as gently as rose petals. They were saved by a miracle, said the Catholics. They were saved by shit, said the Protestants.

The three men escaped, but not without harm. Slavata had a nasty head wound and could not leave the city, so he staggered to the imperial chancellor's house, where the chancellor's wife, Polyxena of Lobkovic, hid him from the mob and tended his wounds. Eventually, a body of men came to her house looking for the administrators, but she stood up to them, berated them, and sent them on their way. The secretary Fabricius had hurt his leg in the fall, but he managed to escape altogether. When he arrived at the emperor's court, he was greeted warmly and was later raised to the nobility. His title included the term "von Hohenfall," making him the lord of "High Fall."

Meanwhile, back in Prague, after news of the Second Defenestration of Prague hit the streets, the city erupted. Protestant burghers roamed the city murdering Franciscan monks and, predictably, invading and pillaging the Jewish quarter. The world that was Prague was changing; fire was on the wind, the Thirty Years' War had begun, and somehow once again the Jews were caught in the middle.[3]

CRUMO

KATHARINA KEPLER was a stubborn woman. She had been accused falsely of witchcraft, and although she knew that she was not a witch, she believed that anyone who truly understood her would know that as well. So why should she be hiding out in Heumaden and Linz? Why should she not be back in Leonberg, proving to them all that they were wrong about her and that the Reinbolds were malicious slanderers? On June 16, 1618, Katharina returned to Leonberg for the first time since October 1616. The Reinbold faction raged. They thought that they had gotten rid of the troublesome old women once and for all. Einhorn was nervous, however. What was she doing back in Leonberg? What could she want among clean Christian people? Would she try to resurrect her lawsuit against them? He thought that he had safely doused the slander case against Ursula Reinbold, but now that Katharina Kepler had returned to Leonberg, she and her family could blow on the embers once again and have the whole thing brought to life. The old woman was altogether troublesome and had to be stopped.

Katharina's house was near the market square, so just standing at her doorstep upon her return, she let loose the hornets' nest of gossip. Still under great suspicion of witchcraft, Katharina sparked fear in many simple people and in some not so simple. While she was away, the fear of her had grown. Instead of dying out, what had been mere suspicion had blossomed into a prejudice, and from a prejudice into a court case. The Reinbold camp, which had been shepherding that prejudice, gathered its forces at once and petitioned to have Katharina arrested.

Up until that point, Einhorn had managed to push through most of what the Reinbolds wanted, but now that Katharina had returned, they didn't have everything their own way and had to split the difference—on June 23, the high court in Stuttgart denied their petition for Katharina's arrest, but at the same time they advised that the witchcraft trial against Katharina Kepler should go forward. So Katharina was allowed her freedom, but the charges stood.

On October 8, the court publicized the witness examinations that they had made on May 7. During this time, Johannes Kepler sought the advice of the well-known jurist Christian Besold, and in April 1619 he received a letter from Besold that frightened him. Besold said that he had studied the forty-nine accusations that the Reinbolds had skillfully gathered for their court action and in his opinion they had the ring of truth. He was not hopeful about Katharina's case and advised Johannes to contact Dr. Bidenbach, one of the duke's legal advisers and a man who was personally close to the Kepler family.

A few months later, on June 10, 1619, Besold met with the Kepler family and their advocates and laid out for them all the behind-the-scenes manipulations that Einhorn had engaged in. Finally, the Keplers began to understand the power of the forces arrayed against them, but it was too late. If they had truly understood what was happening a year before, then they might have taken steps to neutralize Einhorn, but by the time they had figured it all out, Einhorn and the Reinbolds had already curried too much favor in high places, and their whispering campaigns had effectively turned the populace against Katharina. People were already beginning to remember old illnesses, twinges of sciatica and bouts of intestinal parasites, and were assuring each other that Katharina had had a hand in it all. Two months later, on August 16, Dr. Phillip Jacob Weyhenmayer, Jakob Reinbold's advocate, walked into the courtroom and read into the record all forty-nine of the "most terrible and shameful articles":

> Accuser Jacob Reinbolden, in marriage and in the name of his dear housewife Ursula, versus Katharinam Kepplerin:
>
> 1. States in this behalf *(salva priori protestatione)* that it is true, first, that no person shall harm another in body or health in any way, and is forbidden under pain of punishment.
>
> 2. That it is also true that a person who inflicts pain upon another person is responsible and liable for all loss.
>
> 3. And then, thirdly, it is true that the Kepler woman, against the above laws not only took hold of Reinbold's wife but also caused harm.
>
> 4. Fourth, it is true that the Kepler woman in 1613 invited Reinbold's wife into her home and gave her a potion.
>
> 5. That a few footsteps later, Ursula fell ill.

6. And as a result, to this date, she endures unspeakable pain.

7. True that the same Ursula Reinbold has tried numerous proper remedies, at substantial cost to her, but nothing would help.

8. True that the Kepler woman confessed to the magistrate that she gave such a potion to the accuser.

9. True that the Kepler woman tried to hit another person when that person caught her in the unjust act of giving a drink to the glazier's wife.

10. True that Ursula could swear a corporeal oath in good conscience to God that such a potion caused her present unspeakable illness.

11. Then the following is also true and strongly apparent in keeping with the Kepler woman's fiendish misdeeds: The Kepler woman was raised by her aunt in Weil der Stadt, and that same woman was a witch and later burned for it.

12. True that the Kepler woman's own mother (after above fiend was burned and the girl returned to her parents in Ölttingen) was willing to overlook her daughter's upbringing as long as she hadn't learned the fiend's trade at Weil.

13. Further, it is true that the Kepler woman's own son Heinrich states that his mother is out of her mind, since she once rode a calf to its death and then wanted to prepare a roast for him from it.

14. And that this Heinrich declined and wanted to report his own mother and accuse her of witchcraft to the authorities.

15. True that she also wanted to seduce Schützenbastian's daughter to follow the trade of witchcraft, as well as . . .

16. . . . having told the same young woman that there is neither hell nor heaven, but that if one dies, everything would be over, just as it is with the senseless beasts.

17. Turned out to be true then, that the Kepler woman said, if anyone practices the fiend's trade, then that person would have a good life and possess many wonderful things.

18. When the witches gather, they would have the best time, eat the best food and drink the best drink, and enjoy so much voluptuous pleasure that no one could praise it enough.

19. True, that the accused asked the grave digger for the skull of her deceased father, informing him that she wanted to make a drinking cup out of it, and then have the cup bound in silver; when, however, the grave digger let her know that he was not allowed to do this, that he would have to get permission from his superiors, the Kepler woman told him to let it be.

20. True that the Kepler woman injured the brick maker's wife to such an extent that the woman is unable to work even now.

21. True that she gave the schoolmaster [Beutelsbacher], and also another person [the wife of Bastian Meyer], a potion in a pewter cup, so that he became lame and is even now unable to work, and that the other person fell ill and passed away from it.[4]

22. And it is just as true that the infamous Kepler woman gave a similar potion to the barber's apprentice (who had cut her hair), so that he fell seriously ill at once.

23. True that she touched the two-year-old calf of Haussbeckhen, so that it died.

24. No less true that she rode the cow of Michael Stahl, the local saddler, at midnight, so that the cow kicked and raged as if mad and if he hadn't helped the animal right then and infused her, she would have died the same night.

25. True that the butcher Stoffel was injured by the Kepler woman on his foot at the market square, where he sold his meat, and that he has endured great pain for some time.

26. And in turn she remedied his pain.[5]

27. True that she gave the wife of Guldinmann a basket of herbs which, when given to the livestock, made them act as if they wanted to climb the walls.

28. Further true that the above mentioned livestock started to kick and their throats swelled.

29. True that she touched the old brick maker Görge Bretzern's two pigs, so that the same also started kicking and climbing the walls until they finally died.

30. True that during the time the Kepler woman was preparing her slander case, she injured Jörg Haller's girl on her arm so that the girl endured much pain.

31. True that Haller reported her accordingly to the magistrate.

32. True that the magistrate, after listening to the Kepler woman enough said she could not produce a reason for doing so [hitting the Haller girl], which had the ring of truth.

33. That the magistrate hereby advised the Kepler woman that he could not keep quiet about the criminal process and was obliged under oath and office to report the same.

34. True, that the accused ardently implored the magistrate to spare her and not forward a report to the chancellery.

35. As was no less true that the Kepler woman promised to give the magistrate a cup if he would refrain from reporting her.

36. As was further true that the Kepler woman's children tried to corrupt him and keep him away from the truth, but . . .

37. . . . the magistrate accepted neither gift nor offering, but included these things in his report to the chancellery.

38. True that afterward the Kepler woman did not leave to walk home, but walked directly from the court house out of the upper town gate.

39. True that due to the humble reports of the local magistrate, and no less the magistrate of Stuttgart, the court ordered her arrest.

40. As the accused was advised accordingly by her children, it is true that she moved out of the duchy to Austria.

41. And although it is true that the accused party requested judicialiter (by law) to state that she is innocent of any actions causing the unspeakable suffering of Ursula Reinbold, she should still appear before the court in person and defend her misdeeds.

42. True that she never came forward, but ran away and let her kin handle her affairs even to this day.

43. And that by her running away the Kepler woman gave strong evidence that she caused bodily harm to the accuser Ursula.

44. True that the accused after her escape from Leonberg also came to Stuttgart, attacked a young girl, born in Gebersheim, on the open road, so that same was injured to such a degree that she endured much pain for a long time.

45. True, that even some of the Kepler woman's friends consider her a witch and have knowledge of several wicked acts.

46. That she chased her husband out of the house (undoubtedly with her witchly deeds), and thereafter he perished miserably in the war.

47. True that she touched two local children, who also died.

48. True that the general consensus in Leonberg is that the Kepler woman injured the accuser Ursula and caused her the reported bodily injury.

49. True that the accuser Ursula would much rather pay a thousand florins than be subjected to the unspeakable agony she endures daily, and at every moment.

Some of the events reported in these accusations actually happened. Katharina Kepler's aunt had indeed been tried and burned as a witch. Katharina was an amateur apothecary, an herbalist who gathered flowers from the fields and boiled up potions and healing tonics for all occasions. She was not unusual in this, because about half the women in town did the same. She was unusual in that she pushed her concoctions on just about every visitor who happened by. Moreover, it is a real possibility that some of her potions might have been breeding grounds for *Escherichia coli* and perhaps *Clostridium botulinum,* maybe even the virus that causes poliomyelitis. It is also true that Katharina requested her father's skull, so she could turn it into a drinking cup, which she wanted to give to Johannes. She did offer a bribe to Einhorn, and she did leave town at the behest of her children. The rest was codswallop—half vindictive rumor and half superstitious fantasy. The accusations about riding calves to death were part of the folk tradition on witches; if Katharina had been a witch, then midnight death rides on people's livestock would have been a requirement according to the tradition.

The witness accounts bear this out. Jacob Koch's bits of hearsay about what Heinrich Kepler, the unfortunate son, may or may not have said, along with other bits of hearsay about what the saddler told him about noise in his barn late at night are typical:

Jacob Koch, citizen of Leonberg (forty years old):

Some time ago (day and year cannot be recalled by witness), witness ran into Heinrich, the Kepler woman's son, in Palin Theuerer's shop. The men asked each other about how things stood (because at the time both of them were somewhat ill). Heinrich then said that he was the Kaiser's messenger and thought that his condition may have come from the Keltin.

So, the witness said, he personally has it quite good, since his mother could care for him. Kepler then said that his own mother might care for him as well, but she was leading such a scandalous life that he had half a mind to move to his sister's house in Heumaden, and then, when he had returned [to Leonberg], he would report his mother himself. The witness did not recall anything about a calf.

After interrogation: If he spoke out of habit, then one should accept same from the witness out of habit.[6] He knew from the common talk that Heinrich Kepler had said that his mother had once ridden a calf to death and then offered to make him a meal from it. . . .

Also, witness said that Michael Stahl, the saddler, was at Martin Weishaubten's house, where he told the witness that he thought that the Kepler woman had once caused a ruckus in his barn. It happened at eleven o'clock at night, just as the watchman called out the hour—there was such a commotion in his barn, back by the shed, so that he called out to the watchman. His own mother, who had already taken to her bed, lamented. She was afraid and did not want to rise to see about the noise, so he [the saddler] finally worked up the courage to go into the barn himself. There the cows were unruly and angry. He then called his neighbor Hanns Nestler to help him, and the man did him a favor and came (despite his somewhat weak health). Hanns Nestler then asked the saddler and his mother who they thought had caused this disturbance, but they didn't know.

It is amazing to modern readers that most people in the seventeenth century considered such testimony to be quite reasonable when it came to witchcraft trials, for such cases formed a category all their own. If an accused woman—and almost three-quarters of all cases were against women—did not have powerful political influence, she could easily end up being roasted alive.

Johannes, who was still off in Linz, was the most powerful support old Katharina had, but he was less influential than he might have been. His own reputation in the duchy was not very clean. During his struggles with Pastor Hitzler, Kepler had written to his old teacher Matthias Hafenreffer, the theologian at Tübingen, asking him to intercede on his behalf. In typical Kepler fashion, he tried to explain himself, passionately outlining his theological ideas as well as his personal reasons for refusing to sign the Formula of Concord. Behind it all was his desire for reconciliation between the different strands of Christianity, his fear of the coming war, and his belief that God's will was always a force for peace, for harmony between people, even as it was among the stars.

Hafenreffer exchanged two letters with Kepler, and then showed Kepler's letters to the rest of the theological faculty, and from there they went to the consistory. Their response was final. "Either give up your errors, your false fantasies, and embrace God's truth with a humble faith, or keep away from all fellowship with us, with our church, and with our creed." On July 31, 1619, Kepler was excommunicated not just from his local community, but from the Lutheran church. This did not mean that he believed himself to be any less Lutheran, for later, when pressed by the Counter-Reformation, he refused to budge and had to flee Linz, just as he had left Graz rather than convert. It just meant that the Württemberg Lutherans did not want him.

Would Einhorn have had the freedom to persecute Katharina had Johannes obediently signed the Formula of Concord? Perhaps not. He was, after all, merely the lower magistrate of one town in one corner of the duchy. His association with the duke was largely through Kräutlin, the duke's barber and a crony of the young prince. Hidden in the shadows behind Luther Einhorn, therefore, may well have been Matthias Hafenreffer, of the theological faculty of Tübingen, and the Stuttgart consistory, and the sufferings visited upon a half-mad old woman may well have been a punishment for the sins of her famous son.[7]

The duke's consistory had accused Kepler of heresy. Could not his mother also be a witch? How great a jump was there from the first to the second? The duke, of course, had to contend with Kepler's fame throughout Europe. How would it look if he had arrested so famous a mathematician, a friend to dukes and barons throughout Germany, Protestant as well as Catholic, the personal mathematician of three Habsburg emperors, on heresy charges? Kepler himself was nearly invulnerable, but Katharina was another matter. They could go after her, and no one would complain. After all, witchcraft trials had been occurring all over Germany for several hundred years, in Protestant and Catholic territories alike. What was one more old woman?

Most of all, what Katharina Kepler had against her, however, was the fear of shadows, fear that blocked people's minds and strangled their sense of justice. So many people suffered so many mysterious illnesses. Children died unexplained. Cattle went mad. There had to be a reason. More than a reason—there had to be a conspiracy, a conscious group of secret malefactors out for ruin. Just as alienation envelopes people of our time, fear choked the people of the sixteenth and seventeenth centuries. Nighttime visions had plagued Albrecht Dürer all his life. His prayers were screams for help, for the terror of the night, of demons, and of eventual damnation twisted his sleep. So too for the good citizens of Leonberg and the duchy of Württemberg. They were victims of their own imaginations and, tragically, odd little women such as Katharina Kepler paid the price.

In some ways as well, the little town of Leonberg was a Freudian Disneyland. Upwelling sexual fantasies, rejected as sin and too horrible to admit, still bubbled away like mud pots and sometimes ended up in witch-craft testimony:

Dorothea Hanns Klebling, hunter's wife, about thirty-three
years old:
Five years ago, Barbara, the daughter of Schützenbastian, was at
the witness's home for some sewing and told her that she had also
been to the Kepler woman's house for sewing and that the Kepler
woman had asked her to stay the night. Around midnight, the
Kepler woman got out of bed and walked around the room. The
seamstress asked her why she did that and would not lie in her bed.
The Kepler woman then asked her if she might have a desire to become a witch. There would be much joy and fun in it, the Kepler
woman said, where otherwise this world offered neither joy nor
comfort. To this the girl answered that if one had joy and fun for a
long time in this world, one would have to pay for it eternally. The
Kepler woman, however, disagreed. She said there is no such thing
as eternal life, but when a human dies, he dies just as the common
beast. To that the girl responded that the clergy preach that whoever believes and is baptized shall be blessed, but whoever does not
believe, shall be damned. To that the Kepler woman said the reason
we have clergymen is so that we feel safe walking down the
street. . . .
She did not hear anything about a Bülen [possibly a young boy],
but heard that [the Kepler woman] wanted to give the girl [Barbara]
a man so that she would have joy and voluptuous pleasure all the
time.

The forty-nine accusations were based on witness accounts like these, accounts that were hearsay or that waffled or admitted to an ignorance of the facts, but they were read into the record nonetheless. Part of the prob-lem was Katharina herself, for when she was confronted with these wit-ness accounts, her memory failed her and she confused names and dates. She was on the shy side of seventy-five years old and utterly uneducated. It is not really surprising that she could not remember. It is significant, however, that what had started off as gossip and what was admitted to be gossip by some of the witnesses had magically become official court record. What had started off as a dispute between two families and two old friends had become a fight to the death. Leonberg had already burned five other women for witchcraft. Why not one more?

The slander case against Ursula Reinbold had been filed in 1615, four years before, and although the court had gathered witnesses and heard testimony, nothing ever happened. The case simply disappeared into the mist. Not so with the witchcraft trial. On September 9, 1619, the court ordered the trial against Katharina Kepler to go forward and set the date for November 10, 1619. The magistrate examined his twenty-two witnesses, took down their stories, and entered them into the record. But what survives is not verbatim. There are no actual transcripts of what the witnesses said, only Einhorn's summaries. Luther Einhorn, therefore, led the dance. What the witnesses said, believed, or felt about Katharina and how they might have interpreted their experiences were lost, and all that was left was the word of Luther Einhorn.

On December 23, 1619, the court assembled a protocol and began the witchcraft trial. A few weeks later, on January 23, Einhorn sent the protocol, gathered under the directives of April 3 and September 23, 1617, to the duke's chancellery. The chancellery then presented the protocol to the duke, but he found the whole matter distasteful and let the case sit for a time. It is possible that the duke perceived that the charges were trumped up, but, like all men of power, he was caught in a web of his own bureaucratic creation. On February 11, the Reinbold faction contacted him asking for a decision. The Keplers, meanwhile, were still waiting for some closure on Katharina's slander case against the Reinbolds, closure that would never arrive. On March 20, 1620, Johannes Kepler appealed to the duke personally and asked that his mother's trial against Ursula Reinbold be expedited and brought to court as soon as possible. Instead, the duke passed the issue on to the high court, which voted in favor of the Reinbolds and ordered Katharina's arrest.

The days in August around Stuttgart are warm and humid. The low trees along the riversides sip water as if through straws and puff it into the air. Morning rises with a thin mist, which evaporates as the day heats up and becomes invisible, though you can still feel it on your skin. Some days promise cool breezes off the Alps. Other days promise heat and still more flaccid air. On hot days, everything wilts, and people take what time they can to sit in the shade somewhere and drink cool wine. August is a time for peace and for children laughing, with their feet splashing in the streams. With evening, the heat relents, and the air softens. Crickets in the fields. Frogs in the streams. The farms and towns gradually settle into sleep. It was on such an August night in Heumaden that Marx Waltter, the Stuttgart magistrate, appeared at the house of Pastor Georg Binder to arrest his mother-in-law, Katharina Kepler. He arrived long past midnight, in the deep morning, August 7, 1620, when it was still dark but with a hint of sky, long before the town was stirring, before anyone could see what was happening. Marx Waltter had come with a detachment of armed guards, large men with swords, helmets, and grim beards. They pounded on the door, awakening the household, and shouldered their way in. Waltter announced the court order to Pastor Binder, then set about his business, in a hurry to get away. He could not have been proud of what he was doing, or he would not have come in the middle of the night. The men roughly shook Katharina awake, then stuffed her still half asleep into a wooden chest, and carried both the old woman and the chest out of town before anyone noticed.

Because of the general fear of witches, the men may have been as afraid of Katharina as she was of them. Who can say what a witch can do? Katharina sat alone in a jail cell for the next four days. Perhaps at night, when the jailers heard her weeping, they thought that she was weeping over her sins. The presumption of guilt was often so strong in witchcraft trials because the fear of witches was so great—each time they struggled, each time they complained, each time they protested their innocence, it was another proof of their cleverness. They were presumed guilty because the crime as imagined, as with child molesters and serial killers, was so horrible. The point of leaving her in jail for four days, barely attended, alone with her thoughts, was to accelerate her fear, to incite her imagination and inject terror.

On August 11, she was led to the courthouse, possibly still in chains, where the magistrate read her the charges, questioned her fiercely, and confronted her with her accusers and the accounts of witnesses. By law, she would have had the support of an advocate and of her family as well as the support of the war magistrates, since she was a widow, but Luther Einhorn was not particularly fussy about the niceties of law.

Not surprisingly, given the weight of prejudice against her, on August 18, the Stuttgart *Oberrat*, the chief judge, indicted Katharina: "The accused is once more and in all seriousness with warning of the executioners to be examined. If she will not confess and will not speak, all evidence shall be collected, and the accused charged *ad torturam*. If it is brought to light then, to execute the torture and report her confession, i.e., her statement after enduring the torture."

At this point, Christoph, the man whose business dealings with Ursula Reinbold had started all the trouble, suddenly remembered his reputation as an upright citizen and distanced himself from the case. Pastor Binder joined him in this. After all, how would it look for a man of God to be associated with a known witch? On August 26, Christoph asked the town scribe, Werner Feucht, to write a letter for him to the duke, begging the duke to transfer his mother's case to another town, any other town. Three days later, his mother was carted off to Güglingen, where they chained Katharina, now broken and confused, in a musty dark prison cell in the Güglingen Tower.

Margaretha had remained loyal to her mother, however. Soon after the arrest, she wrote to Johannes in Linz and described the entire situation to him. He wrote at once to the Duke of Württemberg and begged him in the name of his "God-given and natural right" to protect his mother, asking him to hold off Katharina's actual trial until he could arrive. The *Oberrat*'s decision was an indictment, but the trial still remained. Soon after, Johannes Kepler left his home in Austria and traveled as fast as he could to his mother's side, to a nasty wet prison cell in Güglingen.

AFTER THE RIOTS HAD DIED DOWN in Prague, the Estates gathered to form an interim government. They had revolted, so they said, to protect their rightful king, Ferdinand, from the insidious influence of the Jesuits. In truth, however, they rebelled in order to overthrow Habsburg power. They elected thirty directors, with ten for the barons, ten for the knights, and ten for the towns. The leadership of the directorate was radical, though the rank and file was moderate. As it turned out, there were moderates and radicals on both sides. The Protestant radicals wanted war, and actively sought alliances with other Protestant nations. The moderate Protestants still believed that they could negotiate and achieve their ends through peace. Emperor Matthias was the moderate on the Catholic side, because he too hoped for negotiations that could avoid a war. Archduke Ferdinand, now king of Bohemia, and the Catholic faction, however, wanted to crush the Protestants wherever they found them, to leave none standing, and to negotiate only with the dead.

Suddenly, Emperor Matthias, by then an old man, died in 1619, just after the defenestration. The most obvious choice for emperor was Arch-Duke Ferdinand, an energetic man of forty who was intelligent, but not to excess, cheerful to those who knew him, and possibly the most virulent anti-Protestant leader in eastern Europe. His election was fairly certain. Even the Protestant electors voted for him, though there was some doubt about Friedrich V of the Palatinate, but in the end even Friedrich voted for him. In fact, just hours after the electors had met in Frankfurt to confirm Ferdinand's election, Ferdinand learned that his own people in Prague had rebelled and rejected him as their king and that they had chosen the same Friedrich of the Palatinate to replace him.

The moderate Protestant faction had watched their hope of negotiation evaporate with the death of Matthias, and they began to gear up for war. It was not a particularly unified rebellion, however. The members of the Estates disliked and distrusted one another and fought over everything. They had no army to speak of and few allies. The Habsburgs, on the other hand, had a deep bench and could call upon the support of Spain, Bavaria, Poland, Tuscany, and even Lutheran Saxony. The Prague Estates, who were new at this game and had not yet established a network of support, approached England, Holland, the Italian Piedmont, and the Republic of Venice, asking for whatever military support they could get, and if not that, then loans to help them raise a local militia. But the Estates were outclassed. They were up against the most successful imperial dynasty in the history of Europe, a dynasty that had lasted longer than the Caesars, the Claudians, the Stewarts, and the Tudors. For hundreds of years, that single family had understood power better than any other on the continent. They knew how to use it and, as a lot, they were mad enough to use it without conscience or restraint. The last two emperors, Rudolf and Matthias, were indecisive and not very effective. That was not true of Ferdinand, however. He had a simple mind, given to seeing the world in blacks and whites, and he had the energy to act. He was, therefore, a dangerous man.

Soon after his election, Ferdinand removed Cardinal Khlesl from his position as the emperor's chief political adviser. This was the same Cardinal Khlesl who had written the notorious response to the Protestant Estates. But that was not why Ferdinand removed him and ordered him confined. The cardinal, who was more intelligent than the new emperor, had advised Ferdinand to negotiate with the Protestants in order to avoid war and because it was the smart political move. But Ferdinand was an ideologue and did not want to make the smart political move, so he sent the good cardinal to his room.[8]

Ultimately, the Protestant Estates were not very successful in gathering international support, and so in the summer of 1619 the Estates in Bohemia and Upper and Lower Austria gathered into a confederation that declared itself to be the enemy of the Habsburgs. It officially banished the Jesuits from the city of Prague forever and gathered a war chest from the sale of confiscated Catholic properties. Then, with great ritual and the blaring of trumpets, it officially threw Ferdinand II off the throne of Bohemia and sent word of its offer of the throne to the young elector palatine, Friedrich, the husband of Elizabeth, the daughter of King James I of England. Friedrich was in Frankfurt at the time, voting for Ferdinand as emperor. As the prince-elector of the Palatinate, Friedrich was the most celebrated Protestant ruler on the continent, and no doubt the offer of the crown to

Friedrich was done with the hope of bringing along more international support.

Friedrich certainly looked the part of the serene monarch. He was a slim, good-looking, elegant young man of twenty-four with a grave and serious demeanor, but one who had no head for war. They said that he would have made a better gardener. He was not particularly intelligent either, though he was not always wise enough to know it. The best that could be said for him was that he dutifully followed good advice when he got it. Had he been born in a time of peace, he would have done well. His education by the French Calvinists had made him into a philosopher, a rhetorician, and a good gardener. He was fascinated with mechanical fountains of any kind. He was a spoiled child married to a spoiled child, neither of whom had any idea about the world they lived in. His wife, Elizabeth, was probably better suited for rule, but she was too busy giving birth and raising her children to take control. Elizabeth, who was a plain woman, had once been courted by numerous princes and kings from all over Europe, from the French dauphin to Gustavus Adolphus when he was the crown prince of Sweden. Understandably, she did not like Ferdinand very much and said that he would make a bad emperor, and so she acquiesced when Friedrich, taken by a sense of divine call, decided to accept the Bohemian offer.

At the end of September 1619, they left Heidelberg accompanied by a thousand soldiers, several hundred servants, and over a hundred and fifty wagons full of goods. As a royal couple, they loved hunting, entertaining, and riding in the forests accompanied by witty young men and women, most of whom spoke only French. The fact that they were going to rule the Czech people, while none of them spoke Czech, did not seem to register. And neither did the fact that they might have to go into battle against the implacable Habsburgs. When they arrived at the Bohemian border, they were met by representatives of the Bohemian Estates, who welcomed them with a great deal of ceremony and at least several Latin speeches. Then, on October 21, they entered Prague to a great celebration, cheering crowds, adoring faces, and troops in review, and there they were crowned king and queen of Bohemia in two different ceremonies, Friedrich first and then Elizabeth, at St. Vitus Cathedral, one of the great religious centers of Habsburg power.

And while Friedrich and Elizabeth held court in the Prague Castle, receiving ambassadors and throwing gorgeous parties, the storm clouds gathered around them. The rebellion that had given them the crown was fraying, because none of the nobles trusted one another enough to levy the taxes that they needed to raise an army. Moreover, for all the grand celebration and the words of welcome, the new king and queen of Bohemia were outsiders, and though they seemed to be oblivious to it, they were treated as outsiders by the very nobility who had hired them in the first place. When Friedrich approached the Prague burghers for a personal loan so that he could gather an army, they refused him, even though it was their own necks on the line as well.

No doubt Friedrich and Elizabeth had no sense of Slavic culture or of the Slavic version of Protestantism. They were both Calvinists, trained in the French style, and could not completely understand the Bohemian Brethren. Elizabeth did not make much of an impression either, with her lavish parties, her French entourage, her fashionable hairstyles, and her plunging necklines. Moreover, it was becoming increasingly apparent to the rebels that Friedrich and Elizabeth had been a bad choice. They were largely chosen with the hope that they would bring along the support of other Protestant rulers, especially James I of England, Elizabeth's father. But Friedrich had never read his father-in-law's signals correctly. King James may have been willing to marry off his daughter to the young prince, to spend a great deal of money on the wedding, and to give them a glorious send-off, but he did not want to involve England in the military quagmire that was central Europe. James could see that fighting the Protestant cause in Germany was a no-win situation, because the Habsburgs were too strong, and besides, the Church of England was never extraordinarily Protestant anyway. Going to war with Ferdinand meant going to war with Spain, and possibly even with France, a bit more than little England was willing to take on.

The rest of the Protestant rulers in Europe were not overly eager to jump onto Friedrich's bandwagon either. The Bohemian nobles had set a bad precedent. The idea of an absolute monarch with the divine right to rule was firmly entrenched in many European courts. The age of constitutional government had not yet arrived, and so few among the royalty of Europe wanted to come to the aid of what looked like an anti-imperial revolution. The Bohemian nobles were not republicans by any means, however, but they had been willing to throw out the king that they had accepted only two years before and to choose a man who was more to their liking. Therefore, feeling the ground shift beneath their thrones, most of the Protestant kings stayed neutral, and although many of them were willing to raise a glass to Friedrich, they weren't willing to do much more.

On the other hand, Ferdinand had to rely on allies to provide his own army. Most of the nobles in the Habsburg lands in Austria and in their attending provincial estates were Protestants and had already sided with the Bohemian rebels. Moreover, he did not have the kind of funding he needed in order to raise a mercenary army. But he was the emperor, and through the judicious doling out of imperial privileges and, of course, of the bits and pieces of Friedrich's old territory of the Palatinate, he could make alliances. Ferdinand gave Maximilian, the Duke of Bavaria and an old Counter-Reformation partner, Friedrich's electoral vote along with a chunk of the Palatinate. The rest of the Palatinate lands went to Philip III of Spain, Ferdinand's cousin, who also sent an army to help. Oddly enough, after being given the promise of the Habsburg territories in Lusatia, the Lutheran elector of Saxony also joined the emperor's side and sent his own army into Bohemia. This showed the fundamental political weakness of the Protestant cause.

On October 21, 1620, Friedrich and Elizabeth were crowned king and queen of Bohemia, but by early November, two Catholic armies and one Protestant Lutheran army had entered Bohemia and were closing on the city of Prague. Suddenly, Friedrich and Elizabeth realized that they needed to be elsewhere. They packed what goods they could into a coach and took along what soldiers and servants they could muster. As they were beginning to set out, Elizabeth suddenly remembered that she had left the baby behind in the castle and sent Baron Christopher Dohna to go and fetch the child. The baron came running back with the infant, wrapped in a bundle, and handed him through the window as the coach was moving toward the gate. Somehow they managed to evade the three armies and to escape Bohemia altogether, but they could not return to Heidelberg, for by taking the crown of Bohemia and thereby usurping Habsburg privilege, Friedrich had made himself an outcast. His old principality of the Palatinate was gone, doled out to the rulers of the invading armies. His home was gone. His gardens were gone. His mechanical fountains were gone. Instead, he and Elizabeth and the children, their few servants, and a few soldiers made their way to Holland, first traveling to Kunstin and then on to Berlin, and from there across Germany to the Dutch lands. There they set up a court in exile. Friedrich retreated further into his fantasy world, styling himself the king of Bohemia until he died in 1632, two years after Johannes Kepler. Elizabeth stayed on in Holland after Friedrich's death and returned to England only one year before she died herself.

⟨ຓຓຓ⟩

ON AUGUST 29, they moved Katharina to Güglingen in response to Christoph's petition to the duke. When it had authorized the transfer, the court also ordered that Katharina be accompanied by a guard. As the two wardens looked on, the jailer shackled the old woman to an iron band with chains, and from that point on, as long as the guards were with her, they gouged her for money, eating their way through her small estate and burning an excessive amount of firewood to keep themselves warm.

The next day Einhorn made it clear that he wanted to examine the accused once again, since he had made this a part of this complaint *ad torturam,* which he had sent along with Katharina to Güglingen. The term *ad torturam* turned the complaint into what we might call a capital crime. It carried with it not only the possibility of torture, but also of execution. Five days later, on September 4, the Güglingen magistrate, Johann Ulrich Aulber, presented the capital charge to the judges of the capital court, asked permission to hold the Kepler woman, and also asked to be given the power to force a reply from her.

On September 26, 1620, Kepler arrived in Güglingen to find his mother sitting alone in a cold, wet, drafty, dark, and nearly unheated prison cell of the Güglingen Tower. The wardens had pressed as much money as they could out of her and heated their own station excessively at her expense. He was furious as well as terrified for her and wrote to the duke at once, appealing for his mercy in the name of God and asking to have her transferred to the house of the city magistrate, where she could at least be warm. She was an old woman after all, in a "dilapidated condition." A broken, dispirited old woman like Katharina was not much of a threat and did not need a guard. The magistrate could keep her at his house "at her own, as minimal as possible, expense." Because of her gender and her old age, guards would not be unnecessary. "To keep the prisoner warm, especially since she has not been convicted and continues to unshakably plead her innocence, but that you mercifully suggest to our adversary, since our mother is kept in such a cold shelter, and was sent to Güglingen in the first place, or perhaps *priori casu,* and not insured with my security (bond), that the guard look out for himself, and that in the interim he should pay [for heat] out of his own pocket."

Kepler also made contact with two of his old friends and advocates on the Tübingen law faculty and asked them to help in her defense. The defense moved into high gear, mounting argument upon argument against the prosecution's case. Johannes was prominent enough in this process that the scribe of Güglingen wrote in the court transcripts that "unfortunately, the prisoner appears with the support of her son Johannes Kepler, the mathematician." Johannes wanted to make sure that his mother got the best defense possible, so he and Katharina's advocate, Johannes Rüff, presented the defense's case in written form. This was expensive and slow going, but it was more certain than courtroom oratory, which could not be referred to later in any exact way.

Suddenly feeling outclassed, the magistrate of Güglingen, Johann Ulrich Aulber, told the court that he could not lead the prosecution without a legal adviser of his own and without additional information. On November 4, the court assigned the duke's own chancery advocate, Hieronymous

Gabelkover, to take over the prosecution of the case and sent the files to him from Güglingen.

Christoph Kepler then accused Johannes of spending too much money on their mother's defense, saying that he was unnecessarily extending the trial through his involvement and opposition. Despairing, Christoph had decided to let the Reinbolds have their way with his mother, perhaps as a way of protecting his own reputation and perhaps, if looked at a little less charitably, as a way to protect his own portion of the inheritance. The Kepler children argued fiercely about the financial situation, about how much money was being paid out for the trial, for the advocate, and for their mother's upkeep. Johannes and Margaretha were willing to protect their mother at all costs, while Christoph had decided that the best thing he could do was to protect the family fortune. Meanwhile, putting his own paddle into the water, Jakob Reinbold requested 1,000 gulden to pay for his dear wife's pain and suffering and asked that this be paid out independently of the outcome of the capital trial, so that Katharina would have to pay even if she were declared innocent.

Suddenly, noting how the wardens continued to consume Katharina's estate, Jakob Reinbold, too worried about her financial condition. In the name of Christian charity, of course. The two guards were eating their way through Katharina's money so fast that by the time the trial ended, there might not be anything left to comfort Reinbold for his trouble. Without any shame at all, he wrote a letter to the duke's chancery, asking the duke that "in the name of God's mercy" he protect Katharina Kepler's funds, saying that she had already spent several hundred gulden since her arrest for food, shelter, heat, and her defense, and hinting that her money was disappearing rather too fast for his comfort. Suddenly, on November 23, out of the blue the Güglingen magistrate, working with the prosecution, included thirty-five new articles in his complaint. It would be a long trial indeed.

On November 24, the high court, after being pressured by Johannes Kepler, denied Reinbold's motion. No doubt even though Einhorn might have enjoyed a special line of communication with the duke's government and had a close relationship with some of the duke's servants, the council was doing its best to follow the law under the circumstances. In witchcraft trials, there was always fear, so that even though the law might be plain, the results of such cases were never that simple. There was always a subterranean assumption of guilt by the very existence of an accusation. After another letter from Johannes Kepler, on December 4, the duke's chancery in Stuttgart decided that the mayor and the court of Güglingen should bear half or at least a third of the cost of heating Katharina's cell. After all, if they were going to insist on wardens, then the wardens would have to be kept warm, and Katharina Kepler should not have to pay for that. Suddenly the wardens decided that they did not need that much heat anyway. One more letter from Johannes, and they reduced her guard by half. But they still expected Katharina to pay for her own upkeep. The jail provided about thirty-five cents a day for a single meal, but that only included a hunk of bread and a bowl of watery broth.[9] Anything beyond that—milk, eggs, meat, beer, or wine—the deputy charged dearly for, especially for any kind of meat, which Katharina "because of her missing teeth could enjoy little of." On top of that, Johannes Kepler had to pay a few coins under the table to keep the magistrate in good humor.

Finally, ten weeks after Katharina's arrest, on December 11, 1620, Katharina Kepler went to trial. On January 8, 1621, the court took more witness examinations in Leonberg under the direction of Luther Einhorn. Even more witness examinations took place in Güglingen on January 13, and on February 23 they published the court protocol. At this point, on April 2, 1621, the court permitted Katharina's advocate to present their defense. Finally, in May, the defense presented its last arguments to the court, written as a legal brief of 126 pages. Kepler had written much of this himself, in his own handwriting, using his own arguments to counter the prosecution's arguments one by one. Gradually the tide began to turn.

On August 22, 1621, the defense presented the final version of their written arguments. The court then sent the trial documents to the Tübingen law faculty, as was normally done in important cases, for the final word. Christian Besold worked behind the scenes. With all that, however, with all Kepler's finely honed arguments, and all of Besold's authority and influence, Katharina Kepler was still convicted. The prosecution had asked for a sentence of "Cognition of Torture," or showing her the instruments and explaining their function. This was not a university lecture, but an exercise in terror.

On September 10, 1621, the judicial faculty of Tübingen ruled as follows: "Learning the detailed facts of torture, but only in the first degree, i.e., in the form of the so-called *territio verbalis,* where the accused is taken to the usual place where one faces the torturer and the latter, under dire warnings, shows his *instrumenta.*" Although the law faculty had accepted most of Kepler's arguments and agreed that there was not enough real evidence to warrant Katharina's death, there was still enough doubt in their minds that they could not acquit her. In the final analysis, the learned doctors too were afraid. At this point, Johannes wanted to get the sentence over with, so he pressured Aulber, who was inclined to hold on to Katharina as long as possible, to set the day of her ordeal. Aulber announced that the sentence would be carried out on September 27 at seven o'clock in the morning. According to the rules, the *Cleemeister,* the Swabian term for the *Wasenmeister,* the one who executes sick animals and buries criminals, needed to be present, but he did not show at the appointed time, so Katharina had to wait one more day.

Tormenting witches and sorcerers was a fine old tradition dating back to the Babylonians. The usual outcome was death by fire, generally by tying the convicted witch (or heretic, since the punishment was the same for both and sometimes the distinction between the two crimes was blurry) to a stake, piling faggots of wood all around, and setting it all on fire. Generally, out of Christian charity, the executioner strangled the witch before the flames could take her, but sometimes that did not work. All too often, the flames burned the executioner's hands, and he was not able to finish the strangling. This happened in the case of Catherine Hayes in London in 1726, where the poor woman screamed for half an hour before she died. In the case of Kathleen Cawches and her two young daughters, also in England, all three women were burned at the same time on the same pyre. Right in the middle of the burning, the younger daughter gave birth to a baby. Someone in the crowd pulled the baby from the fire and gave the child to the priests. After some deliberation among them—

the dean and the bailiffs, the jurists and the provost—the provost commanded that the child be thrown back on the fire, since the infant too was
tainted with heresy.[10] Similar stories occurred all too often on the continent as well. Certainly, on the day of her ordeal, they told Katharina quite
a few such stories in gruesome detail.

One of the favorite ways of determining the guilt of a witch was
through a trial by water, in which they took the accused to the river, tied
her hands and feet together so that she was bent over in a squat, and then
threw her in the water. This was an old Babylonian idea, noted in the
Code of Hammurabi. The idea was that the holy river would decide a person's fate, drowning the guilty and floating the innocent. Thus justice was
something even the waters of the river would attest to. However, the Assyrians turned this on its head. According to their rules, if those thrown in
floated, then the waters rejected them, and they were guilty of witchcraft.
If they sank, then the waters received them, and they were innocent. The
Assyrians had an odd sense of humor. Of course, the onlookers had to be
quick, or the innocents would drown before they could celebrate their ac-
quittal. This Assyrian tradition was the one that traveled into Europe
long before Christianity appeared, and Christian Europe, as it did with so
many other pagan traditions, merely clothed it in Christian robes. Often
when a witch was tried by water, the priest prayed over the water, recalled
details of the Christian tradition, and thanked God just before they
dunked the accused. One example from Germany:

> May omnipotent God, who did order baptism to be made by water,
> and did grant remission of sins to men through baptism: may He,
> through His mercy, decree a right judgment through that water. If,
> namely, thou art guilty in that matter, may the water which received
> thee in baptism not receive thee now; if, however, thou art innocent,
> may the water which received thee in baptism receive thee now.
> Through Christ our Lord.[11]

Of course, all the other usual methods of torture applied to witches as
well—branding with hot irons, the rack, the wheel, and so on, but one
method seemed peculiar to witchcraft trials, the ordeal by pricking. The

belief was that witches had numb spots on their bodies from the devilish growth of extra nipples used to suckle imps and demons of all sorts. These places on the witches body were said to be without feeling, insensible to pain or torment, and so one way to determine whether a woman was a witch was to tie her to a wall and, with sharp needles or sharpened metal rods, "prick" he accused witch over and over in every place imaginable, all in the hope of finding a place where she would not cry out in pain. Apparently, there was a class of witch-hunters called "prickers" who specialized in this form of torment, which had developed it into quite a science. Sometimes the pricking sessions would go on for days and, exhausted from screaming, the accused witches just stopped, too tired to cry out any longer. Which, of course, would be proof that they were indeed witches.

All of these things would have been shown to Katharina Kepler. All the brands, the hot pokers, the rack, the wheel. The executioner would have explained their use in detail and what would happen when someone was burned alive, which, he would have said, was only the beginning of her eternal damnation in hell. He would have explained trial by water and the waters of baptism. He would have explained pricking and showed her the prickers, showed her how they hooked the skin and tore the flesh in little bits. He would have explained everything, but not quietly, not rationally. After each demonstration, he and the magistrate, perhaps even the guards, would demand that she confess her sins, so that even if she could not escape punishment in this life, she would at least escape the wrath of God. Then they would start again and tell her more stories and make more threats, until the old woman was kneeling on the floor, gibbering with fear. Nine times out of ten, by that point the accused would say anything.

At seven o'clock the next morning, September 28, 1621, the executioner carried out the sentence and published the result. According to the Güglingen magistrate, the Kepler woman continued her denial, even after they showed her the *instrumenta*, even after they hounded her to confess. "They may do whatever they wish to me," she said after they read the court protocol and showed her the instruments. "Even if they wanted to pull one vein after the other out of my body, I would have nothing to confess." Then Katharina fell on her knees and prayed an "Our Father" to God, asking God to give a sign to show if she was indeed a witch or a demon or if she ever had her hand in witchcraft. "I know that God will bring the truth to light and will not take his holy ghost from me. The Reinbold woman had brought her illness with her from Ansback to Leonberg, and the schoolmaster's lameness happened because he jumped over a ditch. I know that God will punish the witnesses who brought this upon me, for I am experiencing violence and injustice every day."

The sentence had been carried out, but Katharina had not confessed. According to the verdict, nothing more could be done to her, so Aulber, the Güglingen magistrate, led her back to prison. On October 3, Duke Friedrich of Württemberg commanded that Frau Kepler be released from prison, since she had cleansed herself through her ordeal. The sentence had been carried out, and she was free to go. Einhorn and Aulber refused to let her go and, instead, delayed the final proceedings while holding Katharina in prison, even after the order of her release. It worked once, but this time the duke had less patience. The ducal chancery in Stuttgart finally informed Aulber that if he wanted to hold Katharina after October 7, then any expenses incurred for her care would come from his own pocket. Aulber released Katharina almost at once. From the day that Katharina Kepler was arrested in Heumaden to the day of her release, October 7, 1621, 425 days had passed.

Even with the trial over and the sentence passed, the Reinbold family continued their campaign against Katharina, calling her "evil trash." On November 15, Luther Einhorn, supported by the judges of Leonberg whom he dominated, wrote to the duke, requesting that he forbid the Kepler woman from living in Leonberg. They wanted her to live somewhere else. Einhorn proclaimed his deep concern for the welfare of the town, saying that the presence of the Kepler woman would be a source of terrible contention and continued struggle. He did not, of course, mention that he and his friends would be one of the causes of that contention. Einhorn was terrified that the slander case might finally come to light, bringing his own actions to light with it. Katharina certainly never returned to Leonberg, for aside from Einhorn's petition to the duke, she received word that if she ever showed her face in her hometown again, the people there would beat her to death.

Johannes Kepler, in response to Einhorn's petition, asked the duke to transfer the complaint to Tübingen or Cannstadt. The day before Katharina, destroyed by the long ordeal and weakened by age and abuse, finally died in Heumaden, at the home of her daughter Margaretha, the duke ordered Einhorn to deliver the files to Tübingen. The files, rather mysteriously, disappeared.

൏ඟඟ൏

*Linz. Beginning of 1619. I don't care about the accusation of
boastfulness, even though it has been made by those people
who condemn all that I have written about this matter as silli-
ness. It may manifest itself in words or in the way one con-
ducts his affairs, the scientifically ignorant, only half educated,
feel superior, and fool the people with titles and decorations.
Also the plebeian theologians, as they are called by Pico della
Mirandola. With people of every standing who truly appreciate wisdom, I can easily defend against this accusation by
showing the usefulness for my reader.*

*I know a woman who is very uneasy in her mind. Not only
does she not understand the field of science, which is not surprising for a woman, but she also upsets her entire surround-
ings and, regrettably, makes herself miserable. So to begin
with, we have the constant imagination of a pregnant mother
for medicinal lore, something she shared with her mother-in-
law, my grandmother, who was admirable because of her interest in this, as was her father before her. I received my
mother's build, more suited to studies than to other profes-
sions. Also, my parents didn't have much money, and we
owned no native soil, where I could have grown and stayed.
Finally, there were schools and certain circumstances where
the officials, in good-hearted manner, aided the boys who
were suited to study.*

To examine the secrets of nature, one needs a sharper mind and the talent for discovery, more than for any other profession and the studies that serve them. I want to note furthermore: Jupiter [appearing] in the center of heaven is responsible for my enjoyment of geometry, as it manifests itself in the physical world more than in abstract mathematics, therefore more in physics than in geometry. But when I speak about the success of my studies, what do I find in the heavens that could even remotely point to them?

The scholars confirm that I have examined, improved, or entirely completed some important areas of philosophy. But my stars were Copernicus and Tycho Brahe, without whose records everything I have brought to light so far would remain in darkness. The eminent Kaisers Rudolf and Matthias were my rulers, my refuge Austria above the Enns, the [current] Kaiser's home, and his Stände afforded freedom to me with unusual generosity, just for the asking. Here is the corner of the world where I retreated from the much too restless court, with the agreement of my Kaiser. Here is where I have worked through these years, as they started to lean toward the end of my life, on my harmonic work and on other projects at hand. The astrologist finds in vain, in naiveté, the reasons that I discovered the relationships between the celestial spaces in the year 1596, in the year 1604 the laws of optics, in the year 1618 the reasons why the eccentricity of every planet is just this size and not larger and not smaller, during the time in between the celestial physics including the appearances of the planetary motion, in addition to the true movements themselves, and finally the metaphysical basis for the influence of heaven on our lower world.

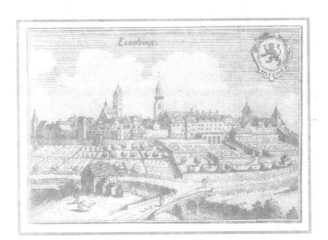

Kepler's handwriting

XIV

To Examine the Secrets of Nature

Where Kepler writes his Harmony of the World
as the Thirty Years' War heats up, and where
he is finally chased out of Linz,
his home for fourteen years.

NOVEMBER 8, 1620. Deep autumn, nearly winter. The air was cold with more than just the time of year. Thousands of men would die that afternoon in a little less than two hours. The noise would be unthinkable—booming cannons, war cries, the screams of dying horses. When it was over, blood soaked into the ground all across the hillside. The Battle of White Mountain lasted from noon until just before two o'clock in the afternoon. The French philosopher René Descartes, in his soldierly days, fought on the Catholic side and witnessed the entire battle. For such a short, decisive conflict, the two sides were fairly well matched. The Protestants occupied the top of White Mountain, a low hill just outside of Prague. They had the better tactical position, commanding what passed for the heights, with about twenty-one thousand men at arms, seven thousand fewer than the Catholics. But their men were tired after days of forced marches, while the Catholic troops were fresh. Moreover, the

Protestants had not adequately prepared for the battle. Their commander, Christian Anhalt, had ordered his men to dig in along the forward curve of the hill, but the men refused, saying they were too tired and that they were soldiers, not peasants. Digging was for farmers and, besides, their shovels were all broken. Instead, Anhalt ordered them to build up five breastworks, three to protect the artillery and two for the infantry.

The Protestants also chose to arrange their troops according to the Dutch fashion in three columns of three lines, interspersing companies of cavalry and companies of infantry. Current military theory claimed that this arrangement was more flexible, which was probably true, but it was also less stable, and in the end that proved fatal.

The Protestant armies were led by Field Marshall Prince Christian Anhalt-Bernstein, the personal adviser to Friedrich V and the man behind the formation of the Protestant Union. He was an ambitious man, fiercely anti-Catholic and anti-Habsburg, who worked tirelessly to advance the Protestant cause, a man of radical temperament who did much to inflame the hatred between Protestants and Catholics. Under him was Field Marshall Count Matthias Thurn, one of the leaders of the Second Defenestration of Prague, the man who had counseled action that day in the Bohemian chancellery building, and then later led an army of mercenaries against Vienna and the emperor. He was the torch that had lit the fuse of the Thirty Years' War, a fearless commander, some said rash, who had been the supreme military leader of the Protestant forces until Friedrich replaced him with Prince Christian Anhalt, a change he fiercely resented. From that point on, Count Thurn complained openly whenever he thought that Anhalt was showing hesitation of any kind.

The men who followed them were Czechs, Austrians, Germans, Moravians, and Hungarians, many of whom were veterans of campaigns against the emperor. Few of them had the fighting experience of the Catholics, however, especially the imperial troops, for they had been fighting the Turks for years. The Catholics also had much better equipment. However, because the Protestants held the high ground, Anhalt believed that the Catholics would avoid a direct attack and would likely wait until the next day anyway. He informed Friedrich of his opinion, and so the new king of

Bohemia, who was not much of a warrior anyway, left the field for the city on the morning of the battle and ordered up his breakfast.[1]

Their enemy, the Catholic League, was founded by Maximilian Wittelsbach, the Duke of Bavaria, a man who was said to be a financial genius, but who could sometimes be stingy and even petty, a man who supported the Catholic cause with all his heart, but at the same time worked constantly to undermine the Habsburgs in favor of his own family. In command of the Catholic army was Lieutenant General Count Jean T'Serclaus, Baron of Tilly, a Walloon from Flanders who had waged war in the pay of the Spanish for years. He was aggressive up to a point and believed that a general could shift the tides of history if he made the right decisions and won the right battles. When he felt that he had corralled enough of an advantage over the enemy, he would grab the opportunity and attack, and attack again. He was willing to break with military fashion when he saw the need and was rarely taken in by the latest theories. A rather conservative, pragmatic commander at heart, he preferred large, deep formations and overpowering force.

His immediate subordinate, Lieutenant General Charles Bonaventura de Longueval, Count of Bucquoy, was a different sort entirely. He supported Tilly, but was not the strategist his commander was; he relied more on personal courage and bravado, much like Jeb Stuart in relation to Robert E. Lee. He believed in the quick thrust, the lightning charge, in wearing down an enemy by feint and maneuver. Their soldiers were generally old campaigners—Bavarians, Germans, Frenchmen, Spanish, and Walloons, men who had fought the Dutch, the Hungarians, and the Turks. They were hard men, men used to death, and their password in the field was something no Protestant would ever say: "*Sancta Maria*," "Holy Mary."[2]

The day before the battle, the two armies faced one another and waited. The Catholic position was the poorer one, for they had to fight uphill all the way. Tilly had organized his army into two wings, with the imperial troops on the right and the army of the Catholic League on the left. They were arrayed in the less flexible but more traditional and more stable configuration of four large square phalanxes of infantry at the center, with columns of more mobile infantry and cavalry protecting their flanks. Anhalt had apparently misread Tilly right from the start. Tilly had been aggressive right off, setting the front of his formation a mere six hundred yards from the Protestant ranks. The Catholic commander had somehow judged that, despite the ground, his army was the stronger one, and he would press that advantage without mercy.

The armies were finally set in order and battle lines drawn, when night fell and the two sides encamped. By next morning, as Friedrich rode off for his breakfast, an autumn mist covered the battlefield. Men moved about in the fog, troops assembled. Suddenly, in the early morning light, Tilly sent soldiers to attack the Protestant outposts along the stream, pushing them back and quickly taking the bridge and a small village nearby named Rep. The Moravian colonel Stubenvoll rode as fast as he could to the Protestant commanders and begged them to counterattack as quickly as possible, to throw the Catholics back across the stream, so that the Protestants could hold on to every advantage they had. The enemy could not be allowed to hold the bridge. Count Thurn agreed, but Anhalt resisted, and finally did nothing. Colonel Stubenvoll, furious, stormed out of the tent. "We are all lost!" he shouted.

Meanwhile, the Catholics had their own problems. Bucquoy disagreed with Tilly's plan and wanted to outflank the Protestants instead. Both Tilly and the Duke of Bavaria, figuring that their strength was sufficient, resolved to attack Anhalt head on, but other commanders agreed with Bucquoy and preferred to nibble around the edges first. A report from one of their scouts, however, from a Colonel Lamotte, said that the Protestants were tired and disheartened and would fall if pressed, but Bucquoy would not change his mind. The disagreement intensified until the war council nearly split. Then a friar in attendance to the duke stood up and preached holy religion to the officers, stirring up their Catholic fire. They belonged to the true faith, while the Protestants were the vilest of heretics. Their side was righteous, and God was with them—that sort of thing. Swept away by their religious zeal, the officers ended the breach and devoted themselves to Count Tilly's bolder plan. They would attack.

The Catholics opened the battle at noon by firing an artillery barrage to soften up the Protestant positions. They had named their cannons the

"Twelve Apostles." But after only fifteen minutes, the apostles stopped spreading the word—the barrage suddenly ended. For a short time, the field was silent, and then Tilly gave the signal by firing all twelve apostles at once. The Catholic forces advanced. A troop of Walloons charged the Protestant lines and drove them into the lines behind them. Imperial Cossacks circled Anhalt's forces on the right, suddenly coming upon them from the rear. At this point, the coordination of the Protestant troops broke down. Count Thurn ordered a counterattack, and his men advanced toward the Catholic League forces, fired a volley, and then broke and ran. Prince Anhalt's son then led a charge against the left flank of the imperial troops and drove them back. Tilly, however, saw this coming and sent cavalry to flank the young prince and capture him.

More and more Protestant lines faltered. Catholic infantry quickly overran Anhalt's entrenchments and captured the artillery. Bit by bit, their army fell away and ran for the city. Anhalt and Thurn tried to rally their men, but it was too late. At this point, the battle was only an hour and a half old, but the Protestant forces had entirely collapsed. All that was left were a few brave pockets of resistance. Four thousand Protestant troops were either captured or killed in that short time, and the Bohemian rebellion was over. But the war had only begun.

After the battle, there was a quick, decisive slaughter, a quick jab to sweep away any resistance, and then the Catholics rested and prepared themselves for the assault on Prague.[3] But the assault never came. The troops' retreating to the city fortified the castle, and the leaders of the rebellion discussed their future into the night. Count Thurn and Johannes Kepler's old friend and patron from Linz, Georg Erasmus Tschernembl, argued that the city should be defended, that they had enough forces remaining to man Prague Castle and die to the last man if necessary. But the rest turned them down. The next morning, Friedrich and Elizabeth fled the castle in shame, nearly leaving their infant child behind. From that time on, Friedrich carried the nickname "Winter King," given to him by Catholic propagandists who satirized him in print: woodcuts showed a post boy riding across the country looking for the king, unable to find him. Most of the Protestant leaders, including Count Thurn, followed after. Soon the victorious imperial army entered the city, and the remaining Protestant leaders, expecting a general amnesty, surrendered the city to the Duke of Bavaria, who accepted it in the name of his emperor, Ferdinand II.[4]

◌〜〜◌

THE EXECUTIONERS OF EUROPE were a people set apart, a tribe of their own, for the sons of executioners married only daughters of executioners, and vice versa. Their position in society was as hereditary as that of kings. They were, like the tribe of Levi, a special people, with rights, privileges, and duties that were passed on from father to son. There was also a certain artfulness that went along with being an executioner, an artfulness and a swaggering machismo. The best executioners could kill with a single stroke and could go on all day long, killing and killing, all blood and showmanship, taking only a few short breaks in between while carrying out the king's vengeance. It was theater.

One myth about executioners holds that somehow they are connected with those they kill and at the last instant before the slaughter, they feel a certain tenderness for those whose heads they are about the lop off. This is unlikely, at least in most cases, though it has made for good poetry. Executioners were professional men who grew up in the life, like show-biz families, and for them to suffer a sudden bout of fellow feeling for those they executed, they would have had to deny everything they knew from childhood on, including the righteousness of their profession and the special position of their tribe. The king, everyone believed, had a right to vengeance; if he were denied that right, the glue that held the social order together would dissolve. For the executioner, the king's right to vengeance was the core meaning of his life, and not only of his life, but of the lives of his family and friends, even his ancestors, going back generations.

The emperor Ferdinand had defeated the Protestant nobles at White Mountain. Although everyone expected him to grant a general amnesty, or at least hoped and prayed he would, few were surprised when he chose vengeance over mercy. He had made promises, of course, to those Protestants who had not fled with the Winter King, but what are promises made to one's enemies? And so Ferdinand called upon his dark angel, the executioner of Prague, Jan Mydlár, to carry out his vengeance. Mydlár was one of the best and prided himself on his work. By all accounts, he slept well at night and, in his cups, boasted how this head fell and that head rolled across the floor.

On June 18, 1621, in spite of all the promises and hints of amnesty, Mydlár received the order from Ferdinand to begin construction of a gallows stage. The order came in the afternoon, and by that evening the executioner and his assistants, his carpenters and metal workers, had begun construction on the scaffold and theater—twenty-two paces long, twenty-two paces wide, and four ells high—in order to execute twenty-seven men, enemies of the emperor and leaders of the Protestant rebellion. Some were nobles, some were knights, and some were burghers. The theater Mydlár constructed was quite an affair. The platform was surrounded by a wooden railing on all sides, for safety's sake, and joined to a balcony on one of the upper floors of the Old Town hall by a wooden bridge. The entire stage was then covered all the way to the ground with a great black cloth.

Imagine the scene. The crowd of burghers, peasants, and soldiers filling the Old Town Square. The conquering lords, generals, and dukes watching the executions from upper windows. And then, with a blast of trumpets, the men were led out one by one by the magistrates to the place of killing. At five o'clock that morning, just as dawn was pinking the eastern sky, the castle cannons fired. Three infantry companies and two companies of light cavalry held the crowd back, while twenty-seven coffins were laid out in rows on the cobblestones beneath the scaffolding. Each coffin had already been assigned—one victim for each coffin.

Mydlár and his assistants dressed themselves in black robes and masks. A large crucifix stood in front of the chopping block so that each of the victims could see the face of Christ just before his head was chopped off. A group of magistrates also dressed in black sat on the balcony of the town hall. Three of the magistrates took on the task of calling for the victims, one by one, then rushing back inside, black robes flapping like crows, to haul out the next one. Trumpets cried, drums rolled as if in some magic show, and a magistrate disappeared into the town hall. He led the next man forward to the place where Mydlár leaned on a bare sword, honed and cared for like a chef's knife. The executioner stood calmly, waiting for the next death.

The magistrates offered each man a chance to speak, to pray, to explain his life, to beg for mercy. The condemned men said good-bye to the world, their friends, and families; to show their superior Christianity, they begged forgiveness for any suffering they may have caused, prayed to God, forgave their killers, and at the instructions of the executioner knelt in place and stretched their necks upon the chopping block, slimy with blood. Each man looked upon the soft face and form of the suffering Jesus as the sword fell and ended his life. Then six *holomci*, special servants of the executioner who carried the dead, all dressed in black masks and cloaks, like riders from the Apocalypse, picked up the headless bodies one by one, the arms and legs still twitching, and carried each one to its appointed coffin.

The execution was a morning's entertainment lasting four hours. Mydlár killed them all, using four different swords to make sure that the edges remained sharp. All in all, he beheaded twenty-four men and hanged the remaining three. Two he hanged from the town hall window, set up specifically for that purpose. The last he hanged from the gallows erected at the center of the town square. When all was over, they gathered the heads separately and chose out twelve special ones that once belonged to the more famous Protestant leaders, jammed wooden pikes through the medulla into the brain, and fixed them to the Stone Bridge so that people could laugh at them and be afraid as they passed the cornice of the bridge tower. At least the skulls were neatly arranged. Six faced the Catholic Church of the Holy Savior, and the other six faced Mala Strana across the river, so that all of Prague, on both sides of the river, could have a fair share of death. As an added entertainment and a grisly message, they nailed the severed hands, the right hands only, of Count Šlik and Dr. Hausenšild to their own skulls.

One of the men executed that day, whose head also ended up on the Stone Bridge, was Johannes Jessenius, rector of Prague University and a friend of Johannes Kepler. Generally a man of peace and Christian virtue, Jessenius was the man who had negotiated a truce between Kepler and Tycho Brahe. Before cutting off his head, Mydlár also cut out his tongue. This was usually a symbolic act done to those who had sinned against the king with their mouths, and most often it was done after the man was already dead. However, in this case Ferdinand specifically reversed the order and insisted that Mydlár cut out Jessenius's tongue while he was still alive. Apparently, the emperor reserved a special hatred for him, because he had tried to bring order to the Protestant cause and had orchestrated several peace negotiations among the distrustful Bohemian Estates. Instead of burying his body, the emperor had his headless corpse cut into four pieces and impaled on posts over a gate leading out of the city. The emperor then ordered Jessenius's head to be one of the twelve chosen for impalement on the bridge and ordered his tongue nailed to his skull.

The heads remained there for nearly ten years until 1631, when the Protestant Saxons entered Prague and a group of Bohemian émigrés, returning to the city, removed the heads from the tower and carried them to Tyn Church, the same church that held the bones of Tycho Brahe. According to legend, twelve ghosts rise from the dead every year on the anniversary of their execution to see if the hands of Master Hanus's astrological clock were still moving, for everyone knew that when that clock ceased to run, the world would end and the Last Judgment would begin.

⟨༄⟩

KEPLER ARRIVED IN LINZ on December 22, 1617, just before Christmas. He had been to Leonberg to try to settle his mother's affairs, but Einhorn had managed to delay Katharina's lawsuit against the Reinbolds one more time. The Christmas season was on, and each family in Linz was preparing itself for the birth of Christ. Families gathered, sang songs, told stories, drank too much, and ate too much. In Kepler's household, however, there was only sadness and death. It was not much of homecoming—two of his little children by Susanna had already died. Now, little Katharina, named for Kepler's mother and just a baby, was deathly ill. As

319

in Graz, Kepler descended into a morbid depression, while thoughts of death choked him. He tried everything to save her—doctors, herbal cures, prayers to heaven, but nothing worked. By February 9, little Katharina's fever had overcome her, and she died. Oppressed by sadness, Kepler could not even find solace in his work. He could not work on the *Rudolphine Tables,* for in his grief he found the calculations too tedious, the work too hard. "Because the Tables require peace," he wrote to a friend, "I have set them aside. I am turning my thoughts back to the Harmony."[5]

By all human reckoning, the previous five years of Kepler's life had been lived in the seventh level of hell. His own church had excommunicated him on a technicality, and Pastor Hitzler's sinful misuse of Kepler's confidentiality had threatened his reputation and even his livelihood. His mother had been vilely accused of witchcraft, and when she sought justice, she received the worst kind of bureaucratic manipulation. The Kepler family was beginning to suspect that old Katharina was in serious danger. Kepler too had been accused, but they couldn't make it stick. And finally his children, one after the other, had died of some traveling pox. By all human reason, he should have been a broken man, but, instead, he turned back to his *Harmony of the World,* a work dedicated to the belief that the universe is finally good and ultimately beautiful.

Loss and suffering, however, dressed him in the morning and tortured him at night. If only their church had not set them adrift, for even if God had not failed them, the Lutheran church of Württemberg had and failed them deeply.

The Lutheran consistory, clouded with orthodoxy, had denied his appeal after Pastor Hitzler's excommunication. On his trip to Swabia, Kepler had tried to reconcile himself with Matthias Hafenreffer and the church, but failed. His enemies, hidden in shadows, had conducted a whispering campaign, accompanied by secret letters between them and members of the consistory. Their opinion of Kepler was not very high. In 1619, a year after the consistory's decision, while the witch trial was still under way in Leonberg, Erasmus Grüninger, one of the members of the council, wrote a letter to Lukas Osiander, then superintendent of the Tübingen seminary, defending their handling of the case, saying: "Touching on this Kepler, we have long had dealings with this crazy man, but to no avail, because he would not listen to what anyone else says. We [of the consistory] did not want to keep silent with the worthy faculty of Tübingen, about what the consistory had written to him years ago on this subject [of his excommunication], because we felt that the faculty would likely have dealt with him the same way, since not one of us would want to change his beliefs just to please Kepler's sick head."[6]

Was this pique at some former student who had dared to deviate from orthodox Lutheranism? No, it was more than that, far more, for Kepler had committed the unforgivable sin—he had warned them of the fire to come. He had played the prophet against them, played Jeremiah predicting disaster to Jerusalem, disaster for them and for all of Germany because of their refusal to treat other Christians with Christian courtesy.

Württemberg was a provincial place, a big small town. Although some of the church leaders had traveled outside the duchy and some were even cosmopolitan in their way, few had ever seen Europe as Kepler had. Few had stood in the halls and courtyards of Prague Castle or attended the emperor—three emperors to be sure—or had kept up a steady correspondence with kings, dukes, princes, and bishops, as Kepler had. Few of them had ever suffered eviction for their faith or struggled with the dark night as Kepler had. For them, the events in Graz or Prague or Linz were the substance of reports from far-off mission lands, abstract in a way, foreign and dreamlike, something to be clucked at over beer and sausages. Kepler, however, had seen the reality from the front, had watched the storm clouds of war piling ever higher, starting small in Graz and then spreading with the ever rising power of Ferdinand of Habsburg, the arch-Catholic and protector of the Counter-Reformation. Kepler had seen it all, and knew what was coming.

The threefold split in Christianity, like a festering wound, would have to be cleaned of putrefaction, he wrote to Hafenreffer, and cauterized by the fire of war. He could see it coming, but they could not. In his letter to Hafenreffer, Kepler had written: "Then suffering will give you insight over many things, things that you conceal from your students who are even now growing up in the service of the church."[7] War is coming. Be prepared.

He was right, but they could not see. Not wanting to play the heretic before his family and his numerous friends, Kepler finally wrote down all of this, setting forth his beliefs about the relations between the churches, explaining his theological opinions, and defending himself against all the talk that surrounded his excommunication. The pamphlet, published anonymously in 1623, and titled *A Profession of Faith and Defense Against Numerous Unjust Rumors Which Arose Because of It*, circulated among his friends and relations. Things quieted down for a while then. Two years later, however, the conflict caught fire once again, and the consistory, never squeamish about privacy, published Kepler's last letter to Hafenreffer to show that they were of one mind against him. Kepler then revised his pamphlet and republished it, this time for the public. He responded in footnotes to each of the points raised by his letter and set out his true beliefs and precisely reasoned opinions.

Meanwhile, the Counter-Reformation stood nearby like carrion birds, with the Jesuits ready to gather in another convert. But they too would find Kepler a difficult fellow. His beliefs were his own, and he would not surrender them to either pope or consistory. His faith was in God, divinely unaffected by human theology, while the stars danced across the sky, waltzing to their own music, stately and perfect, far above the discords of earth.

In his life, Kepler wrote three great cosmological works—the *Mysterium Cosmographicum, The Mystery of the Universe*; the *Astronomia Nova, The New Astronomy*; and the *Harmonice Mundi, The Harmony of the World*. The last of these, his book on world harmony, he began soon after being forced out of Graz, while he was still working on the *Astronomia Nova*. The idea of harmony was a perfect blend of philosophy, astronomy, music, and theology.

But the *Harmony* had been coming a long time and was the culmination of a number of lesser works. During the years he lived in Linz, Kepler suffered so much from his struggles with the Lutheran church that he often had difficulty jumping back into his astronomical studies. Instead, he finished a series of projects that had been sitting in the back storeroom of his brain for a number of years, starting as far back as 1600, the terrible year he fled Graz to take up employment with Tycho Brahe in Prague. Among these was a series of short works on odd questions, most notably his method to measure the inside of a wine cask. He also wrote his *Epitome Astronomiae Copernicanae,* his textbook on Copernican astronomy, modeled quite consciously on the *Epitome Astronomiae,* the famous textbook on Ptolemaic astronomy by his old master Michael Mästlin.

The story starts as far back as 1613, the year that Ursula Reinbold claimed to have taken her fateful drink of potion. During the heat of the summer, Kepler had traveled from Linz to Regensburg to meet with Emperor Matthias and to answer several astrological and astronomical questions. Coming home, he traveled by boat down the Danube River, as he often did, and along the way he noticed that there were wine casks lashed together in various places along the shore. Each cask was a different size and shape, because barrel making was a craft and each piece was made one at a time. Typical of Kepler, his mind quickly recognized in this a new mathematical problem. How can one calculate the interior volume of a wine cask? Remember that integral calculus had not yet been invented, and so Kepler was trying to a find a way to describe a curved surface with precalculus mathematics. In doing so, he created the basic theory behind calculus by imagining a number of thin discs of varying sizes, stacked in such a way that they would exactly recreate the interior shape of the cask. If you could calculate the area of each disk and calculate how that area changed as you moved down the length of the barrel, then you could add up the areas and have the volume of the cask.

Again typical of Kepler, he tried to find some way to generalize this new method. He started by describing shapes produced by conic sections—the circle, the ellipse, and the parabola—that were rotated around a single line on a plane, thus creating a solid shape. In taking slices out of each conic section, he could thereby reproduce the shape of the rotated three-dimensional object. What Kepler had done was to set the imaginative beginnings of calculus, which was later formulated separately by Newton and Leibniz. He developed his new method into a short book—the *Nova Stereometrica Doliorum Vinariorum,* or *A New Stereometry of Wine Casks.*

Kepler couldn't find a publisher for his new work, so he took a lesson from Tycho Brahe, invited the printer Johannes Planck to come to Linz, and sent the manuscript to him. The book finally came out two years later, in July 1615. Even though the *Stereometria* was the first book ever published in Linz, Kepler's bosses at the school and among the representatives of the Estates were not overly thrilled by it. A book on measuring wine casks was not quite the work they had in mind for him to do. It didn't seem as noble as the *Rudolphine Tables,* which were based on the great work of Tycho Brahe and dedicated to the emperor. And well, they just didn't understand what Kepler was doing. Measuring wine casks was mundane, to be sure, but finding newer and simpler ways to measure curved surfaces was not.

In the process, however, Kepler learned something about publishing. He was not as rich as Tycho Brahe, who could afford to run his own press and could therefore produce a small print run of his esoteric works simply to publicize his ideas. But Kepler could still do something of the same sort, if on a lesser scale. He purchased his own set of printer's type and even ordered some mathematical symbols to be cast to round it out. This type set was the most precious thing he owned, and since he was sliding sideways into the publishing business, he had to try to make some money along the way. Kepler had already learned, as far back as the *Astronomia Nova,* that his books were not best-sellers, because few people could understand his complex mathematics or appreciate his methods. He was not the satirist that Galileo was; he just did not have the flair for controversy. But he had learned that he could make money by printing presentation copies to be given as gifts to important members of the nobility, for which they always returned a comfortable stipend. Kepler found that he could make more money on the presentation copies than on the regular sales.

With his *Stereometry,* he had ordered both a Latin and a German version printed, so that he could cover a wider circle of the nobility and thereby collect more stipends. When all was done, after he had given

away all his gift books, he had made enough money to pay for the printing costs and even pocketed an extra 40 florins. After listening to the three
emperors' empty promises for so many years, Kepler had finally decided
to go commercial, and in the case of small-print-run books, that meant
getting them into the hands of the people with the money.

Holding his intellectual nose, he also began to publish his yearly astrological calendars once more, after a hiatus of eleven years. The first of
these new prognostications came out in 1616, and though Kepler looked
down on his own astrological work, saying that it was "only a little more
honest than begging," he had learned that if he wanted to do great works,
he had to support them with his astrological penny stinkers, which were
much more lucrative.

Kepler had also discovered, as far back as 1611, that terrible year, that
if he ever wanted his ideas to reach a larger audience, then he would have
to write his own textbook, his own *Epitome*. In this book, he would not
only explain Copernicus's theories to the educated masses, but his own as
well. His old teacher Michael Mästlin had been the rock star of astronomical textbook authors in Germany with his *Epitome Astronomiae* and had
made some decent money in his time. Though Mästlin had introduced
Kepler to the Copernican universe, his own textbook had done one of the
best jobs of explaining the Ptolemaic universe. Kepler's own textbook, the
Epitome Astronomiae Copernicanae, or *Epitome of Copernican Astronomy,* took off from Mästlin's book and did for Copernican astronomy
what Mästlin had done for the Ptolemaic system. He gave Planck the
manuscript for the first volume of the *Epitome* in May 1616.

Then at the end of 1617 he wrote another prognostication for the year
1618, and then another for 1619. His calendar for 1619 took a shot at the
Württemberg church. "I know a gelding animal that sits among the roses,
covered in majesty, and stares out at its enemy, that other animal, without
fear at all, though its enemy will soon cause its death. Be careful, therefore, get ready for the stroke, stop shoving and remember that you are
here for the sake of the milk, not for your own benefit."[8] Suddenly Kepler
was a prophet. Everyone wondered who the gelding animal was; some
wrote to Kepler, some stopped him on the streets, begging him to let them

in, to tell them the secret. Many of them thought he was talking about the pope; others about the House of Habsburg; others about the Jesuits or even the Rosicrucians. Kepler never explained himself, however, for he knew that those he addressed his message to, the consistory in Württemberg, already understood him well enough.

That year of 1617, however, was another terrible year. His daughter Margareta Regina, the little one, had already died on September 8, of consumption and epilepsy, before Kepler arrived home. And his step-daughter, Regina, the little girl that Barbara had brought into the marriage, a girl whom Kepler had raised as if she were his own flesh, a woman now fully grown with a husband and children of her own, died in Walderbach, near Regensburg. She and her husband, Philip Ehem, a son of a prominent Augsburg family, had just moved there with their children. Ehem had a good job too, as Friedrich V's representative from the Palatinate to the imperial court. Suddenly, Regina took sick, and then all too quickly died. Ehem was lost in his grief for his wife and fear for his children. He begged Kepler to send his eldest daughter, fifteen-year-old Susanna, to help care for the children. Kepler agreed, but new worries had piled on top of the old.

He was still suffering from the news of Regina's death when he received the terrible letter from his sister, Margaretha, the wife of Pastor Binder, about the strange and terrifying accusations of witchcraft against their mother. So Kepler began his journey from Linz, up the Danube to Regensburg, passing through Walderbach, on the way to Leonberg, to see Philip and to leave his daughter Susanna with him. Along the way, he read a little book on harmony in music by Vincenzo Galilei, the father of Galileo Galilei, called *Dialogo della musica antica e moderne, A Dialogue on Ancient and Modern Music*. Kepler's Latin was strong enough that he could pick his way through Galilei's Italian. In this book, Vincenzo had returned to the original theory of harmony based on Pythagorean mathematics, an idea that piqued Kepler's interest because it appealed to his own astronomical, mathematical, and theological ideas. He had long been convinced that the best way to synthesize all three, to get that peek into God's cosmographic mind, was through the idea of harmony as it applied to the motion of the planets.

Ultimately the trip to Leonberg was a disaster. He was never able to silence the accusations against his mother, nor was he able to reconcile himself with the Lutheran church, no matter how hard he tried. His visit to Hafenreffer in Tübingen had ended in his final excommunication. After a short stop in Walderbach on the way back to Linz, when he returned home, his children started dying. One after another, the world dropped stones on him. Weighed down by all the troubles of his life, he could not return to the grinding calculations of the *Rudolphine Tables,* and so he started his work on the *Harmony.* Like Mozart, who would follow him two hundred years later and who would write his sweetest music during the blackest times, Kepler was at his best when things were worst.

So what was this "harmony"? It is a complex word for us, meaning just about anything good depending on the predilections of the hearer. For Kepler, however, the word had a precise mathematical meaning. Each instance of harmony was a regular mathematical pattern that Kepler found in the world, a comparison between two or more things the coming together of which created something beautiful. Harmony was a geometric as well as an arithmetic idea and turned on everything from the complex shapes of snowflakes to the motions of the planets in the heavens. It was not a single abstract experience, like "humankind," but a complex array of individual harmonies, something more like "people." The harmonies were arranged in phalanxes of ever more complicated patterns coalescing into a great cosmic symphony, a music so profound that it harrowed the heart and set fire to the soul.

At the core, harmony corralled all that Kepler believed about science as well as all he believed about God, for the two could not be easily separated. Divine harmony was his answer to the troubles of the world, troubles that he was all too well aware of. Over the years, he collected tidbits about the multitude of harmonies from writings in philosophy, theology, astronomy, and mathematics, until he gradually formed a new synthesis. Geometry was at the heart of it, for the secret structure of the universe was geometric, and therefore geometry was the blood of harmony, the marrow of God's thought. Kepler's researches led him back to the roots of mathematics.

For some time, he had tried to get his hands on a copy of Claudius Ptolemy's *Harmony* in the original Greek, for Ptolemy had tried to identify the qualities found in certain things that made them beautiful. But the book was hard to find, and Kepler failed at every attempt until Herwart von Hohenberg loaned him a copy to read. It amazed Kepler that even with several thousand years of difference between them, he and Ptolemy had contemplated the same ideas. This was reassuring to Kepler, because it intimated that he was onto something divine that lurked in the human mind. "The same idea about the construction of harmony has emerged from the minds of two very different men (and separated by so many centuries), simply because they were two men who had dedicated themselves to the contemplation of nature."[9]

As his little girl lay dying of pneumonia, Kepler stood at her bedside, watching as her tiny chest rose and fell with every ragged breath, and prayed almost without hope, while the ideas of harmony he had collected over the years almost imperceptibly jelled in his mind. When little Katharina died, Kepler locked himself away in his library, choked with grief, and as a refuge plodded on with his last great work. The *Harmony* was his fortress. He could enclose himself there, inside the perfection of mathematics, inside the transcendental beauty of geometry, and bolt the door of his mind to keep the screaming world outside. In those few stolen hours, he transported himself to that place of perfection and beauty, that place where God's will alone ruled the world.

In the process, while he wrangled with Luther Einhorn over the accusations against his mother, he remembered a sudden time when, in a momentary contemplation of harmony, he had discovered the third of his planet laws:

Eighteen months ago, the first light of dawn hit me; three months ago, the light of morning; and then, only a very few days ago, the complete light of the sun has revealed this remarkable spectacle.

Now, nothing holds me back. Indeed, I live in a secret frenzy. I sneer at mortals and defy them by the following public proclamation: I have pillaged the golden bowls of Egypt, to decorate a holy tabernacle for my God, far from the lands of the Egyptians. If you will forgive me, then I am happy. If you are angry with me, I will survive it. Well then, I will throw the dice; I will write a book, if not for the present time, then for posterity. To me, they are one and the same. If the book must wait a hundred years to find its readers, so what. God has waited six thousand years to find a true witness. [10]

This third law, his "harmonic law," set the relationship between the mean distance of a planet from the sun and the period of its orbit. This meant that one could calculate the time each planet would take to travel around the sun by knowing its distance, something that carried Kepler's first two laws and his application of physics to astronomy one step further. It was a cosmic regularity, one that had deep implications for Newton's law of gravity, for it showed the relationship between the distance from the sun of a body in orbit and the time it took that body to complete its cycle.

This was, for Kepler, another peek into God's mind. Kepler's mysticism orbits around this single idea, for he was no plodding empiricist, no earthbound pragmatist. His joy was in the perfect beauty of mathematics, especially geometry, which he always expressed in mystical terms. He was a mystical rationalist, a man who found transcendence by embracing reason rather than by abandoning it. In this, he was the inverse of Blaise Pascal, the French mathematician from Rouen who was born just a few years after Kepler had died and who experienced mystical insight by breaking through the veil of rationalism to find there a God of the burning bush, of the sacred fire, the God of Abraham, Isaac, and Jacob, the God of revelation. Although Kepler believed in the Bible and accepted its teachings, he saw within the Book of Nature a mysticism equally grand, equally vital, and equally important.

In this simple statement, Kepler also identified himself as a thoroughgoing Platonist, with earthbound harmonies rising and converging into the perfect harmonies of the mind. In doing so, he distinguished between sensual harmonies and perfect harmonies, or the harmonies of mathematics. Music, of course, was the living heart of sensual harmony, for it not only had the power to enrapture the soul, but also to engorge the mind with perfect order. Harmony in music, as in all things, is a matter of comparison. Two things are brought into contact—two tones, two colors, two objects, two ideas, and they either blend together to conjure a greater, higher experience or they do not. In other words, they are harmonious or they are not. This is not first a matter of calculation, but of immediate experience. The eye does not need to calculate to see the harmony of color, nor does the ear need to do sums to hear the harmony of music. The harmony is there in the experience, in the first astonishing encounter, emerging like the sun from the ocean out of the byplay of individual things.

However, harmony is not created by the eyes or by the things themselves, but by the soul, which resonates with the interaction between sounds, colors, things, and ideas and builds a new beauty for itself. It resonates because God has planted even deeper harmonies, and the ability to recognize them, into each human person at birth. The structures of the mind are harmonious themselves, which, like the Ideas of Plato, are implanted inside us, and it is up to us to remember them. It is from the comparing of sensual experience that these deeper harmonies emerge, as the human person lives in the world, feels it, weighs it, and measures it, with each experience opening up to new experiences beyond that.

This is not a matter of bare imagination, however. The mind does not make things happen. The colors, sounds, and objects exist in the exterior world and will not disappear if the observing human person walks away. If a tree falls in the forest, it makes a sound, but whatever sound it makes cannot form a harmony without someone to hear it. Thus, "harmony" is a primary category of existence.

Discovering harmonies is essential to discovering the world. In discovering harmonies, one makes comparisons between the two things that exist in the world—two sounds, two colors, two chairs, two people—in order to produce a harmony, and then goes on to compare that harmony and the internal prototype that exists inside the mind. This grand procession of harmonies has one ultimate function: "to reveal, to understand,

and to bring to light the resemblance of the proportion in matters of sense with that exact prototype of a true harmony, that prototype which abides inside the mind."[11] Such prototypes are the perfect harmonies, the Platonic ideas. These perfect harmonies are what the soul recognizes, and in that recognition are aroused the feelings of beauty and joy. I recognize them as I recognize the land of my birth, my neighborhood, my home. I am like an Alzheimer's patient who has been given a wonder drug and suddenly remembers everything. All education is a matter of remembrance, of bringing the harmonies to light. Mathematics is therefore the ultimate education, for it raises into the conscious mind those innate harmonies that we see projected into the world, in color, light, and joyful sound. And because these harmonies are innate, even the simplest people—the poor, the peasants, the barbarians, the uneducated—can recognize them, though they may do so unconsciously, for they emerge in the very act of perception.

Kepler rejects the idea that the human mind is an empty slate, a *tabula rasa,* and instead embraces Plato's notion that the harmonies are born in us, along with a secret storehouse of knowledge that is best expressed in mathematics. Our entire flesh, our bones, our eyes and ears are built for the discovery of this knowledge. The senses that we possess do not exist by themselves, but exist to serve the mind. The mind is not a sifting device, not a computer calculating numbers. The mind is the human soul listening to the universe and finding there a resonance with the mind of God: "If the mind had never benefited from an eye, then it would, in order to understand the world outside itself, demand an eye and invent its own laws for its creation."[12]

All of this for Kepler the mathematician, as it did in its own way for Einstein, comes down to geometry. "Geometry, being part of the Divine mind from time beyond memory, from before the origin of things, has provided God with the models for creating the world, models that have been implanted in human beings, together with the image of God. Geometry did not arrive in the soul through the eyes."

Kepler's reasoning, therefore, begins with the circle, which he accepts as fixed in the mind. One can inscribe a vast array of polygons inside a circle. Those polygons, which can be constructed with a straight edge and compass, like triangles, squares, and octagons, Kepler says are entities and he considers them to actually exist. Those polygons that cannot be inscribed in a circle—those with seven, eleven, or thirteen sides—Kepler says are nonentities, for they do not exist at all. They do not exist in the mind except as words, for they cannot even be imagined. One could draw a seven-sided polygon, of course, but such a figure could not be inscribed inside a circle. For Kepler, therefore, the world is ordered between those things that are possible and those that are not.

All of this has metaphysical importance. His thinking here becomes more medieval than modern, for his geometric speculations take on emblematic significance within an Aristotelian hierarchy of perfection. If the circle is the perfect two-dimensional shape and a symbol of infinity, then the sphere is the perfect three-dimensional shape and is therefore the symbol of the most complete form of infinity. Kepler instantly identifies it with the Trinity. He places God the Father at the center of the sphere, for the Father is the center of all that lives and moves and has being. God the Son is the surface of the sphere—as the round, dimensionless center point explodes outward into space, so too the Son is the presence of the Father expanding out into the world. The Holy Spirit is the radii from the center to the surface, joining the center with the surface, Father to Son, Son to Father, and remaining constant at each point of the sphere. This is a strange bit of speculation to the modern mind, which does not work like this at all, but it made perfect sense to the intellectuals of the seventeenth century.[13]

Kepler further identifies the created mind of the human being with the two-dimensional circle, which expresses a two-dimensional level of infinity and projects onto a flat surface some of the very qualities of the sphere. Therefore the polygons that could be inscribed inside a circle Kepler said actually existed in the mind. The human mind is an image of God's divinity projected into bodily flesh.

Kepler finds three essential expressions of harmony in the world—in geometry, music, and astronomy. Geometry is the greatest of them, because it is the bridge between those harmonies found in the senses and those found innate in the soul. It is in music, however, that those perfect harmonies best touch the senses. Here we feel them, see them, listen to them. In astronomy, finally, these harmonies express themselves in the structure of the universe. The motion of the stars is an expression of them, so that the human mind, which touches the shapes of circles, triangles, and the like, can also caress the universe itself. It is here that reason reaches its peak, for it is here that human beings shake hands with the Creator God. It is mathematics at its most profound, its most mystical. But like all mysticism, it requires a struggle, a dark night of the soul, a tramp through twisted jungles that clears the mind and prepares for those explosions of insight. In those moments when a person has wrestled with Reason itself, "then he awakens to the true light; he is taken by an astonishing rapture, and, rejoicing, he inspects the whole world and all its different parts, as though from a high tower."[14]

⚬〰〰〰⚬

KEPLER RETURNED HOME soon after his mother's trial had ended, in November 1621, but by then Linz had become a conquered city. Old Katharina had recently died, a broken woman, after all of Kepler's attempts to save her. And then he returned home to the flotsam of war. The Catholic League's success at White Mountain had crushed the Protestant revolt in eastern Europe and had left occupying armies all over the territory of the rebellious Estates. Bavarian troops still marched through the city as Kepler's river barge landed, a bit more than a year after the battle. As with all occupying armies, the Bavarians had made it clear who was in charge of the city, and Ferdinand, now firmly in power, had made sure that his rebellious subjects could see which way the religious world was turning. For a few years he waited, consolidating his power, but everyone knew that he intended to reestablish his program of Catholicization, begun in Graz twenty years before, that would sooner or later choke the Protestant schools and churches until they died. Pastor Hitzler, Kepler's old antagonist, was one of the first men thrown into prison.

Little happened for the next two or three years after Kepler returned to Linz. The city, like Kepler himself, was depressed. The Battle of White

333

Mountain had solved the problem of open rebellion by Protestants with a great iron boot, but from that time on all of Upper Austria held its collective breath. The Protestant Estates had participated in the rebellion, and some of their most prominent leaders, even a few friends and supporters of Kepler's, had been executed. Some were continuing the fight, but the fight was not going well. And so the presence of Bavarian troops was a running sore for the people of Linz, for these men were Germans, not Austrians, an occupying army intended by the emperor to send the people a message—the time of the "new doctrines" was coming to an end.

Still, the battle raged on in other parts of Germany. Friedrich V of the Palatinate began holding mock court in the Netherlands, but Christian von Braunschweig and Ernst von Mansfield, both followers of the former king of Bohemia, had swallowed their shame at Friedrich's cowardice and taken to the field to continue the fight. The emperor, who had eradicated all the new doctrines in Prague and in the rest of Bohemia, would soon turn his eye toward Upper Austria. Just as in Graz, Lutherans found it harder and harder to attend services. Protestant leaders, even some of the most prominent of the nobility, either converted or were exiled. Protestant ministers were either jailed like Pastor Hitzler or sent out of the region.

Generally, the people ground their teeth at the emperor's new restrictions, but what could they do? There were troops in the city. Foreign troops. Catholic troops. But surprisingly, nothing happened—no explosions, no outbursts, no demonstrations, no sudden reign of terror. In those two or three years, nothing much happened at all, except for subtle things. There were new restrictions on marriage and on burial, the old Graz story all over again, only this time it was unfolding more slowly. The gloom gathering over the city darkened bit by bit every day. The Lutherans waited for the terrible punch line, for the Counter-Reformation to arrive in force. They knew how it would begin: one edict, then another, coming faster and faster, until the Protestants would be surrounded by them. Then one day the emperor would simply tell them to convert or leave.

Reports of war filtered down to Kepler almost weekly: a bloody skirmish here, another battle there, someone denouncing the emperor, someone being denounced. Kepler hated it, calling the whole warlike noise "barbaric neighing." Worst of all for Kepler was the loss of intellectual freedom. No one could think honestly; no one could argue; no one could search for the truth in mutual fellowship. The slightest disagreement had become a cause for debate, and the slightest debate had then become the cause for denunciation. "What execrable stupidity forces these people to run into the middle of new fires when trying to escape old ones!"[15] The world had become chaos all around him, with people running as if they were crazed from one battle to the next, and Christianity itself had become the cause for war.

Kepler's solution to this chaos was typically Platonic. Everyone should study mathematics and philosophy, he said, "so that the contemplation of these things would lead the mind away from desire and the other passions, out of which emerge wars and all other evils, to a love of tranquility and temperance in all things." This was a sweet idea, if a bit naïve. But it fit the rest of his mysticism—if one could not find peace on earth, one could find it in the contemplation of the heavens. But Kepler also understood that the world had gone beyond good advice. The people of his time had become beasts, driven mad by religion, which was a state of affairs that, in the long run, only God could cure. "The more anyone falls in love with mathematics, the more fervent will be his dedication to God, and the more he himself will make every effort to practice gratitude, the crown of virtues, so that he will join me in prayer to the merciful God that much more sincerely: let him crush the warlike confusion, eliminate devastation, snuff out hatred, and venture forth to discover that golden harmony once again."[16]

Ultimately, Kepler was a man of hope. He believed that the war would end and that sooner or later Christian Europe would come to its senses. When that happened, he believed, God's grace would shine in the world, for those who committed the sin of war would repent in dust and ashes, and Europe would be reborn. But no one wanted to listen to such optimism. This was partly because of their stubbornness and partly because in this religious war Kepler was sitting out alone in no-man's-land. His own church, the church of his parents and grandparents, the church that had

335

raised him and had given him his education, the church of his own con-science, had rejected him. It had rejected him because, like his mother, he would not bend, he would not be the obedient believer that they demanded he be, because he would not believe in the authority of the consistory to overthrow his reason. It had rejected him, but he had not rejected it, for no matter what they thought, Kepler was a Lutheran—in his heart a member in good standing. The fact that they would not include him in Communion hurt him deeply, but in his own mind it did not place him outside the church. How the church leaders wanted to behave was their business.

Meanwhile, the Catholic Counter-Reformation was waiting to pluck Kepler like a pear. Because of his desire to treat all Christians with cour-tesy and to befriend anyone who shared his love of astronomy, Kepler had built many friendships among the Jesuits and often stayed in Jesuit houses as he traveled across Europe. Jesuits Johannes Deckers in Graz, Albert Curtius in Dillingen, and Paul Guldin in Vienna were all regular correspondents and often debated the fine points of theology with him. Each of these men was deeply interested in astronomy, and all harbored a not so secret desire to bring Kepler to the Catholic faith. They would fail. In his letters, Kepler repeatedly reminded his Catholic friends that it was their church that created laws to oppress their Protestant fellow Chris-tians, laws that Kepler too had suffered from. They were the ones using the power of the state to silence opposition and in that were one of the worst offenders. Kepler reminded them that he was still a Lutheran and would remain so. He proved this when Pastor Hitzler, the Lutheran min-ister who had caused him so much trouble and done so much damage to him, was finally released from prison and exiled from Upper Austria. Kepler regularly wrote to him in Württemberg to send along his greetings and good wishes.[17]

In 1625, however, things began to move in the wider world. Duke Albert von Wallenstein had convinced the emperor to put him in command of an army, so that he could fight under the general command of Count Tilly, the victor at White Mountain and also the commander-in-chief of all the forces loyal to the Catholic League. The emperor was not disap-pointed. Wallenstein soon defeated the remaining followers of Friedrich, the onetime king of Bohemia, and consolidated Ferdinand's power over his territories. Ferdinand was now more powerful than ever, which was fine for Ferdinand, but the rest of Europe worried. It was good that Ferdinand had managed to put down his little rebellion, and everyone agreed that the Catholic church's success with the Counter-Reformation in Austria was laudable, but having a Habsburg emperor with real power behind him frightened everyone.

Cardinal Richelieu began scheming at once. He had been running the French government for one year when Wallenstein took command and, typical of Richelieu, the arch-Frenchman, he did everything he could to undermine Ferdinand's position. Admittedly, he was a cardinal of the Roman Catholic church and a supporter of the Counter-Reformation and, admittedly, Ferdinand was doing the work of the church, but Richelieu was a Frenchman first and gave his primary efforts to the ascendancy of France, even at the expense of the faith.

Secretly, he contacted the Danish king Christian IV, who was always looking for good fight, and offered him money to take on the Habsburgs. Christian invaded, and Tilly and Wallenstein quickly made plans to coordinate their armies in the fight against the Danes and then to take on the princes of northern Germany along with them, thereby bringing the Counter-Reformation into another Protestant stronghold.

Then, unexpectedly, in the spring of 1626 a peasant revolt exploded in Upper Austria. The pressure that Kepler could see building after his return to Linz 1621 had finally burst. The forced conversions, the loss of a proper Lutheran ministry, the insulting presence of the Bavarian soldiers had all slowly enraged the Austrian peasants, and, as if with a single unified shout, they rose up and formed an army. Amazingly, they organized themselves into troops and burned and plundered the homes and castles of the powerful. Monks and nuns fled into the night as peasant battalions gathered around their cloisters and burned them to the ground. A new leader emerged among them—Stephen Fadinger—who not only formed them into an army, but a successful army. They quickly conquered the city of Wels, occupied it, and then laid siege to Linz. That day was June 24, a day when farmers would ordinarily be planting and growing crops, and so the quick drop in agriculture promised a lean winter. Many starved in the city, and some took to eating horsemeat. The peasants besieged Linz for nearly two months, until August 29, when the peasant army withdrew at the appearance of imperial forces. Eventually, by the end of that year Count Peppenheim defeated the peasants and slaughtered them by the thousands.

During the siege, the peasants started a fire that destroyed the house of Johannes Planck, Kepler's printer. Planck was in the middle of printing the *Rudolphine Tables* at the time, and his printing press and all that he had printed up to that time was destroyed, with the exception of those parts that Kepler had taken home. Kepler would have to begin again with another printer. To do that, he would have to leave town.

At the time, Kepler was living in a country house near the city walls. The battle burned on all around him, and the city had garrisoned troops in Kepler's house. Every time the battle caught fire one more time, men rushed out to fight, no matter what time of the day or night. Guns fired off all the time, the air was filled with the smell of gunpowder, and the noise of the war was all about him. All in all, Kepler was lucky. He was one of the few who did not starve during the siege. He had enough food to keep his family healthy, so none of them were forced to eat horsemeat or rats.

When the battle was finally over and the peasant rebellion put down, Kepler knew that he could no longer print the *Rudolphine Tables* in Linz, so he wrote to the emperor to ask permission to leave Austria and move to Ulm. Ferdinand agreed, for there was nothing else the emperor could do if he wanted the tables printed. Once again, Kepler packed his bags, his possessions, his family, and his precious set of printer's type, loaded them onto a barge, and traveled up the Danube to Regensburg. He would complete the *Rudolphine Tables*. He must. He was committed to it.

ᏇᏆᏇ

*My second homeland, that I left behind, is in much danger
with the yoke of tyranny around its neck. If the noose tightens,
my return will not be possible. But I won't despair just yet, I
still hesitate to extend my thanks to the* Stände *and continue
my duty under danger, if I am able to and they will not dismiss
me. Shall I go overseas, where Wotton invites me? I, a German
who loves the mainland and fears the closeness of an island. . . .
I, with a young wife and a flock of children. The maternal inheritance of my children and my assets are deposited with the
Austrian* Stände. *If the* Stände *will be dissolved, which I see
coming, all that will remain is that the people forgo their property. Recently it became illegal in Bavaria to pay even a penny
to a creditor or wage earner. Even if this is a temporary law, I
fear it can become a permanent one.*

FROM KEPLER'S JOURNAL
1623

*The year 1623. A son, Fridmar, was born to me on January 24
in Linz. He was baptized in the country home of the honorable
Johannes Reu (?), first preacher. Godparents: Dr. Abraham
Schwarz, Herr Sebastian Paumaister, Baron Jörger with wife
Katharina, maiden name Hagen, from Kärnten. My son Sebald,
completely exhausted by smallpox, died on Corpus Christi Day.
I published the "Glaubensbekenntnis" ["Confession of Faith"].*

ALBERT. DVX. FRITLAND. COM. WALLEST. ETC.

Albert von Wallenstein, Duke of Friedland

XV

My Duty under Danger

*Where Kepler seeks a home for his last few years
after fleeing Linz, argues with the Jesuits,
finds patronage with Wallenstein,
and dies in Regensburg.*

KEPLER TOOK TO THE ROAD once again, but this time there was no Tycho Brahe waiting at the other end to welcome him. Where would he find a home this time? How would he live? How would he raise his children? Where would he find a place to work in peace? All of Europe was at war it seemed, and wherever he tried to run, the war was already there. Kepler's greatest joy was in the ecstasy of perfect order, in the astonishing beauty of God's mind, and that is where he preferred to stay; and yet, as if by some irresistible force of gravity, he was forever pulled back to the mud and the blood of the earth, to live out his life in the battlefield that was Germany.

The most immediate task before him was to find a publisher for the *Rudolphine Tables,* and the best place to do that was in Ulm, on the Danube River, a few days' walk from Tübingen. He had many connections there, old friends and acquaintances, correspondents of many years standing, even relatives. Ulm was the home of his distant cousins the Ficklers, the city where the great Benigna Fickler, a woman of learning, had once held court. With great sadness, Kepler gathered his family about him once again and bought passage for the journey upriver, away from their home in devastated Linz. It was the middle of winter, however, and the river was frozen above Regensburg, so they could go no farther together than there.

After spending a few days finding a safe residence for his wife and children, Kepler packed his precious letter type into a wagon and continued the journey on to Ulm alone, fighting cold and snow all the way. Once in Ulm, he found residence with an old friend of his from Prague, Gregor Horst, who had once been a professor of medicine at Wittenberg and was now the city physician. Horst owned a small house adjoining his own larger house near the Cathedral Place, but facing a back alley; there Kepler, exhausted from his trip, unpacked his bags and settled in.

Horst immediately sent out feelers to his friends and acquaintances in town to scout out a printer for the *Tables.* The word on the street was that one local printer, Jonas Sauer, could do the work quickly and cheaply—especially cheaply. Horst spoke well of him, which was good enough for Kepler. He was ready to begin printing. He had the ink. He had plenty of paper, already sent to him in Ulm—two bales of the good stuff for the presentation books and two bales of the ordinary stuff for the general market. And he had his set of letter type, including the special type blocks he ordered cast in Linz to print the astronomical symbols.

The work carried on, but Kepler was soon disappointed with the personality of the printer. The man was proud, rash, and unpleasant. At one point, Sauer ran into money problems and tried to extort more money out of Kepler, money that Kepler didn't have. Then, as his money problems increased, Sauer began arguing with Kepler over the contract and even threatened not to finish the job unless Kepler helped him with his problems. Furious, Kepler decided to take the printing job away from him and find someone else in Tübingen. Because Kepler was feeling cornered by his situation, his moods had become more erratic, and he did his best to fend off depression, that dark beast always lurking inside him, waiting like a leopard for him to slip and fall. In a sudden fit of pique, he set out for Tübingen to check on alternate printers. He couldn't ride a horse, though, or sit in a rough wagon for long, for he had developed an inflamed abscess on his buttocks, and the gunshot of pain he felt with every trot of a horse or bounce of a wagon had become intolerable. Instead, he set out on foot, and made it all the way to the town of Blaubeuren, when the deep February weather stopped him. He was not as young or as healthy as he used to be, so he turned right around and walked back. Sauer got to keep the job after all and, surprisingly, completed it without too much more trouble. Kepler, however, fought through the blizzards of his own changing moods. The world had once again become too heavy for him, and eventually the depression avalanched over him.

Meanwhile, his son Ludwig had reached university age and was studying medicine in Tübingen. The cost of a university education, then as now, was enough to drain a family's resources, and Kepler had to shepherd his money carefully to be able to pay for his part of Ludwig's education and the printing of the *Tables* at the same time. Ludwig, of course, always needed more money in order to live the life that every university student believes he deserves, and so he wrote to his father asking for help. Kepler would not give it, for he had given his son "expense money" enough, he said. He had given him the clear understanding that the Kepler family was poor and would have to live accordingly. Ludwig had enough money to survive and to complete his education. He would get no more from his father. Kepler had only enough money to complete the printing of the *Tables,* a task he considered his sacred duty. "To use this money, even a bit of it, for my son would delay the work, and that for me would be a sacrilege." Although he loved his son and wanted to provide for him, the proper time for the production of the *Tables* was rapidly ending. The emperor expected the work to be done. His sense of responsibility to his old master Tycho Brahe hung over him. "This is the way the matter stands: either I finish the work now, or I don't finish it at all."[1]

Finally, after months of work, the printing was nearly done. Kepler contacted the Brahe family to inform them of this so that they could write a dedication on behalf of their long-dead Tycho. The old master's two sons, Tycho and Georg Brahe, composed a title page and a dedication, which they sent to Kepler, who then sent it on to the imperial commissioners, saying that he had no disagreement with what they had written. Kepler, however, had written another dedication to the emperor, which he wanted to include, a longer one, in which he explained the reasons for the long delay. With all this, the printing, almost miraculously, was finished.

The print run was fairly large for Kepler—a thousand copies. Since he had paid for the entire printing himself, he needed to recoup its expenses, so he decided to send the entire run to a book dealer named Tampach in Frankfurt for the book fair. On September 22, 1627, Kepler arrived with a few copies of the *Tables*. But because of the legal issues raised by his complicated relationship with the Brahe family, Kepler could not set a price for the book, so the whole question went into arbitration. The imperial commissioner sent copies to various experts at the fair to ask their opinion about the price. The Jesuits he sent one to set the price high, at 5 gulden. Some of the other experts set it at only 2, so the commissioner split the difference and set the price at 3.

It was a decent price, but not enough to make any real money. Kepler figured that the amount of book sales of astronomical tables would never be very high, since mathematical works were rarely popular, especially in wartime. "There will be few buyers, as with all mathematical works, especially in these times." He decided to pursue alternative markets, including international sales, so he secretly sent a number of copies to a friend of his in Strasbourg, Johann Bernegger, with the request that he spread the word about the *Tables* among the university crowd in that part of France.

Then once again the Brahe family decided it was not happy and made a fuss. More than likely, they saw this as a duty to the memory of their father. A little rain, a little tempest, and no one would be allowed to forget Tycho the Dane. Kepler had broken their agreement, they said. He had not sent them a copy of the finished tables for the family to review before the book went to print. This was true. Kepler did not want to send them a copy of the *Tables* before the printing because they were such fussbudgets, and because it would give them another chance to harangue him about following the Tychonic cosmology rather than the Copernican. Kepler didn't want to go through that again. They complained they didn't like

Kepler's dedication because it took too much attention away from their illustrious parent. They didn't like this phrase, they didn't like that phrase.
They wanted a new title page. Finally, to please them, Tampach the book-
seller had to replace the signature page on all the copies, and then they
nitpicked about that too. Needless to say, when they were done, Tampach
wasn't very happy either.

But for Kepler, the great work was accomplished, and he felt the wind
shifting in his life, a sea change bringing along something new. Kepler realized that after printing the *Rudolphine Tables,* an important phase in his
life had ended. He had been working on the *Tables* ever since the death of
his old master. For twenty years, he had burrowed into the mathematics
of Tycho's observations and, with antlike labor, built the jumble of data
into a set of tables that would revolutionize astronomy. It was over. He
had done his duty to the emperor. So where would his life go from there?
What he wanted, what he had always wanted, was a nice, comfortable
position somewhere far away from the war and the endless bickering between Christians, a little position somewhere that would allow him to
give lectures on things astronomical and astrological, where he could
gather his family about him without having to leave them in some other
city, and where he could live out his remaining years in peace and security.
After years of struggle, he was willing to go anywhere, anywhere at all, in
Germany, Italy—even England.

He was fairly certain by that time that his days as imperial mathematician were coming to a close. After all, Emperor Ferdinand had already re-
instated his Counter-Reformation measures in Upper Austria and was
busy extending them throughout the empire. In the summer of 1627,
Ferdinand had decreed that all non-Catholic officials in Upper Austria
should be removed from their positions and encouraged to emigrate.
Sooner or later, Ferdinand would remember that his imperial mathematician was also a Lutheran and would put him to the question as well.
Kepler would not give in, no more than he did in Graz or Linz. Drowning
in his depression—so much death, so many of his children lost!—Kepler
imagined the worst.

Nevertheless, he was required to bring the *Tables* to Prague to present them to the emperor, but he was afraid. That February, he had written to his friend Schickard that he had a "heart swarming with anxiety and stung by fear of the future, but a heart that could still find new hope in a single word of encouragement."[2] Fearing that Ferdinand would receive him badly, both for his Protestantism and for the tardy publication of the *Tables,* he decided he needed help, so he visited an old friend, Philip, Landgrave of Hesse, in October 1627 and asked the prince for his assistance.

The landgrave was sympathetic and understood Kepler's insecurity quite well. He knew how shaky Kepler's position was and had always been. Philip referred the matter to his nephew Landgrave Georg, who had succeeded him as ruler of the principality. Georg had remained Catholic, loyal to the emperor and the Counter-Reformation, while his uncle Philip had become a Calvinist. Nevertheless, the two men were both well disposed toward the imperial mathematician and agreed to help him in any way they could. They advised him to go on to Prague, while they would write letters to people they knew.

With Landgrave Georg's help, Kepler gathered the remaining copies of the *Tables* and set out for Prague. He left Ulm on November 25, 1627, and on the way stayed with the Jesuits, especially Father Albert Curtius in Dillingen for two days. Oddly enough, for all his unmovable Protestantism, Kepler had an excellent relationship with the Jesuits, the chief architects and ideologists of the Counter-Reformation. As a Lutheran, he was expected to avoid them as the very spirit of evil, just as they were expected to see him as a heretic and an apostate. But the Jesuit motto to "find God in all things" was similar enough to Kepler's own view of mathematics and astronomy that Kepler and the Fathers of the Society of Jesus found that they were kindred souls, that they were all men of scholarship and faith. His letters to them were always cordial, and so were his visits. Kepler enjoyed the intelligent conversation he found among them and often looked forward to his short visits to their houses. After a few days, however, he took leave of Curtius and traveled on to Regensburg, and from there went on to Prague.

He found the city in quite a state of celebration. The first phase of the Thirty Years' War was winding down, and it seemed as if Ferdinand had triumphed. The peasant revolt had been put down, and the last remnants of the Protestant rebellion had been cornered and their leaders killed. Wallenstein had defeated both Ernst von Mansfield and Christian von Braunschweig, the last two supporters of the exiled Winter King, Friedrich, and both men had died during the battle. Wallenstein was the man of the hour; his star was shining. He was the most successful general in the empire and had proven it. He was more successful than Tilly, more successful than anyone. He had fought the Danish king and pushed him back all along the northern part of Germany, until the Danes finally had retreated to a series of islands. A few weeks before Kepler arrived, Wallenstein had come to the city, set himself up in his opulent, some said decadent, new palace in the shadow of Prague Castle.

The general had already been appointed the General Colonel Commander-in-Chief and the General of the Baltic and Oceanic Seas. He had risen so fast and had become so wealthy so quickly, that he was roundly hated in Protestant circles all across Europe and held in suspicion and fear in Catholic circles. Both France and Sweden, one country Catholic and the other Protestant, fretted about Wallenstein and schemed to bring him down. He was simply too successful, and his successes were too brutal and too final. The Swedes worried that the emperor might turn his eye northward, might convince himself that the Catholic church needed to return to Scandinavia. The French thought what the French always think, that the rest of Europe would be a better place if it were French. Nevertheless, as Kepler entered Prague, Wallenstein was the most celebrated man in the empire. People waved to him in the streets, knelt before him, kissed his hand, while plots within plots, schemes within schemes, swirled around the General Colonel Commander-in-Chief as the mist swirled above the Vltava River. The fact that he would one day be assassinated would have come as no surprise to the people of Prague.

Kepler, of course, had no idea about any of this. He was not much of a politician and disliked intrigue in general. He saw in the love of power that drives so many great men the evil that had brought death to the world. Likely, all he noticed the day Wallenstein marched through the streets in triumph was a city in full celebration.

He had work to do himself, however. He needed to present himself to the emperor. Putting it off as long as he could, he finally dragged himself up the Steep Stair to the castle and waited in the great hall to be announced to the emperor. Standing there, Kepler may have thought of his friend Jessenius and the headsman's ax. But surprisingly, Ferdinand was in a splendid mood. His son, another Ferdinand, a boy who would later become Emperor Ferdinand III, had just been named king of Bohemia, an important stepping stone to the imperial throne, and was crowned at St. Vitus Cathedral with every bit of trumpeted glory that the Habsburg family could gather into one place. Which was considerable. The whole world was there, and Ferdinand was luminous with glory. He had gotten everything he wanted—power, fame, the ascendancy of the Catholic church—and then suddenly, standing before him, was his old imperial mathematician, nearly forgotten after years of war, a rarity in the imperial court, a man who had actually completed something. He had brought a book to present to his emperor, a copy of the *Rudolphine Tables,* named after Ferdinand's unfortunate uncle Rudolf, that was a compilation and summary of the greatest work of astronomical observation of the time.

Kepler thought he was about to be fired, but—*mirabile dictu!*—the emperor welcomed him kindly and with some ceremony. This fearsome Emperor Ferdinand, who had earlier cast him out of Graz, whom Kepler had dreaded for so long, was unaccountably gracious. And the court, whose halls he had haunted for so long waiting to be paid, was suddenly full of admirers. Kepler was another man of the hour, or as much of one as a mathematician could be. He was certainly no Wallenstein, but he had done his part, and the emperor praised him.

Kepler's depression, which had been partly fed by fear of this meeting, suddenly lifted like a morning fog burned off by afternoon sun. He was once again in Prague, the center of intellectual life in the empire, and he was the imperial mathematician, praised and admired by the court. The emperor was unaccountably pleased with him, and it seemed that, at least for now, his job was not in jeopardy.

But so much had changed. Where were Kepler's old friends? Tycho, his old master, was dead. Jessenius the peacemaker was dead. And where were the others? Where were the Lutherans, the Augsburgians, the Utraquists, and the Bohemian Brethren? All had disappeared. In Rudolf's day the city had boiled with arts and arcane sciences, but in Ferdinand's empire Prague Castle had become a true fortress. Soldiers were everywhere. The Old Town, which had once been a glorious hodgepodge of Catholic and Protestant and Jew, was now Catholic to the core. The Jews were still there, for they had been loyal to the emperor, which is one of the reasons so many of the Protestant uprisings quickly rampaged through the ghetto. But now, the Protestants were all gone, and the city seemed to have lost something of its feisty soul.

Still, Kepler had to admit that the emperor was kind to him. Ferdinand bestowed on Kepler 4,000 gulden and commanded that the cities of Ulm and Nuremberg split the cost of it between them. Kepler never saw a penny of this money, but it was a nice thing for Ferdinand to say. Kepler asked to stay in the service of the emperor and to continue to live in Habsburg lands, and Ferdinand was more than happy to oblige. Ferdinand told him that he would find a position for Kepler at the imperial court and then perhaps a teaching position at one of the universities.

But, of course, there was this little matter of Kepler's religion. A trifle, really. Ferdinand let it be known quietly that Kepler would have to convert to Catholicism for all of these wonderful things to happen. After all, his own church had abandoned him, so where else could he go? Why not do it? He wasn't really a Calvinist at heart—everyone knew that—and the position the emperor would find for him would be a good one. Why not convert? Some of his best friends were Catholic. Some of his relatives too. There were the Ficklers, who would be glad to welcome him back to the true faith. And even his own long departed father-in-law, Jobst Müller, had converted.

At this point, the Jesuits appeared on stage. The Jesuits had befriended Kepler. He and Father Albert Curtius in Dillingen had been writing back and forth for years. Father Paul Guldin in Vienna, who had been raised a Protestant, but who as a young man had converted to Catholicism and joined the Jesuits, was a longtime friend. The last shoe finally dropped when Guldin, in a letter to Kepler, asked if he might be interested in joining the Catholic church. After all, there were great advantages in it, not only in this world but in the next. Kepler wrote back to him on February 1, 1628: "Just as I entered my life," he wrote, "my parents initiated me into the Catholic church, sprinkled me with the holy water of baptism, and thus endowing me as a child of God. Since that time until now, I have never left the Catholic church."[3] Kepler, however, meant these words differently than the Jesuits meant them. Kepler believed that the Catholic church was not limited to those who accepted papal authority, but extended to everyone baptized in the name of Christ. Baptism, anointing with chrism, and faithful adherence to the teachings of Jesus were the marks of a true Catholic. Some Catholics followed Rome, while others followed the Augsburg Confession. "If you tell me that the church is that group of people united under one pope to spread the errors once cast off by the Augsburgians, and to rule them in matters of conscience, then you set forth the single characteristic which would prevent me from accepting the church you speak of, if this were offered by itself."[4] The rule of one pope could not define the church for Kepler. Only the rule of one's conscience could do that. "Just as quarrels occur among the burghers and political factions of a city, so too, because of human frailty, mistakes happen among the citizens of the one church, separated by time and space."[5]

Kepler's idea of the church was similar to the ideas expressed by St. Augustine in his *City of God*. There exists a mystical union among all believers, and differences between Christians, for all their violence, cannot destroy the Body of Christ. The divisions within the church are the result of human error, perhaps even human sin. They are not what God intended for us. This was the same belief that Kepler had already expressed to the members of his own church, a belief that had finally gotten him excommunicated. Who was this Kepler to council peace in the time of war? In some ways, the Jesuits were true friends at least, and for all their desire to convert him, they were willing to accept him when the Württemberg

Lutherans would not. But he was still a potential convert, a possible victory for them and for the Roman Catholic church, and neither he nor they could forget that.

Because of his baptism, Kepler believed, and therefore because of his membership in the Catholic church, he did not need the religious authority of the pope to teach him what the Scriptures and the fathers of the church had already taught him. Like a good Lutheran, he denied magisterial authority to any office within the church and reserved that authority to the texts of Scripture and tradition, and to the Holy Spirit that was granted to him in his baptism. He knew the religious practices and beliefs that he could not accept, ideas that in his view had built up over the centuries and that he considered to be dark innovations. He denied the adoration of Jesus in the Eucharist. Following Luther, he rejected the veneration of religious icons, pictures of saints, and even pictures of Jesus. He rejected the Mass as a sacrificial act, and the sharing of Communion only in the form of bread. The early church, he said, did not practice these things or hold these beliefs. He was not a proud man, he said, and recognized that the Catholic church was the mother church of the West. But he must obey Christ, whose rule commanded the church itself. "Therefore, do not think poorly of me, best friend, for I abide in the Catholic church. In order to reject what I do not consider to be Catholic or apostolic, I am prepared, not only to surrender the rewards which his Imperial Majesty has graciously and generously consented to offer me, but also to give up the Austrian lands, the entire kingdom, and even—and this weighs heavier than all the others—astronomy itself."[6]

Kepler would give up everything he loved and treasured in this world to maintain his grasp on his belief in God and Christ. He had to follow his conscience, which he could not surrender without surrendering his soul. For Kepler to give up his beliefs, beliefs that he had suffered so much over with his own church, would be to give up Christ himself.

As in so many other times in Kepler's life, the forces around him began to congeal all because he would not give up his conscience. Soon after he sent his letter to Guldin, he received a letter from the imperial representatives in Linz. The Reformation Commission (named as such, but actually a product of the Counter-Reformation) had reconvened, and Kepler, appearing on the list of non-Catholics, received a fairly standard letter commanding him to reveal himself ready or not ready to "accommodate" himself to the work of the commission. In other words, he would have to become Catholic or he would lose his position as district mathematician for Upper Austria. He had already told the Jesuit Paul Guldin that he was not willing to convert to Catholicism, which meant that sooner or later, he would lose his position as imperial mathematician. The fire was already lit; the Reformation Commission in Linz had merely added a bucket of coals to it. Kepler wrote again to Guldin, saying:

> I keep hold of the Catholic church. Even when it goes mad and
> thrashes about, I remain faithful to it with heartfelt love, as much
> as any frail human can do. Should the church accept me with my
> few small reservations, then I will persist in the perfection of my
> science under the leaderships of the dominant party. I am willing in
> silence and with full patience to carry on: I will abstain from all
> insults, mockery, hatred, hyperbole, calumny, and ridicule of any
> people who are of goodwill. I shall take to heart those sermons in
> which I see the light of divine grace; I shall make it my practice to
> avoid processions and the like, for I do not wish to give offense to
> anyone, and not because I am passing judgment on those who take
> part, but because such events mean two different things for two
> different people. Yes, I can also attend Mass and join my prayers
> with the prayers of the faithful, with one condition—if you accept
> my objection and that of all my family, to the degree that we will
> not accept those things that our convictions tell us are in error—but
> are only asked to accept the general and final, sacred and Catholic
> purpose of the Mass, which is to raise to God our prayers and the
> offering of our praise and good works, always bearing in mind that
> unique sacrifice accomplished on the altar of the cross, that this
> sacrifice may be of good use for us, that the church may be taught
> by these historical acts about the memory of the death of the Lord.[7]

In effect, Kepler was asking no more than what Thomas More had asked of Henry VIII. He would do none harm, he would think none harm. What he would do, to the limits of his conscience, would be to practice the faith within a Catholic milieu, if he could do so without causing scandal or division. But like Thomas More's, Kepler's own fame had worked against him. He could not be allowed to slink into the corners of the church, to spend his life in quiet and peace. He was famous, a man of science, a man whose conversion would be a great victory for the forces of Catholicism and the Counter-Reformation. He was an honest man, and was known to be honest. He was nonpolitical, and known to be nonpolitical. He was also known to have suffered on this account. If they could turn him, then their victory would have been that much sweeter. For the same reasons, then, that the Lutherans excommunicated him, the Counter-Reformation could not allow him to camp out in the middle between the great armies.

Before Guldin responded to Kepler's letter, he sent it to an unnamed brother in the order. He wanted to make sure that Kepler got the best advice from the most proper sources, but this backfired. The implication was that Kepler's friendship had not mattered as much as the push to convert him, and Guldin had referred the matter to a higher authority. His letter back to Kepler was riddled with heavy-footed theological argument. When Kepler received Guldin's response, he smelled betrayal. He had already experienced enough of clergymen who could not keep confidences in the Lutheran church, and he considered his letters to Guldin to be confidential, between friends, and not a matter for public discussion within the Jesuit order. He understood quite well what the Jesuits, and indeed the emperor, wanted from him, but he could not give it. After reading Guldin's response, he sent a short letter back restating his personal reasons for holding the beliefs he held without trying to muster any theological arguments to support them. And that was that. The debate was over.

So Kepler was alone. He held to his beliefs and would not change them. His beliefs kept him outside of all the churches, and that was the way it had to be. But even so, he felt the loneliness deep inside him. True believers on both sides, Protestant and Catholic, friends on both sides, fretted about his salvation. During the long struggle with his own church, Lutheran friends had often written to him and worried about his apparent fall from the true faith. And what would this mean for the salvation of his soul? Now Guldin was doing much the same. "I assure you seriously that I have never been farther from toying with my salvation," Kepler wrote to him.

As before in Graz, Kepler's faith had put him in an uncomfortable position. He had rejected the Jesuits' advances, and therefore the emperor's. Württemberg no longer accepted him. Many of his former Protestant patrons had been executed or exiled, so his list of powerful friends was growing shorter. Suddenly, as if from heaven, Kepler found himself with a new patron—Wallenstein himself, the new general colonel commander-in-chief of the emperor's armies. But it was an odd relationship; both men were interested in the comings and goings of the heavens, but for very different reasons. Wallenstein was an astrology addict. Like every egomaniac, he believed that he had a destiny that the stars would sooner or later reveal to him. To a certain degree, Kepler was responsible for this.

The story began in 1608, while Kepler was at the height of his career as the imperial mathematician. He was visited by a Dr. Stromair, a well-known physician, who had come to Prague partly to ask Kepler to cast a nativity, a birth horoscope, for an important man. This important patron would not give his name, however, and preferred to remain in the shadows. This was not too unusual in astrological circles, but one can still imagine the doctor providing Kepler with just enough information to do the horoscope, and then assuring him that the man was indeed a powerful lord and would be most grateful for Kepler's cooperation, and his discretion. Of course, the man wished to remain anonymous for political reasons.

Kepler responded to this request, as he always did, with his usual caveat: "My work is intended for people who understand philosophy, not for those whose minds are infected with credulous ideas, those who think that an astronomer should be able to predict particular events in a person's life and pluck future eventualities from the heavens."[8] For Kepler, astrology revealed more about a man's character than about his future. He well understood its limitations and warned everyone who came to him for a peek into the future that the astrological arts were of little use in predicting the particular events of life. For Kepler, the heavens said more about God than about destiny, and he would allow no one to go away mystified on this account. To do less than this would be dishonest. However, Dr. Stromair was a man with a reputation for learning, a man who understood philosophy and the limitations of the astrological arts, and therefore, Kepler agreed to a cast the horoscope on the basis of the doctor's recommendations. Taking the birth information, Kepler wrote a long description of his shadowy patron, a description that was quite accurate in the end:

One could honestly describe this man as alert, quick, industrious, of an impatient disposition, with a passionate craving for innovation, who dislikes the common run of humanity and human relationships, but who struggles for new, unproven, or otherwise abnormal means of human endeavor, yet is much more thoughtful than he lets anyone see. Saturn ascendant gives rise to a profound, yet melancholic nature, and carries with it a penchant for alchemy, magic, sorcery, communion with the spirits, contempt for the human laws and human customs, and of all religion, which infuses in him a suspicion of everything which either God or humans do, as if everything was a confidence game, and much more has been hidden from people than is generally believed.

For with this man, it can also be seen that he possesses a great eagerness for glory and a need to pursue worldly honors and worldly power, and because of this he will make many dangerous enemies, both hidden and revealed, but he will be able to overcome them. It can also be seen from this nativity that this man has many similarities to the former Polish chancellor, the English queen, and people like them, for these people have many planets standing around the horizons in positions of rising and setting. For such reasons, one cannot doubt, if he pays proper attention to the events in the world, that he will attain the high honors he seeks, along with great wealth, and after making friends at court for himself, he will also find a high-ranking lady, and make her his wife.[9]

Wallenstein was then only twenty-five years old, but was already hungry for news of his future and believed that he possessed some magnificent destiny. Despite the cloak and dagger, it is possible, even likely, that Kepler knew the identity of his shadow patron. Wallenstein had the look about him of a man whose stars were happy. If Kepler knew who was behind the request, then he read him well and gave him just enough detail with just enough obscurity to make him think that the horoscope had come from the gods themselves.

But then suddenly, in 1624, as Kepler's life seemed to run into a box canyon once again, Wallenstein appeared once more. Sixteen years after casting the original horoscope, Kepler received a letter from Gerhard von Taxis, an officer working for Wallenstein and also the man who originally sent Dr. Stromair to visit Kepler. That letter contained his original manuscript of the horoscope along with a cover letter written by von Taxis. Von Taxis was now an important man himself, the captain general of Friedland. But once again, he kept the name of his famous patron a secret, and, if he also knew that Kepler had known all along who is patron was, he kept that a secret as well. The letter from von Taxis asked that Kepler revisit his original horoscope for his secret patron, because much had changed in sixteen years, and the patron wanted to know more specific details about his destiny. Kepler found a series of notes scribbled onto the manuscript in the margins, notes that had been penned by Wallenstein himself, responding to the predictions and making comments about what he believed was true about Kepler's predictions.

What was Kepler to do? His service with the emperor was coming to an end, all because of his most inconvenient conscience, and, well, there was money in doing the horoscope. Wallenstein had offered him a sizable stipend for the work. Kepler took on the task, because he had nothing better going at the moment. Once again, he warned Wallenstein that only a fool is led around by the wild predictions of an astrologer. Only a fool would "let himself be used like an entertainer, actor, or a mountebank. There are many young astrologers who are perfectly willing to play such a game, and anyone who likes to be tricked while still awake, may seek them out and enjoy their theatrics. Philosophy and therefore real astrology testify to the works of God, and are therefore sacred and not frivolous, and I for one do not wish to dishonor them."[10]

Perhaps it was Kepler's honesty that drove Wallenstein to pursue him. Certainly, he could have picked any one of a thousand young astrologers to tell him exactly what he wanted to hear. But Kepler was honest and only half believed in astrology, but then again, his predictions, for all of his doubts, all too often came true. He was a very good soothsayer, whether he believed in it or not. No one could deny that. Even so, although Wallenstein evidently trusted Kepler, he did not really want to hear his warnings and exceptions. He wanted to know other things. He wanted to know if, when he died, would he die of a stroke? Would he die in some foreign land far away from his homeland? If he chose to leave his native land, would he find wealth and power in this new land? Should he stay in the military? If so, where would he serve? Would he continue to have good luck as a soldier? Who were his enemies, and what would *their* astrological signs be?

Kepler answered Wallenstein's questions rather harshly. Anyone, he said, who turned to the heavens to find answers for daily problems or to peer at his future destiny without reference to his own behavior and personality never learned anything in school. Still, Kepler did try to answer Wallenstein's questions. He did so, however, with demands that his patron keep Kepler's original warnings in mind and on the basis of reason. "A ruler who believed in astrology as much as this person does, and knew all this, would, without any hesitation, send a commander with such a remarkable constellation against his current enemies, if he were sure of the man's faithfulness."[11]

By looking at the celestial particulars of his patron's birth, Kepler was able to make certain predictions that Wallenstein found useful. Because of the complexities of their different stars, Kepler predicted that Wallenstein and Emperor Ferdinand, while able to work together for the most part, would often find their relationship difficult. Finally at the end, Kepler's predictions for Wallenstein, as they had sometimes done in the past, got spooky. Kepler stopped his horoscope at 1634, stating that "a horrible disorder" would assail him during the month of March. This was close enough, for Wallenstein was murdered by assassins on February 25, 1634. The general colonel commander-in-chief at once offered his patronage to Kepler for all of the work he had done on the horoscope. His patronage, he said through von Taxis, could be a great advantage to the gentleman.

Kepler, with really no other prospects except to leave the empire entirely and go off to Italy, or even England, accepted Wallenstein's patronage. The emperor, as Kepler expected, was not satisfied with Kepler's modifications on his Catholicity, modifications that sounded suspiciously Lutheran. One of the honors that Ferdinand had lavished on his most successful general was the duchy of Sagan in Silesia, and Wallenstein quickly offered Kepler a comfortable residence there, a quiet place to work, and the use of a printing press in a place where Protestants were still permitted to practice. What more could Kepler ask for? Wallenstein promised him a salary of 1,000 gulden a year, and by taking residence there in the general's territory he could practice both his religion and his astronomy without question. No one would bother him there. Kepler had been right about Wallenstein—he was loyal to his emperor, but tepid in matters of religion. The only qualm that he felt was about Wallenstein himself. The man had risen so far so fast, that he was bound to fall sooner or later. Wallenstein was almost a character in a fable, he fit the type so well. Sooner or later, one of his enemies would get him, and that would leave Kepler a pilgrim once more.

But Sagan never quite worked out. Kepler had met up with his family in Prague, and they traveled the rest of the way to Sagan, carrying much of what they owned in a wagon, though they had left a good portion of their household items in Regensburg, including parts of Kepler's library, a few globes, and his astronomical instruments. They arrived in midsummer, July 20, 1628. The house was comfortable enough, but very quickly they realized that Silesia was a foreign country. Austria had been much like Swabia, but Silesia was far different from either. None of them spoke the Silesian language adequately, and the people in Sagan often made them feel like barbarians. As for an intellectual life, Kepler had to rely on his correspondents to supply that. Not for the first time in his life, Kepler felt lonely and depressed.

Moreover, it soon became clear that Kepler was not going to receive any of the money owed him by the imperial treasury. The emperor had handed over the responsibility of getting Kepler his money to Wallenstein, something that the general did not want to handle. He certainly did not want to pay Kepler the 12,000 accumulated gulden the treasury owned him, and so he offered the astronomer an estate and a title, none of which materialized. Then he offered Kepler a position at the University of Rostock, but Kepler did not want that. Rostock was in the middle of the war zone. Then, in the middle of all that, war broke out in Sagan between Protestants and Catholics, as it had done everywhere else in the empire. Wallenstein was generally an easy ruler on matters of religion, but this time he had to act, and suddenly the Counter-Reformation arrived in Sagan. The Jesuits soon followed. Grabes von Nechern, Wallenstein's captain general, took charge of the conversions. He had been raised a Protestant himself, but what was that in relation to his lord's command? In 1628, he published the order that everyone in the duchy would convert to Catholicism or emigrate—that was the only choice.

In some ways, the persecution of Protestants took a nastier turn in Sagan than it had in either Graz or in Linz. Von Nechern, following the Jesuit lead, required that everyone attend the Corpus Christi processions. Lutheran churches had to pay the salaries once given to their Lutheran pastors to the Jesuit college instead. Obeying the new rules, they gathered all the books they considered heretical and burned them in the public square.

Kepler's position with Wallenstein kept him free of the effects of these new rules, but he had suddenly become a suspicious character. People who once greeted him in the street suddenly kept away from him, and even his neighbors avoided him in fear. As the only tolerated Protestant in the city, Kepler was watched by von Nechern's agents, and those who befriended him had the same shadow of suspicion cast over them. Although Kepler was not in danger himself, his neighbors were.

There were, however, a few sweet occurrences in Kepler's life at this time. The first was that he was finally able to set up his press, the one that Wallenstein had promised him. He had been searching for an appropriate one for several years and, after a few false starts, found one in Leipzig. To accomplish this, Kepler had to travel from Sagan to Görlitz, and then on to Friedland to make all the arrangements. Wallenstein then offered Kepler twenty fresh bales of paper each year, a treasure, and an extra 20 gulden a week for miscellaneous expenses. Kepler was in business. Within a short time, he published his *Ephemerides*, astronomical tables, for 1629. On top of that, he began work on a piece that had been sitting on his desk since his university days in Tübingen. This was the *Somnium seu Astronomia Lunari*, or *The Dream, or Astronomy of the Moon*. Really a work of science fiction, this was a little fable, with massive commentaries, about a man whose mother introduced him to the demons of the air, who carried him to the moon, where he described what he saw from there. This little fabulous exercise began with a debate that Kepler wanted to stage during his university days, in which he would describe what the earth would look like from the moon within a Copernican universe. But the debate never went anywhere, because the virulently anti-Copernican professor Vitus Müller refused to let Kepler stage his debate. Tales of this little story had even come up during his mother's witch trial, as evidence of both Katharina's and Johannes's odious associations with the devil. But in 1629, Kepler picked up the fable once again and organized it into a little work of scientific contemplation.

The other sweet event was the wedding of his daughter Susanna to Kepler's assistant Jakob Bartsch, the young man who would, after Kepler's death, help collect his papers and publish the *Somnium*. Bartsch married Kepler's daughter Susanna after some romantic negotiation at the instigation of Kepler himself. He admired the young man and thought him a good match for his daughter. At the time, Bartsch was studying mathematics in Strasbourg with Philip Müller, an old correspondent of Kepler's. The only reservation Kepler had about him was that he was studying to become an astrologer. After some deliberation, they decided that even if Kepler and his wife, Susanna, would not be able to make the wedding, because they would have to travel through several war zones, the wedding should then take place in Strasbourg, far from the conflict. The very afternoon of the wedding, on March 12, 1630, young Bartsch also graduated as a doctor of medicine. The next month on April 18, Kepler's wife,

Susanna, who was still in her child-bearing years, gave birth to another daughter, Anna Maria. It was possibly the happiest time in Kepler's life.

Then the world collapsed once again. The emperor promulgated the Edict of Restitution, which commanded that all territories taken by the Reformation under the control of the Habsburg emperors should immediately introduce the Counter-Reformation into their territories. This meant that Württemberg, so long exempt from Ferdinand's program, had finally come under the ax, and what Kepler had long before written to Mästlin and Hafenreffer had finally came true. The Thirty Years' War, which seemed to have been winding down, suddenly caught fire once again, and as the fire burned on Württemberg burned along with it.

In response to some political encouragement by Richelieu, Gustavus Adolphus of Sweden decided to weigh in on the side of the Protestants, and in June 1630 he invaded Germany at Pomerania. Wallenstein saw at once how much danger the empire was in and readied himself for war. Oddly enough, it was at this moment that Ferdinand removed him from command, the one moment when he needed Wallenstein the most. Ferdinand had become involved with his attempts to get his son Ferdinand set up to succeed him as emperor, and so he called a congress of imperial electors to meet in the traditional city of Regensburg. Young Ferdinand had already received the thrones of Hungary and Bohemia, and all that was left was his election as imperial heir. However, the electors balked. They were unhappy with Ferdinand, unhappy especially with Wallenstein, who levied outrageous taxes in his territories in order to support the war, and also because, well, he was simply too successful.

Maximilian, the Duke of Bavaria, who had been rewarded with Friedrich V's electoral privileges after supplying Ferdinand with the army he needed to crush the rebellious Bohemians after the Defenestration of Prague, was in secret contact with Richelieu. Jealous of Wallenstein's many successes, he intrigued at the congress to form an anti-Wallenstein resistance. They refused to vote for young Ferdinand unless the emperor broke with his favorite general. The emperor had had his own difficulties with the general colonel and agreed. Wallenstein was out, and there was no one strong enough to fight Gustavus Adolphus.

In the middle of this, Kepler decided to travel to the congress, and from there to Linz. He owned several government bonds from there, one for 2,000 gulden and one for 1,500, and wanted to ensure that they paid the promised interest, which until that time they had avoided doing. After some negotiation, the representatives in Linz told him to appear before them on St. Martin's day, which was November 11, and they would see what they could do about paying him.

Silesia was in the midst of battle at the time, as was much of the empire. Kepler left Sagan with little hope of his return. Whether he had a premonition or was suffering from depression once again is uncertain, but Bartsch later wrote that Kepler expected that his family would see the Day of Judgment before they would see him again. For the trip, he took a number of books, some clothes, and documents that contained "all his wealth." Before he left, he wrote to his friend Johann Bernegger in Strasbourg, and gratefully accepted his invitation to bring his family to that city as a refuge from the endless war. "Hold tight with me to the only true anchor of the church—pray to God for our church and for me."

On November 2, 1630, after a long, exhausting ride through winter weather, Kepler rode a half-dead horse into Regensburg across the stone bridge spanning the Danube. He came to a house, now Keplerstrasse 5, at that time owned by an acquaintance of Kepler's, an innkeeper named Hildebrand Billj, who put him up. Kepler was there for only a few short days, exhausted, drained, and depressed, when he caught a cold that turned into a fever. Much of his life, he had suffered with fevers, agues, and bouts of fatigue, possibly from an array of recurring viruses that Kepler, in his often depleted state, had become vulnerable to. The medical arts being what they were at the time, Kepler believed that his fevers had come from fire pustules, *sacer ignis,* hidden in his body. Very quickly, however, the fever spiked, and Kepler's life was in danger. Billj sent for a doctor, who bled Kepler to balance his humors.

Supposedly, as Emperor Ferdinand was about to leave on his ship, he heard of Kepler's illness and sent a group of his retinue of gentlemen to visit him with the emperor's prayers and good wishes, along with 25 Hungarian ducats for the doctor. Kepler became delirious after that and drifted in and out of consciousness. Several Lutheran pastors came to visit him in his last hours, though they would not share Communion with him. According to one report, Kepler told one of the pastors that all he wanted to do was bring peace between Lutherans, Catholics, and Calvinists, and the pastor responded that one might as well try to reconcile Christ with the devil. The last clergyman to come, Pastor Christoph Sigmund Donauer, sat by his bedside and asked Kepler, moments before he died, about how he hoped to claim salvation. Kepler said to him, his voice weak and trailing away, that "his hopes were solely in Christ, from whom comes all solace, welfare, and protection." A few minutes later, his breath, ragged and weak, gave way, and Kepler died.

Three-quarters of a mile inland from the river and from the Billj house on what is now Keplerstrasse, there was the Protestant Cemetery of St. Peter, and on November 17, 1630, they buried in that place Johannes Kepler, the boy his own family did not want, the boy who had grown to become the great man who proved Copernicus and helped set the course of science. The funeral was crowded with great men, all come to the city for the congress, many who knew him, all who knew of him. Pastor Donauer preached the sermon: "Blessed are they who hear the Word of God and keep it" (Luke 11:28). They then carved on his tombstone, which was later lost, a short poem that Kepler himself had written only months before. Was it depression or premonition?

Mensus eram coelos,
nunc terrae metior umbras.
Mens coelistis erat,
corporis umbra jacet.

I used to measure the heavens,
But now I measure the shades of the earth.
Although my soul was from heaven,
The shade of my body lies here.

Some claimed that they saw balls of fire fall from heaven the night they buried Kepler. Perhaps earth's war had touched heaven at last. Even dead and buried, Kepler could not escape the war. A few years later, Gustavus Adolphus attacked Regensburg. Many Protestants had come to that city as refugees, and they were caught between the forces of their emperor and the forces of their fellow Lutherans. A battle took place, in which cavalry overran the Cemetery of St. Peter, and destroyed Kepler's gravestone. No one really knows where he is buried today, though there is a monument.

☙❧

I HAVE BEEN TO THAT PLACE. After staying in Weil der Stadt, I traveled on to Prague, where I had words with the German student in the train and made peace with him while talking about Kepler. On the way back, I stayed over in Regensburg for a day, and after visiting Hildebrand Billj's house and the room where Kepler died, I walked up to the park where the cemetery once stood. The memorial that reminds visitors of Kepler's grave stands in an open glade sheltered by elm trees and Japanese pine. An old oak, gnarled and twisted at the base, as if the wood has melted over time, stands nearby. It looks as if it has a beer belly. Across the bicycle path is a fountain, and lovers sit on the orange-painted benches, nuzzling one another in the sunlight. Kepler is alone, however, for his memorial is off to one side, untended and partly forgotten. An old man stops by and we speak in the no-man's-land between the languages. He asks me whose bust is on the pedestal, and I tell him Kepler's. He says, "Ah, Kepler!" as if the two were old friends.

The bust is under a concrete cupola supported by eight columns. Someone has marked black dots on his eyes, making him look like a sorcerer. All around are beer bottles—Löwenbrau—crushed cans and cigarette butts, as if one more recent army has camped out nearby. Two graffiti written on one column, in German and English, are *"Antipolitische Volkspartei, Freiburg"* ("Antipolitical People's Party, Freiburg") and "Never Forget the White Rose of Resistance!"[12]

Somehow, this seems appropriate. I say a prayer and leave.

Notes

INTRODUCTION

1. I. Bernard Cohen, "Kepler's Century," in *Kepler: Four Hundred Years*, ed. Arthur Beer and Peter Beer (Oxford: Pergamon Press, 1975), pp. 11–12.

CHAPTER 1

1. One of the most complete accounts in English of this trial and the events leading up to it can be found in Max Caspar, *Kepler*, trans. and ed. C. Doris Hellman (New York: Dover, 1993), pp. 240–58. This book is perhaps the best biography of Kepler.
2. "Terrorization by word" consisted of showing the accused the instruments of torture and explaining their function, accompanied by verbal abuse and accusation.
3. Berthold Sutter, *Der Hexenprozess gegen Katharina Kepler*, 2d rev. ed. (Weil der Stadt: Kepler Gesellschaft, Heimatverein Weil der Stadt, 1984), p. 37.
4. Sutter, *Der Hexenprozess*, p. 111.

CHAPTER 2

1. The German *"Pfui Teufel"* is a common expression when tasting or seeing something disgusting. Anyone would have used the expression, but it has ironic significance since the Kepler woman is accused of lending a hand to the devil's trade.
2. Christian Frisch, *Joannis Kepleris Astronomi Opera Omnia*, 8:670–71.
3. Dave Gavine, "James Melvill Sees a Great Comet in 1577," *Astronomical Society of Edinburgh Journal* 40.
4. Berthold Sutter, *Der Hexenprozess gegen Katharina Kepler*, 2d rev. ed. (Weil der Stadt: Kepler Gesellschaft, Heimatverein Weil der Stadt, 1984), p. 36.

CHAPTER 3

1. The affectation of writing in the third person was merely the proper form for horoscopes, which were often anonymous. The astrologer was supposed to

see what was to be seen in the stars without reference to outside sources of information. The subject of the horoscope was often, therefore, not named. Because Kepler is writing his own horoscope, he cannot help slipping back and forth between first and third person. In this case, Kepler's horoscope is in fact a thoroughgoing self-criticism, a listing of his faults in the service of moral growth.

2. Johannes Kepler, *Selbstzeugnisse,* ed. Franz Hammer (Stuttgart: Frommann, 1971), p. 26.

3. Charlotte Methuen, *Kepler's Tübingen* (Brookfield, VT: Ashgate, 1998), p. 46.

4. The first thing that Ulrich did in his education reform in 1535 was to restructure the university in Tübingen, which meant weeding out those professors who were loyal to the pope and emperor rather than the duke himself and those who were even suspected of such loyalties. But the reform went too far, and many of the faculty left, even some the duke wanted to keep. Moreover, reforming the faculty did not build up the university in general; it did not attract more students, nor did it help fill the ranks of much needed Lutheran pastors. To solve this problem, Ulrich set up the *Herzogliches Stipendium,* or *Stift,* modeling it on a scholarship system already established in Hesse. On February 5, 1536, the duke instructed Tübingen's town council, along with the councils of every town in Württemberg, to set aside 25 gulden to support three students at the university the next year, and 25 gulden for each year thereafter.

But Ulrich's son Christoph knew that no reformation of the *Stift* in Tübingen could survive without a good system of feeder schools to select out those few with the right qualities to move on to service in the duchy, and so he set up the system of German and Latin schools. Through this system, Christoph wanted to offer education to "children of poor, pious people, of hard-working, Christian, God-fearing character and background, and suited for study." Such children would then study to become Lutheran pastors or perhaps teachers or sometimes clerks. These children of the poor would then grow the ranks of the new middle management of Württemberg.

5. We must take care when bringing the words *astrology* and *science* together. Astrology was losing its scientific credibility at this precise moment in history. Few today would acknowledge astrology as a science, but nearly everyone did in the seventeenth century, and it was Kepler and Galileo who finally ended this.

6. The word "monastery" here is not to be read figuratively, for the schools were in fact monastic in structure and tradition. While there, Kepler was be-

ginning his preparation to be a Lutheran pastor, and though he never landed in the ministry, he spent most of his youth studying for it. Modern prejudice would expect that Kepler was educated as a scientist, in the scientific method, in a laboratory or observatory somewhere, but there was no such thing as a "scientist" per se in Kepler's day, for it was Galileo, Kepler himself, and later Newton who eventually set the pattern.

In Kepler's time, then, the primary course of study for a young intellectual was either law or medicine or the church, and the oldest schools were the monasteries. In strict monastic order, during the summer prayers started at four in the morning, as the deep night thinned into the coming day. In winter, prayers started at five, not because the boys needed to sleep, but because the sun rose later and without electric lights the school hallways must have seemed like tunnels. The boys attended Communion at least six times a year and took turns reading aloud the daily lesson. After prayers, they marched off to *lectio theologica*, or theological study, largely the reading of biblical texts, stressing theological and grammatical analysis. Each boy sat upon a backless wooden bench at a tilted wooden desk. Paper was too precious for note taking, so most of the learning was by rote and recitation. Then, of course, the principal would preach to the entire school and give the boys the proper Lutheran interpretation of the text at hand. Often these sermons were contentious, handling one at a time the latest theological conflict between Protestant and Catholic or sometimes between Protestant and Protestant. In the 1580s, during the time Kepler studied at Adelberg, these sermons would have targeted the Calvinists' doctrine of predestination and their doctrine on Holy Communion, a question that would later be the cause of Kepler's excommunication from the Lutheran church.

7. Max Caspar, *Kepler*, trans. and ed. C. Doris Hellman (New York: Dover, 1993), p. 41.

8. Caspar, *Kepler*, pp. 38–39.

9. John Hudson Tiner, *Johannes Kepler: Giant of Faith and Science* (Fenton: Mott, 1999), p. 35.

10. Martha List, "Kepler as a Man," in *Kepler: Four Hundred Years*, Vistas in Astronomy, vol. 18, ed. Arthur Beer and Peter Beer (Oxford: Pergamon, 1975), p. 97.

CHAPTER 4

1. Carolly Erickson, *The First Elizabeth* (New York: St. Martin's Press, 1983), pp. 364–77.

Notes

2. The ritual was a prime source of money for the university library, so no one wanted to abandon the rite until it got just too silly, which happened around 1819. Walter Jens, *Eine deutsche Universität* (Munich, 1977).
3. Jürgen Sydow, "Kepler's Homeland—Württemberg," in *Kepler: Four Hundred Years,* ed. Arthur Beer and Peter Beer (Oxford: Pergamon Press, 1975), pp. 136–37.
4. Goethe, *Faust,* part 1, lines 828–31.
5. Sydow, "Kepler's Homeland—Württemberg," pp. 136–37.
6. Peter Matheson, *The Imaginative World of the Reformation* (Minneapolis: Fortress, 2001), pp. 42–43.
7. This translation was taken from Robert E. Smith, who wrote it as part of Project Wittenberg. The versification is my own. *http://www.iclnet.org/pub/resources/text/wittenberg/prayers/morning.txt.*
8. John Hudson Tiner, *Johannes Kepler: Giant of Faith and Science* (Fenton: Mott, 1999), p. 47.
9. Anthony Grafton, *Cardano's Cosmos: The Worlds and Works of a Renaissance Astrologer* (Cambridge: Harvard Univ. Press, 1999), p. 6.
10. Tiner, *Johannes Kepler,* p. 50.
11. Max Caspar, *Kepler,* trans. and ed. C. Doris Hellman (New York: Dover, 1993), p. 44.
12. Tiner, *Johannes Kepler,* p. 39.
13. Kepler, *Mysterium Cosmographicum,* GW i, 9:11–23; GW viii, 23:11–23.
14. Ioan P. Couliano, *Eros and Magic in the Renaissance* (Chicago: Univ. of Chicago Press, 1987), p. 23.
15. Couliano, *Eros and Magic,* p. 24.
16. Couliano, *Eros and Magic,* p. 4.
17. Nicolaus Copernicus, *On the Revolutions of the Heavenly Spheres,* trans. Charles Glenn Wallis, Great Minds Series (Amherst: Prometheus, 1995), p. 3.
18. G. J. Toomer, *Ptolemy's Almagest* (Princeton, NJ: Princeton University Press, 1998), p. 41.
19. Tiner, *Johannes Kepler,* p. 45.
20. Ferguson, Kitty, *Tycho and Kepler: The Unlikely Partnership That Forever Changed Our Understanding of the Heavens* (New York: Walker, 2002), p. 64.

CHAPTER 5

1. Berthold Sutter, "Kepler in Graz," in Arthur Beer and Peter Beer, eds., *Kepler: Four Hundred Years* (Oxford: Pergamon, 1975), p. 140.

2. Letter from Kepler to the school inspector in Graz, April 19, 1594. In Justus Schmidt, *Johann Kepler: Sein Leben in Bildern und eigenen Berichten* (Linz: Rudolf Trauner, 1970), p. 223.
3. Schmidt, *Johann Kepler,* p. 223.
4. Johannes Kepler, "*Revolutio anni,* 1594" in Christian Frisch, *Joannis Kepleris Astronomi Opera Omnia,* viii, 2. S. p. 677.
5. Darwin Porter and Danforth Prince, *Frommer's Austria* (New York: Hungry Minds, 2001), p. 438.
6. James R. Voelkel, *Johannes Kepler and the New Astronomy* (New York: Oxford, 1999), p. 25.
7. Kepler's journal for 1595, in Frisch, *Opera Omnia,* viii, 2, nr. 683.
8. Voelkel, *Johannes Kepler,* p. 27.
9. Voelkel, *Johannes Kepler,* p. 27.
10. Letter to Michael Mästlin in Tübingen, Graz, October 30, 1595. In Schmidt, *Johann Kepler,* p. 223.
11. A *Werkschuh* was an ancient German linear measure somewhere between a foot and a yard long, which varied according to the use and was far from precise. Surprisingly, Kepler uses the unit here, even though he didn't have much trust in it.
12. Schmidt, *Johann Kepler,* p. 224.
13. *http://www.mathacademy.com/pr/prime/articles/platsol/index.asp.*
14. Johannes Kepler, *Mysterium Cosmographicum,* trans. A. M. Duncan, intro. and comm. E. J. Aiton (Norwalk: Abaris, 1999).

CHAPTER 6

1. Kepler, *Mysterium Cosmographicum,* GW i, p. 53.
2. Kepler, *Mysterium Cosmographicum,* p. 53.
3. Cf. Max Caspar, *Kepler,* trans. and ed. C. Doris Hellman (New York: Dover, 1993), pp. 72–73. Caspar's description of the marital arrangements is delicious, but he doubts that Jobst Müller insisted on proof of Kepler's nobility. I tend to disagree, mainly because I think that, given the kind of man Müller was, there would have been little other reason for him to agree to the marriage and, frankly, little other reason for Barbara to have been interested in Kepler in the first place. The Müllers were largely clueless about Barbara's suitor's potential for greatness and never understood it even after he became the emperor's mathematician.
4. Oddly enough, the Müllers achieved nobility on their own soon after Jobst died in 1601. In 1623, the emperor knighted Jobst's son Michael for the

many great deeds and services—mostly financial—that the Müller family had rendered to the empire and to the ruling house of Austria. From that time on, he was permitted to sign himself *"von und zu Mühleck"* and to use red wax to seal his letters. The title didn't last long, however. Michael had no male heirs, and so nobility died with him.

5. J. Hübner, *Die Theologie Johannes Keplers,* pp. 165–75. In Charlotte Methuen, *Kepler's Tübingen* (Brookfield, VT: Ashgate, 1998), pp. 205–6.
6. Kepler, *Mysterium Cosmographicum, GW* viii, 9–20. In Caspar, *Kepler,* p. 63.
7. Quoted in Michael Walter Burke-Gaffney, S.J., *Kepler and the Jesuits* (Milwaukee: Bruce, 1944), pp. 3–4.
8. Caspar, *Kepler,* p. 65.
9. Kepler's journal for 1597, Frisch, *Opera Omnia,* 8, 2, p. 689. In Justus Schmidt, *Johann Kepler: Sein Leben in Bildern und eigenen Berichten* (Linz: Rudolf Trauner, 1970), p. 226.
10. Burke-Gaffney, *Kepler and the Jesuits,* pp. 6–9. The letter to Mästlin was also quoted here. I have followed Burke-Gaffney's succinct account fairly closely.
11. Letter to Mästlin, in Schmidt, *Johann Kepler,* p. 226.

CHAPTER 7

1. Some in Weil der Stadt refer to Charles as "Banana Charlie" because of the floppy hat on his statue, which stands in the market square; the hat looks like a peeled banana.
2. Christopher Hibbert, *Rome: The Biography of a City* (London: Penguin, 1985), pp. 153–62.
3. Max Caspar, *Kepler,* trans. and ed. C. Doris Hellman (New York: Dover, 1993), p. 77.
4. Letter of Kepler to Mästlin, June 11, 1598, *GW* xiii, nr. 99:379–82. Quoted in Caspar, *Kepler,* p. 78.
5. Caspar, *Kepler,* p. 78.
6. Kepler's journal, 1598, in Frisch, *Opera Omnia,* 8, 2, p. 699.
7. Letter from Kepler to Mästlin, June 11, 1598, *GW* xiii, nr. 99–360f. In Caspar, *Kepler,* p. 77.
8. Letter from Kepler to Georg Friedrich von Baden, October 10, 1607, in *GW* xvi, nr. 451. Quoted in Caspar, *Kepler,* pp. 81–82.
9. Letter to Herwart von Hohenberg, December 9, 1598, in Justus Schmidt, *Johann Kepler: Sein Leben in Bildern und eigenen Berichten* (Linz: Rudolf Trauner, 1970), p. 227.

10. Letter to von Hohenberg, December 9, 1598, in Schmidt, *Johann Kepler,* p. 227.
11. Caspar, *Kepler,* p. 82.
12. *GW* xix, 337, nr. 7.30. In Caspar, *Kepler,* p. 82.
13. Letter to von Hohenberg, December 9, 1598, in Schmidt, *Johann Kepler,* p. 227.
14. Caspar, *Kepler,* pp. 80–81.
15. Letter from Mästlin, July 4, 1598, *GW* xiii, nr. 101.
16. Letter from Tycho, April 1, 1598, *GW* xiii, nr. 92.
17. Ferguson, Kitty, *Tycho and Kepler* (New York: Walker, 2002), p. 233. The original letter has been lost, except as a copy that Kepler had also sent to Mästlin.
18. Letter to Mästlin, February 26, 1599, in Schmidt, *Johann Kepler,* p. 229.
19. Letter to Herwart von Hohenberg, April 10, 1599, *GW* xiii, nr. 117:174–79.
20. *GW* iv, 308:9–10.
21. Letter to Herwart von Hohenberg, August 6, 1599, *GW* xiv, nr. 130–226f.
22. Letter to Herwart von Hohenberg, July 12, 1600, *GW* xiv, nr. 168:109–11. In Caspar, *Kepler,* pp. 102–3.
23. Letter to Mästlin, September 9, 1600, *GW* xiv, nr. 175:52–56.

CHAPTER 8

1. Eduard Petiška and Jan Dolan, *Beautiful Stories of Golden Prague* (Prague: Martin, 1995), pp. 5–7.
2. Ferguson, Kitty, *Tycho and Kepler* (New York: Walker, 2002), p. 267.
3. Letter to Mästlin, December 16, 1600, *GW* xiv, nr. 180:6f, in Max Caspar, *Kepler,* trans. and ed. C. Doris Hellman (New York: Dover, 1993), p. 118.
4. Letter to Mästlin, December 20, 1601, *GW* xiv, nr. 203:24–26, in Caspar, *Kepler.*
5. *Tychonis Brahe Dani Opera Omnia,* 10:3/Rosen, 312, 313. Quoted in Ferguson, *Tycho and Kepler,* p. 283. After Tycho's funeral, the talk on the streets in Prague proposed that he had been poisoned. This was not entirely unreasonable, since some of his symptoms could have been caused by an overdose of heavy metals. Certainly, Tycho had his enemies at the court, since many feared that as a Lutheran he had far too much influence over the unfortunate Rudolf. The Catholic council was not above a bit of murder, to be sure. However, the most likely case is that Tycho, who practiced homeopathic medicine, which was considered part of an astrologer's overall function, overdosed on some of his own medicine. Many of the remedies he concocted were high in mercury, and if his bladder infection proved stubborn

against his cures, he may have simply kept taking them in larger doses until he poisoned himself.

Forensic studies of Tycho's DNA, found in a small box containing bits of Tycho's beard and a small piece of his shroud, were done in 1991. Any heavy metals in the body would be present in the hair strands, and because hair grows at a regular rate, concentrations could be graphed over time by taking different points on the strand of hair as time points. Tycho's beard showed a high concentration of lead, which meant he could have died from lead poisoning, but that by itself was not all that conclusive, because many people who lived in cities at the time may well have had high lead concentrations for environmental reasons, most notably the use of lead in plumbing. There was a little arsenic in the hair, but there was a much larger than normal concentration of mercury. Also uremia may be caused by mercury poisoning.

In 1996 another test was conducted, using the PIXW (Particle Induced X-ray Emission) method on a piece of Tycho's beard that included some of the root. This test showed that mercury had been ingested shortly before Tycho's death, which meant that he probably died of a mercury overdose. More than likely Tycho did this to himself in an attempt to cure the urinary disorder, perhaps prostatic hypertrophy (or possibly bladder stones), that had been troubling him. It was this failed cure that caused the uremia that killed him. See Aase R. Jacobsen, *Planetarium* 30 (December 2001):4.

CHAPTER 9

1. Letter to Herwart von Hohenberg, July 12, 1600, GW xiv, nr. 168:102–4.
2. Max Caspar, *Kepler*, trans. and ed. C. Doris Hellman (New York: Dover, 1993), p. 182.
3. Angelo Maria Ripellino, *Magic Prague*, ed. Michael Henry Heim, trans. David Newton Marinelli (New York: Picador, 1995), p. 93.
4. Peter Demetz, *Prague in Black and Gold: Scenes from the Life of a European City* (New York: Hill and Wang, 1997), p. 198.
5. Demetz, *Prague in Black and Gold*, p. 198.
6. Caspar, *Kepler*, pp. 154–56.
7. Ripellino, *Magic Prague*, p. 131.

CHAPTER 10

1. Peter Demetz, *Prague in Black and Gold: Scenes from the Life of a European City* (New York: Hill and Wang, 1997), pp. 220–23.
2. Johannes Kepler, *Dissertatio cum Nuncio Sidereo*, GW iv, 281–311.

3. Max Caspar, *Kepler*, trans. and ed. C. Doris Hellman (New York: Dover, 1993), p. 200.

4. Caspar, *Kepler*, p. 201.

5. This little bit can be found in a note in Dava Sobel's wonderful biography of Galileo, *Galileo's Daughter: A Historical Memoir of Science, Faith, and Love* (New York: Penguin, 2000), p. 39.

6. Caspar, *Kepler*, p. 201.

7. Letter from Kepler to Tobias Scultetus, from Prague, April 13, 1612, in Justus Schmidt, *Johann Kepler: Sein Leben in Bildern und eigenen Berichten* (Linz: Rudolf Trauner, 1970), p. 245.

8. Caspar, *Kepler*, p. 207.

CHAPTER 11

1. A wonderful account of early modern witchcraft is Robin Briggs, *Witches and Neighbors: The Social and Cultural Context of European Witchcraft* (New York: Penguin, 1996).

2. J. Widdowson, "The Witch as a Frightening and Threatening Figure," in V. Newall, ed., *The Witch Figure* (London, 1973), p. 208. Quoted in Briggs, *Witches and Neighbors*.

3. Kepler to an unknown member of the nobility, October 23, 1613, GW xvii, nr. 669:84–88.

4. Letter of Kepler to Hoffmann, April 26, 1612, GW xvii, nr. 715:8–9.

5. Kepler, *Glaubensbekenntnis*, GW xii, 28:44–47.

6. Kepler, *Glaubensbekenntnis*, GW xii, 28:3–5.

7. Kepler to Hafenreffer, November 28, 1618, GW xvii, nr. 808:55–56.

8. Kepler, *Glaubensbekenntnis*, GW xii, 28:17–21.

9. Kepler to Mästlin, December 12/22, 1616, GW xvii, nr. 750:260–66.

10. Letter from Kepler to an anonymous nobleman, October 23, 1613, GW xvii, nr. 669:6–7.

11. Letter from Kepler to an anonymous nobleman, October 23, 1613, in Justus Schmidt, *Johann Kepler: Sein Leben in Bildern und eigenen Berichten* (Linz: Rudolf Trauner, 1970), pp. 246–47.

12. Letter, October 23, 1613, in Schmidt, *Johann Kepler*, pp. 246–47.

13. Letter, October 23, 1613, in Schmidt, *Johann Kepler*, pp. 246–47.

14. Letter, October 23, 1613, in Schmidt, *Johann Kepler*, pp. 246–47.

15. Letter, October 23, 1613, in Schmidt, *Johann Kepler*, pp. 246–47.

16. Letter, October 23, 1613, in Schmidt, *Johann Kepler*, pp. 246–47.

Notes

CHAPTER 12

1. Berthold Sutter, *Der Hexenprozess gegen Katharina Kepler,* 2d rev. ed. (Weil der Stadt: Kepler Gesellschaft, Heimatverein Weil der Stadt, 1984), p. 21.
2. In Germany, *Hexenschuss* is still a popular term for lumbago.

CHAPTER 13

1. My original term for "religious concessions" was "religious liberty," which, on reflection, was anachronistic. Religious liberty was hard to come by at the time. Religious groups, Catholic to Protestant, Protestant to Catholic, Protestant to Protestant, that wanted liberty for themselves rarely offered it to others. Religious toleration was yet an undiscovered idea and was problematic for that time. One historical prejudice that has come down to us is that intolerance was a Catholic practice and tolerance was Protestant, but the historical record does not bear that out. In the seventeenth century, religious intolerance was the game everywhere.
2. William P. Guthrie, *Battles of the Thirty Years' War: From White Mountain to Nordlingen* (Westport: Greenwood Press, 2002), p. 48.
3. Peter Demetz, *Prague in Black and Gold: Scenes from the Life of a European City* (New York: Hill and Wang, 1997), pp. 224–26.
4. There are several possibilities for the death of Frau Meyer. The potion that Katharina Kepler gave to her may well have been tainted not on purpose, but out of mishandling. Perhaps she died of botulism toxin. Or possibly she died of one of the numerous pestilences that swept through the region, and the two illnesses, hers and Beutelsbacher's, were not connected. Katharina Kepler told the jury that Beutelsbacher tried to jump over a ditch that day and possibly broke his leg, and that was why he went lame. Why an old school chum of Johannes's would want to accuse his mother of witchcraft is lost to us. He may have resented Johannes's success or may have actually connected his illness with the death of Bastian Meyer's wife. He may also have been lying.
5. According to Max Caspar, the butcher's name was Christoph Frick, not Stoffer.
6. This may have been a cocky answer under interrogation, possibly indicating some hostility toward Einhorn.
7. Admittedly, I am going out on a limb speculating about this. It is unlikely that Hafenreffer played a direct part in the trial of Katharina Kepler, and unlikely that he did much more than cluck over these terrible accusations from a distance, but his actions and the actions of the consistory certainly gave Einhorn power that he would not have had otherwise.

8. Rolbert A. Kann, *A History of the Habsburg Empire: 1526–1918* (Berkeley: Univ. of California Press, 1974), p. 49.

9. The German term *Wassersuppe* implies a not very nourishing soup and is often used metaphorically to represent "poor man's food." It is a soup made from water rather than more nourishing ingredients like meat, milk, or wine.

10. Michael Kerrigan, *The Instruments of Torture* (New York: Lyons Press, 2001), p. 81.

11. Kerrigan, *Instruments of Torture*, p. 89.

CHAPTER 14

1. William P. Guthrie, *Battles of the Thirty Years' War: From White Mountain to Nordlingen* (Westport, CT: Greenwood Press, 2002), p. 64.

2. Guthrie, *Battles of the Thirty Years' War*, pp. 59–60.

3. Peter Demetz, *Prague in Black and Gold: Scenes from the Life of a European City* (New York: Hill and Wang, 1997), p. 227.

4. Demetz, *Prague in Black and Gold*, p. 228.

5. Kepler to Wackher von Wackenfels, *GW* xvii, nr. 783:46–48.

6. Grüninger to Osiander, July 1, 1619, *GW* xvii, nr. 843:10–19.

7. Kepler to Hafenreffer, November 28, 1618, *GW* xvii, nr. 808:65–67.

8. Calendar for 1619, Frisch, *Opera Omnia*, i, 486–87.

9. *Harmonice Mundi*, *GW* vi, 289:35–39.

10. *GW* vi, 480. No citation given by Caspar. Kepler's third law of planetary motion is best expressed in modern notation as:

    $p^2/a^3 = k$

    or, the period (p) of a planet's orbit, that is, the time it takes to make one revolution, squared, i.e., multiplied by itself, divided by the mean distance (a) that the planet is from the sun, cubed, i.e., multiplied by itself two times, is equal to a constant (k). This means that the relationship between (p^2) and (a^3) remains constant throughout the motion of the planet in its orbit. This is Kepler's harmonic law because as Kepler saw it, this relationship possessed the same kind of harmony that one could find in musical chords or in colors that work together. This harmony is innate in the human soul, placed there by God as a key to understanding God's mind.

11. *Harmonice Mundi*, *GW* vi, 215:30–33.

12. *Harmonice Mundi*, *GW* vi, 223:26–35.

13. I refer the reader to Edwin A. Abbott's wonderful little science fiction book *Flatland: A Romance of Many Dimensions* (New York: Penguin, 1998), in which different shapes are given different personalities and the sharper the points that a polygon has in two dimensions, the nastier its disposition.

14. *Harmonice Mundi, GW* vi, 16:35–38.
15. *Mysterium Cosmographicum, GW* viii, 33–35.
16. *Epitome Astronomiae Copernicanae, GW* vii, 9:10–12.
17. Around that time, the authorities in Graz publicly burned Kepler's astrological calendar for 1624, his last ever, even though it had been dedicated to the representatives of Styria. One of Kepler's friends asked him if his calendar had offended the religious sensibilities of the people there, for it predicted that forcing the people to attend "hated divine services" would lead to great hardship, so "that the ordinary people would be willing to bargain a golden cup for a simple slice of bread." Kepler knew better, however, and informed his friend of the true reason. Apparently, the people in Graz did not like the fact that Kepler gave Styria second billing under that other province—Austria above the Enns. Local pride trumps religious sensibilities any day.

CHAPTER 15

1. Kepler to Schickard, April 19, 1627, GW xviii, nr. 1042:42–48.
2. Kepler to Schickard, February 10, 1627, GW xviii, nr. 1037:60–62.
3. Kepler to Guldin, February 24, 1628, GW xviii, nr. 1072:41–44.
4. Kepler to Guldin, February 24, 1628, GW xviii, nr. 1072:45–49.
5. Kepler to Guldin, Spring 1628, GW xviii, nr. 1083:85–86.
6. Kepler to Guldin, February 24, 1628, GW xviii, nr. 1072:104–10.
7. Kepler to Guldin, February 24, 1628, GW xviii, nr. 1072:114–33.
8. Frisch, *Opera Omnia*, viii, 348.
9. Gerhard von Taxis to Kepler, December 14, 1625, GW xvii, nr. 704.
10. Frisch, *Opera Omnia*, viii, 351–52.
11. Gerhard von Taxis to Kepler, September 25, 1625, GW xviii, nr. 1016:18–20.
12. The "White Rose" was a student-based, anti-Nazi movement that suffered terrible martyrdom in the 1930s.

Kepler Time Line

1527   Sack of Rome; end of the Renaissance; beginning of the Counter-Reformation.

1571   Johannes Kepler born in Weil der Stadt on December 27.

1573   Brother Heinrich born.

1575   Kepler family moves to Leonberg.

1577   Kepler's parents attain citizenship; Kepler becomes "Burgerssohn of Löwenberg."

          Kepler sees the great comet with his mother.

1577–83   Kepler attends school in Leonberg (with interruptions).

1578–79   Kepler attends the Latin school.

1579   At year end, Kepler's education interrupted due to family move to Ellmendingen, near Pforzheim.

1580–82   In Ellmendingen, "heavily burdened by farming chores."

1582–83   During the winter Kepler back at Latin school in Leonberg; probably lived with the Guldenmann grandparents in Etlingen.

1584   Family returns to Leonberg.

          Sister Margaretha born on June 26.

          Kepler attends lower cloister school of Adelberg.

1586   Kepler promoted to upper cloister school in Maulbronn.

1587   Brother Christoph is born on March 5.

          Kepler matriculates at Tübingen University.

1589   Father Heinrich leaves family for good on January 5.

          Sixteen-year-old Heinrich runs away from home.

1591   Kepler receives baccalaureate degree and begins theological studies.

1596   *Mysterium Cosmographicum* printed in Tübingen.

1597   Marriage to Barbara Müller von Mühleck on April 27.

1598	Due to Counter-Reformation measures, exiled from Graz for a short time.
1600	First meeting with Tycho Brahe in Prague at Benatky Castle.
	Final exile from Graz on August 7.
	Arrival in Prague as a refugee on October 19.
1601	Death of Tycho Brahe; Kepler appointed imperial mathematician.
1604	*Astronomiae Pars Optica* published.
1605	At Easter time, Kepler discovers that the orbit of Mars is elliptical.
1609	*Astronomia Nova* published.
1611	Death of wife, Barbara Kepler nee Müller.
1612	Excluded from Communion by Pastor Hitzler and the Württemberg consistory.
	Death of Rudolf II; ascension of Matthias to imperial throne.
1613	Marriage to Susanna Reuttinger on October 30.
1614	Brother Heinrich returns to Leonberg.
1615–16	Six witches are executed in the jurisdiction of Leonberg.
1615	*Nova Stereometria Doliorum Vinariorum* published in Linz, first book ever published there.
1617	First volume of the *Epitome Astronomiae Copernicanae* published in Linz.
1618	On May 23, the Second Defenestration of Prague and the beginning of the Thirty Years' War.
1619	Battle of White Mountain and the flight of Friedrich V, the Winter King.
	Harmonice Mundi, Book V, published in Linz.
	Death of Emperor Matthias; five months later Ferdinand II ascends to the throne.
1620	Second volume of *Epitome Astronomiae Copernicanae* published in Linz.
1602–21	Kepler assists his mother, who is released after fourteen months of imprisonment on the charge of witchcraft.
1621	Third and final volume of *Epitome Astronomiae Copernicanae* published in Frankfurt.

1626   Under pressure from the Counter-Reformation, Kepler and his family leave Linz.

1626–27   The *Rudolphine Tables* are printed in Ulm.

1626   Kepler declines to convert to Catholicism.

             Wallenstein offer his patronage and Kepler moves to Sagan; four months later the Counter-Reformation begins in Sagan.

1630   Death of Johannes Kepler in Regensburg on November 15.

1633   Galileo Galilei is tried by the Inquisition in Rome.

1647–50   Peace conference in Westphalia, slow end to the Thirty Years' War.

Source Readings

COLLECTED WORKS OF PRIMARY SOURCES

Caspar, Max, and Walther von Dyck, Franz Hammer, and Volker Bialas, eds. *Johannes Kepler Gesammelte Werke*. 22 vols. Munich: Deutsche Forschungsgemeinshaft and the Bavarian Academy of Sciences, 1937–.

Dreyer, John Lewis E., ed. *Tychonis Brahe Dani Opera Omnia*. 15 vols. Copenhagen: Libraria Gyldendaliana, 1913–29.

Frisch, Christian. *Joannis Kepleris Astronomi Opera Omnia*. Frankfurt and Erlangen, 1858–71.

INDIVIDUAL WORKS

Baroja, Julio Caro. *The World of the Witches*. Translated from the Spanish by Nigel Glendinning. London: Phoenix, 1964.

Baumgardt, Carola. *Johannes Kepler: Life and Letters*. London: Gollancz, 1952.

Behringer, Wolfgang. *Witchcraft Persecutions in Bavaria: Popular Magic, Religious Zealotry and Reason of State in Early Modern Europe*. Translated by J. C. Grayson and David Lederer. Cambridge: Cambridge Univ. Press, 1997.

Brauner, Sigrid. *Fearless Wives and Frightened Shrews: The Construction of the Witch in Early Modern Germany*. Edited by Robert H. Brown. Amherst: Univ. of Massachusetts Press, 1995.

Briggs, Robin. *Witches and Neighbors: The Social and Cultural Context of European Witchcraft*. New York: Penguin, 1996.

Burke-Gaffney, S.J., Michael Walter. *Kepler and the Jesuits*. Milwaukee: Bruce, 1944.

Caspar, Max. *Johannes Kepler*. Stuttgart: GNT-Verlag, 1995.

———. *Kepler*. Translated by C. Doris Hellman. New York: Dover, 1993.

———, and Walther von Dyck. *Johannes Kepler in seinen Briefen*. 2 vols. München: A. Oldenbourg, 1930.

Copernicus, Nicolaus. *On the Revolution of Heavenly Spheres*. Translated by Charles Glenn Wallis. Great Minds Series. Amherst, MA: Prometheus, 1995.

Couliano, Ioan P. *Eros and Magic in the Renaissance*. Translated by Margaret Cook. Chicago: Univ. of Chicago Press, 1987.

de Bourgoing, Jacqueline. *The Calendar: History, Lore, and Legend*. Translated by David J. Baker and Dorie B. Baker. New York: Abrams, 2001.

Demetz, Peter. *Prague in Black and Gold: Scenes from the Life of a European City*. New York: Hill and Wang, 1997.

Donahue, William H. *The Dissolution of the Celestial Spheres, 1595–1650*. New York: Arno, 1981.

Dudák, Vladislav. *Prague Castle: Hradèany*. Translated by Jiøí Trojánek. Prague: Baset, 1998.

Duncan, David Ewing. *Calendar: Humanity's Epic Struggle to Determine a True and Accurate Year*. New York: Bard, 1998.

Dunn, Richard S. *The Age of Religious Wars: 1559–1715*. 2d ed. New York: Norton, 1975.

Einstein, Albert. *The World As I See It*. Translated by Alan Harris. New York: Kensington, 1984.

Erickson, Carolly. *The First Elizabeth*. New York: St. Martin's, 1983.

Ferguson, Kitty. *Tycho and Kepler: The Unlikely Partnership That Forever Changed Our Understanding of the Heavens*. New York: Walker, 2002.

———. *Measuring the Universe: Our Historic Quest to Chart the Horizon of Space and Time*. New York: Walker, 1999.

Field, J. V. *Kepler's Geometrical Cosmology*. Chicago: Univ. of Chicago Press, 1988.

———. "A Lutheran Astrologer: Johannes Kepler." *Archive for History of Exact Sciences* 31 (1984): 189–272.

Gerlach, Walther, and Martha List. *Johannes Kepler*. München: R. Piper, 1966.

Gingerich, Owen. *The Eye of Heaven: Ptolemy, Copernicus, Kepler*. New York: American Institute of Physics, 1993.

Gleick, James. *Isaac Newton*. New York: Pantheon, 2003.

Goethe, Johann Wolfgang von. *Faust*. Translated by Walter Kaufmann. New York: Doubleday, 1961.

Grafton, Anthony. *Cardano's Cosmos: The Worlds and Works of a Renaissance Astrologer*. Cambridge: Harvard Univ. Press, 1999.

Grünter, Doebel. *Johannes Kepler: Er veränderte das Weltbild*. Graz: Styria, 1983.

Guazzo, Francesco Maria. *Compendium Maleficarum: The Montague Summers Edition*. Translated by E. A. Ashwin. Reprint. New York: Dover, 1988.

Hausenblasová, Jaroslava, and Michal Stronek. *Das Rudolfinische Prag*. Prague: Gallery, 1997.

Hollingdale, Stuart. *Makers of Mathematics*. New York: Penguin, 1989.

Holton, Gerald. "Kepler's Universe: Its Physics and Metaphysics." *American Journal of Physics* 24 (1956): 340–51.

———. *Thematic Origins of Scientific Thought: Kepler to Einstein*. Rev. ed. Cambridge: Harvard Univ. Press, 1988.

Janz, Denis R., ed. *A Reformation Reader: Primary Texts with Introductions*. Minneapolis: Fortress, 1999.

Jardine, Nicholas. *The Birth of History and the Philosophy of Science: Kepler's A*

Defense of Tycho Against Ursus with Essays on Its Provenance and Significance. Cambridge: Cambridge Univ. Press, 1984.

Kann, Robert A. *A History of the Habsburg Empire, 1526–1918.* Berkeley: Univ. of California Press, 1974.

Kepler, Johannes. *Epitome of Copernican Astronomy, Bks. IV and V, & Harmonies of the World, Bk. V.* Translated by Charles Glenn Wallis, 1952. Reprint. New York: Prometheus, 1995.

———. *The Harmonies of the World.* Translated by E. J. Aiton, A. M. Duncan, and J. V. Field. Philadelphia: American Philosophical Society, 1997.

———. "Selbstcharakteristik." In *Johannes Kepler Selbstzeugnisse.* Edited by Franz Hammer. Translated by Esther Hammer. Stuttgart-Bad Cannstatt: Frommann, 1971.

———. *Kepler's Conversation with Galileo's Sidereal Messenger.* Translated by Edward Rosen. New York: Johnson Reprint, 1965.

———. *Kepler's Somnium: The Dream or Posthumous Works on Lunar Astronomy.* Translated by Edward Rosen. Madison: Univ. of Wisconsin Press, 1967.

———. *Mysterium Cosmographicum: The Secret of the Universe.* Translated by A. M. Duncan. New York: Abaris Books, 1981.

———. *New Astronomy.* Translated by William H. Donahue. Cambridge: Cambridge Univ. Press, 1992.

———. *The Six-Cornered Snowflake.* Translated by Colin Hardie. Oxford: Clarendon Press, 1966.

———. "Rudolphine Tables: Introduction." Translated by Owen Gingerich and William Walderman. *Quarterly Journal of the Royal Astronomical Society* 13 (1972): 60–73.

———. *Kepler's Somnium.* Translated by Edward Rosen. Madison: Univ. of Wisconsin Press, 1967.

Kerrigan, Michael. *The Instruments of Torture.* London: Lyons Press, 2001.

Koestler, Arthur. *The Sleepwalkers: A History of Man's Changing Vision of the Universe.* New York: Macmillan, 1959.

Kors, Alan Charles, and Edward Peters. *Witchcraft in Europe, 400–1700: A Documentary History.* Philadelphia: Univ. of Pennsylvania Press, 2001.

Koyré, Alexandre. *The Astronomical Revolution: Copernicus—Kepler—Borelli.* Translated by R. E. W. Maddison. Reprint. New York: Dover, 1992.

Kozamthadam, Job. *The Discovery of Kepler's Laws: The Interaction of Science, Philosophy, and Religion.* Notre Dame, IN: Univ. of Notre Dame Press, 1994.

Lombardi, Anna Maria. "Johannes Kepler: Einsichten in die Himmlische Harmonie." *Spektrum der Wissenschaft* 4 (2000).

Luther, Martin. *The Bondage of the Will.* Translated by J. I. Packer and O. R. Johnston. Grand Rapids, MI: Fleming H. Revell, 1957.

———. *Luther's Large Catechism: A Contemporary Translation with Study Questions.* Translated by F. Samuel Janzow. St. Louis: Concordia, 1978.

———. *Martin Luther's Basic Theological Writings.* Edited by Timothy F. Lull. Minneapolis: Fortress, 1989.

Marešová, Soòa, trans. *The Prague Golem: Jewish Stories of the Ghetto.* Prague: Vitalis, 2002.

Methuen, Charlotte. *Kepler's Tübingen: Stimulus to a Theological Mathematics.* Brookfield, VT: Ashgate, 1998.

Midelfort, H. C. Erik. *Mad Princes of Renaissance Germany.* Charlottesville: Univ. of Virginia Press.

———. *Witch Hunting in Southwestern Germany, 1562–1684: The Social and Intellectual Foundations.* Stanford, CA: Stanford Univ. Press, 1972.

Parker, Geoffrey, ed. *The Thirty Years' War.* 2d ed. London: Routledge, 1984.

Petiška, Eduard, and Jan M. Dolan. *Beautiful Stories of Golden Prague.* Translated by Norah Hronková. Prague: Martin, 1995.

Polišenský, P. V. *History of Czechoslovakia in Outline.* Prague: Bohemia International, 1991.

Ptolemy, Claudius. *Ptolemy's Almagest.* Translated by G. J. Toomer. Princeton, NJ: Princeton Univ. Press. 1998.

Ripellino, Angelo Maria. *Magic Prague.* Translated from the Italian by David Newton Marinelli. Edited by Michael Henry Heim. London: Picador, 1995.

Rosen, Edward. *Three Imperial Mathematicians: Kepler Trapped Between Tycho Brahe and Ursus.* New York: Abaris Books, 1986.

Schmidt, Justus. *Johann Kepler: Sein Leben in Bildern und eigenen Berichten.* Linz: Rudolf Trauner, 1970.

Stephensen, Bruce. *Kepler's Physical Astronomy.* New York: Springer-Verlag, 1987.

———. *The Music of the Heavens: Kepler's Harmonic Astronomy.* Princeton, NJ: Princeton Univ. Press, 1994.

Taub, Liba Chalia. *Ptolemy's Universe.* Chicago: Open Court, 1993.

Tester, Jim. *A History of Western Astrology.* Woodbridge: Boydell, 1987.

Thomas, Keith. *Religion and the Decline of Magic.* New York: Oxford Univ. Press, 1971.

Tiner, John Hudson. *Johannes Kepler: Giant of Faith and Science.* Fenton: Mott, 1977.

Tracy, James D. *Europe's Reformations: 1450–1650.* Lanham: Rowman, 1999.

Voelkel, James R. *Johannes Kepler and the New Astronomy.* New York: Oxford Univ. Press, 1999.

———. *The Composition of Kepler's Astronomia Nova.* Princeton, NJ: Princeton Univ. Press, 2001.

Wheatcroft, Andrew. *The Habsburgs: Embodying Empire.* New York: Penguin, 1995.

Wilson, Curtis. "How Did Kepler Discover His First Two Laws?" *Scientific American* 226 (March 1972): 92–106.

Yates, Frances A. *The Rosicrucian Enlightenment.* London: Routledge, 1972.

Index

Achilles, Friedrich: hunting party of, 236; Katharina Kepler's arrest and, 265; Luther Einhorn and, 260

Adelberg (Germany), 44–46

Adolphus, Gustavus: attack on Regensburg, 364; invasion of Germany by, 361

ad torturam (capital crime), 298

alchemists: distinction between scientists and, 178; fate of, 180–181; in Prague, 177–178

alchemy: Edward Kelley and, 184; Kepler's suspicions about, 203

Analytica Posteriora (Aristotle), 57

angels, 182–183

Anhalt-Bernstein, Christian, 312, 314, 315

anima movens (spirit of movement), 94

Anna Marie, Archduchess of Austria: death of father, 162; marriage to King Philip II, 161

Aquinas, Thomas, 44

archpresbyter, 117

area law, 184

Aristotle: Kepler's study of, 57–58; model of universe, 60, 61–62, 63–65; planetary orbits and, 137, 173–174

art, 111, 171. *See also* museum (*Kunstkammer*)

Assyrians, 303

Astonomia Nova (Kepler): Kepler's work on planetary motions and, 173–177; laws of planetary motion in, 172–173; as modern text, 178

astrologers: as meteorologists, 233; in Prague, 177–178

astrological calendars, 325–326

astrology: astrological calendars of Kepler, 325–326; astronomical clocks, 51–52; horoscope for Heinrich Kepler, 115; Kepler as astrologer in Tübingen, 56–57; Kepler's advice to rulers, 224–225; Kepler's beliefs about, 178–180; Kepler's horoscope, 31–33; Kepler's predictions in Graz, 79–80; Kepler's readings for Albert von Wallenstein, 354–358; Kepler's view

of, 42–43; at Lutheran school in Graz, 78; medicine and, 136

astronomers: accounting for appearances, 173; practiced medicine, 179; protectiveness of observational data, 153–154

Astronomiae Pars Optica (Kepler), 172

Astronomia Nova (Kepler): contents of, 185–187; obstructions to publication, 205–207; publication of, 211; quote from, 47

astronomical clock: importance of, 51–52; in Prague's Old Town Square, 34; at Tübingen, 48

astronomy: *Astronomia Nova* (Kepler), 185–187; Copernican model of universe and, 124–125; Galileo's discoveries, 214–222; God and, 40–42, 43, 91–93; harmony and, 332, 333; Kepler and Brahe collaboration, 130–131, 150, 152–154; Kepler and Galileo, 4; Kepler on inclination towards, 47; Kepler on stars, 167; Kepler's book published, 93–96; Kepler's polyhedral hypothesis, 80–84; Kepler's preoccupation with, 88–89; models of universe, 59–68; motion of moon, 136–138; planetary orbits, 172–177

Auersperg, Dietrich von, 203

Augsburg Lutheran, 54–55

Augustine, Saint, 246–247, 350

Aulber, Johann Ulrich: capital crime charge and, 298; Katharina Kepler's sentence and, 302, 304; Katharina Kepler's trial and, 299; release of Katharina Kepler, 305

aurum potabile (liquid gold), 178, 184

Austria. *See also* specific cities of: Counter-Reformation in, 333–338; Kepler sent to, 68

autopsy, 177–178

Babylonians, 302, 303

Baden, Margrave Georg Friedrich, 117

baptism: Catholic church and, 350; children baptized in Catholic faith, 127; Kepler and, 351

Index

Bartsch, Jakob: on Kepler, 362; marriage of, 360
Barvitius, 163–164
Battle of White Mountain: described, 311–316; solved Protestant rebellion, 333–334; Thirty Years' War and, 241
Bede, Venerable, 97
Benatky Castle, 128, 129–133
Bernegger, Johann Matthias: invitation to Kepler, 362; letter from Kepler to, 339; *Rudolphine Tables* and, 345
Besold, Christian: accusations against Katharina Kepler and, 280–281; work on Katharina Kepler's case, 301
Besold, Christopher, 67
Bessel, Friedrich Wilhelm, 125
Beutelsbacher, Benedict: Katharina's potions for, 28; testimony of, 20–21
Bible, 37, 40–41
Bidenbach, Dr., 281
Billj, Hildebrand, 362, 364
Binder, Georg: arrest of Katharina Kepler and, 291; distance from Katharina, 292; Katharina's estate and, 267; marriage to Margaretha Kepler, 204
Binder, Margaretha. *See also* Kepler, Margaretha: letter to Kepler about Katharina's situation, 7, 9, 326; loyalty to mother, 292; marriage of, 28, 204; protection of Katharina, 300
biographies, Johannes Kepler, 6
birth: of Anna Maria Kepler, 361; of Christoph Kepler, 45
birth of: of Friedrich Kepler, 202; of Johannes Kepler, 23; of Katharina Kepler (daughter of Johannes Kepler), 254; of Ludwig Kepler, 202–203; of Sebald Kepler, 254; of Susanna Kepler (daughter of Johannes Kepler), 116, 202
Bocskay, István, 208, 209
Bohemia, 293–298
Bohemian Brethren: banned, 208; struggle for power in Prague, 157–159
Bohemian Brethren in the Bohemian Confession, 158
Bohemian Estates: choice of Friedrich V for king of Bohemia, 293–298; rebellion of, 333–334; Rudolf II and, 209, 210, 211; Second Defenestration of Prague and, 275–279

Book of Concord, 55
Book of Nature, 88, 91
Bourbon, duke of, 108, 109
Bragadino, Marko, 181
Brahe, Georg, 343–345
Brahe, Tycho: collaboration with Kepler, 152–154; comet of 1577 and, 25, 26, 49–50; death of, 155–156; feud with Ursus and Kepler, 122–125; fight with Kepler, 132–134; house in Prague, 169; illness of, 154–155; as imperial mathematician, 148; invitation to Kepler, 128–129; Kepler as assistant to, 150–151, 152; Kepler's collaboration with, 129–133; Kepler's eulogy on death of, 141–142; Kepler's letter to, 149; meeting with Rudolf II, 163–164; model of universe, 60, 186, 187; motion of moon and, 137; *Mysterium Cosmographicum* and, 115, 118; portrait of, *144*; rights to publish work of, 205–206; *Rudolphine Tables* and family of, 156–157; *Rudolphine Tables*, printing of, 343–345; work to secure Kepler's position, 134–136; wrote horoscopes, 79
Brahe, Tycho, Jr., 343–345
Braunschweig, Christian von, 334, 347
bread and wine, 44. *See also* Communion
Brengger, Johan Georg, 167
Brenner, Martin, 138
bride, 74–75
Bruno, Giordano, 167, 214
Bucquoy, Charles Bonaventure de Longueval, Count of, 313, 314
Budova, Václav, 276, 277
Buonarroti, Michelangelo: painting of Sistine Chapel, 111; Paul IV and, 112
Bürgi, Jost, 198
burial, 116

Caesar, Julius, 97
calculus, 323–324
calendar: astrological calendars by Kepler, 325–326; debate over, 97–98
Calvinists: doctrine of Communion, 213; Friedrich V and Elizabeth as, 296; Kepler accused of being Calvinist, 242, 248–249; Kepler and, 44–45; Kepler's ideas as heresy, 225; Kepler's refusal to condemn, 58; Lutherans and, 55; sermon against, 39
camera obscura, 138

Index

Kepler, Cordula, 254
Kepler, Fridmar, 254, 339
Kepler, Friedrich: birth of, 202; death of,
223, 224; father's love of, 204; godparents
of, 203
Kepler, Heinrich (brother of Johannes
Kepler): father attempts to sell, 26; in
Prague, 204; problems of, 28; statements
against Katharina Kepler, 282, 286
Kepler, Heinrich (father of Johannes
Kepler): abandonment of Katharina, 16;
absence of, 26–27; Kepler's boyhood and,
39; Kepler's education and, 38
Kepler, Heinrich (son of Johannes Kepler),
115–116
Kepler, Hildebert, 254
Kepler, Johannes: accusations against
Katharina and, 280–281; accusations/
court action against Katharina, 13–18; at
Adelberg school, 44–46; Albert von Wallenstein's patronage of, 358–360; appeal
to duke about Katharina, 290; area law
formulation, 184; arrival in Prague,
150–151; Astonomia Nova work of,
185–187; astrological readings for Albert
von Wallenstein, 354–358; astrology and,
177, 178–180, 181; Astronomia Nova,
obstruction to printing, 205–207; Astronomia Nova publication, 211–212; Astronomia Nova quote, 47; astronomy and
God, 91–93; attack on Prague and, 210;
birth of, 23; birth of Anna Maria Kepler,
360–361; boyhood of, 36–38; calendar
debate and, 96–99; children of, 202–204;
collaboration with Tycho Brahe, 152–154;
Copernican model of universe and,
59–68; Counter-Reformation in Prague
and, 158–159; courtship of Barbara
Müller, 89–90; death of Barbara Kepler
and, 226; death of daughter Katharina
Kepler, 319–320; death of Friedrich
Kepler and, 223–224; death of Margaret
Regina Kepler and, 326; defense of
Katharina Kepler, 299–301; enemies of,
35–36; excommunication from Lutheran
church, 287–288, 320–322; exemption/re-
turn to Graz, 118–122; family of, 24–29;
Ferdinand's mission and, 114; fight with
Tycho Brahe, 132–134; Galileo's discover-
ies/telescope, 214–222; gossip about

Katharina and, 260; handwriting of, 310;
Harmonice Mundi, 125–127, 307–308;
harmony work of, 326–333; horoscope
of, 31–33; house on Leonberg market
square, 12; illness of Barbara Kepler,
222–223; illness of/death of, 362–364; as
imperial mathematician, 156–157; journal
excerpt, 229, 339; Katharina Kepler's release and, 306; Katharina's move to Linz
and, 265; Katharina's trial/sentence and,
301–302; Kepler v. Reinbold, 268–269; at
Latin school, 38–43; leaves Graz, 140;
leaves Prague, 239; in Leonberg for
Katharina's defense, 268–269; letter from
Kepler to Johann Matthias Bernegger,
339; letter from Kepler to the Senate of
Leonberg, 7–10; letters to Johann Georg
Brengger, 167; letter to duke of Württemberg, 268, 271–272; letter to Michael
Mästlin, 85; letter to Tobias Scultetus,
189–190; letter to unknown nobleman,
227–229; Lutheran persecution in Graz,
116–117; Lutherans banished from Graz,
117–118, 127–128; marriage/family of,
114–116; marriage of daughter, 360; marriage portrait of, 86; marriage problems
with Barbara Kepler, 196–202; marriage
to Barbara Müller, 96, 99–100; marriage
to Susanna Reuttinger, 254; as mathematics teacher in Graz, 71–84; motion of
moon theories of, 136–138; move to Linz,
239–240; Mysterium Cosmographicum
dedication, 87–88; Mysterium Cosmographicum publication, 93–96; planetary
orbits and, 172–177; Prague Castle and,
165–166; Prague residences of, 169–170;
presentation of Rudolphine Tables to
Ferdinand II, 345–349; pressure to convert to Catholicism, 349–354; problems
with Lutheran church/accusation of
heresy, 242–251; Protestants forced to
convert to Catholicism, 138–140; publishing of, 324–325; Rudolphine Tables publication, 341–345; search for new
position, 212–213, 225–226; search for
new wife, 251–253; travels of, 204–205;
travel to Güglingen, 292; travel to Prague,
147–149; at Tübingen University, 49–59;
Tycho Brahe and Nicholas Reimarus
Ursus, 122–125; Tycho Brahe, hopes to

Photographic and Art Credits

All images in this text were used by permission of the Kepler Museum, Weil der Stadt, with the exception of:

Peter Paul Rubens, Equestrian Portrait of Cardinal-Infant Ferdinand of Austria during the Battle of Noerdingen, against the Swedish army, 1634. Painted 1636. 355 x 258 cm. Museo del Prado, Madrid. Photo credit: (c) Erich Lessing/Art Resource.

Astronomy/Cosmological model. "Planisphaerium Copernicanum." (Heliocentric planetary system of Copernicus, 1510). Copper engraving, coloured. From: Christoph Cellarius, Harmonia Macrocosmica, 1660. Photo: akg-images.

Prague Defenestration, 23 May 1618 (Protestant opposition in Bohemia to the Counter-Reformation. Two imperial lords-lieutenant were thrown out of a window of Prague Castle). "Prague Defenestration of 1618." Copper engr., subsequently coloured, by Matthaeus Merian (1593–1650). From Theatrum Europaeum, 1635 ff. Photo: akg-images.

The photographs of the astronomical clock in Prague and in Tübingen belong to the author.

CPSIA information can be obtained at www.ICGtesting.com
Printed in the USA
LVOW10s1202111214

418311LV00007B/115/P